T0212349

An Introduction to
Computational
Systems Biology

Chapman & Hall/CRC
Computational Biology Series

About the Series:
This series aims to capture new developments in computational biology, as well as high-quality work summarizing or contributing to more established topics. Publishing a broad range of reference works, textbooks, and handbooks, the series is designed to appeal to students, researchers, and professionals in all areas of computational biology, including genomics, proteomics, and cancer computational biology, as well as interdisciplinary researchers involved in associated fields, such as bioinformatics and systems biology.

Introduction to Bioinformatics with R: A Practical Guide for Biologists
Edward Curry

Analyzing High-Dimensional Gene Expression and DNA Methylation Data with R
Hongmei Zhang

Introduction to Computational Proteomics
Golan Yona

Glycome Informatics: Methods and Applications
Kiyoko F. Aoki-Kinoshita

Computational Biology: A Statistical Mechanics Perspective
Ralf Blossey

Computational Hydrodynamics of Capsules and Biological Cells
Constantine Pozrikidis

Computational Systems Biology Approaches in Cancer Research
Inna Kuperstein, Emmanuel Barillot

Clustering in Bioinformatics and Drug Discovery
John David MacCuish, Norah E. MacCuish

Metabolomics: Practical Guide to Design and Analysis
Ron Wehrens, Reza Salek

An Introduction to Systems Biology: Design Principles of Biological Circuits
2nd Edition
Uri Alon

Computational Biology: A Statistical Mechanics Perspective
Second Edition
Ralf Blossey

Stochastic Modelling for Systems Biology
Third Edition
Darren J. Wilkinson

Computational Genomics with R
Altuna Akalin, Bora Uyar, Vedran Franke, Jonathan Ronen

An Introduction to Computational Systems Biology: Systems-Level Modelling of Cellular Networks
Karthik Raman

Virus Bioinformatics
Dmitrij Frishman, Manuela Marz

For more information about this series please visit:
https://www.routledge.com/Chapman--HallCRC-Computational-Biology-Series/book-series/CRCCBS

An Introduction to Computational Systems Biology

Systems-Level Modelling of Cellular Networks

Karthik Raman

CRC Press
Taylor & Francis Group
Boca Raton London New York

CRC Press is an imprint of the
Taylor & Francis Group, an **informa** business

A CHAPMAN & HALL BOOK

First edition published 2021
by CRC Press
6000 Broken Sound Parkway NW, Suite 300, Boca Raton, FL 33487-2742

and by CRC Press
2 Park Square, Milton Park, Abingdon, Oxon, OX14 4RN

Library of Congress Cataloging-in-Publication Data

Names: Raman, Karthik, author.
Title: An introduction to computational systems biology : systems-level modelling of cellular networks / Karthik Raman.
Description: First edition. | Boca Raton, FL : CRC Press, 2021. | Series: Chapman & Hall/CRC computational biology series | Includes bibliographical references and index.
Identifiers: LCCN 2020050614 | ISBN 9781138597327 (hardback) | ISBN 9780429486951 (ebook)
Subjects: LCSH: Systems biology. | Biological systems--Computer simulation. | Molecular biology--Data processing. | Computational biology.
Classification: LCC QH324.2 .R344 2021 | DDC 570.1/13--dc23
LC record available at https://lccn.loc.gov/2020050614

ISBN: 978-1-138-59732-7 (hbk)
ISBN: 978-0-367-75250-7 (pbk)
ISBN: 978-0-429-48695-1 (ebk)

Access the Support Material: https://www.routledge.com/9781138597327

To my parents and teachers

In memory of
Sunder Mama & Prof. E. V. Krishnamurthy

Contents

PART III **Constraint–based Modelling**

CHAPTER   8 ■ Introduction to constraint-based modelling   173

PART IV Advanced Topics

CHAPTER   11 ▪ Modelling cellular interactions   249

CHAPTER   12 ▪ Designing biological circuits   275

Preface

MATHEMATICAL APPROACHES to model very large biological systems and networks form the foundation of *computational systems biology*. The main objective of this book is to deliver an insightful account of many such modelling approaches. The field in itself is highly multi-disciplinary, and the intended audience for this book, therefore, includes biologists and engineers alike. The idea is to present a sufficiently compelling account of the modelling strategies, but at the same time, retaining simplicity so as to not ward off interested biologists and others from less quantitative backgrounds. Although the area is relatively nascent, some good books have already appeared to date; the emphasis in this text will be to present a balanced account of a broad spectrum of modelling paradigms. We will adopt a hands-on approach, with exercises, codes, and, importantly, practical insights into troubleshooting modelling projects.

Organisation of this book. This book is divided into four parts. Following a short introduction to modelling, Part I focusses on static networks, studied using graph theoretic approaches. Part II moves on to dynamic modelling, covering classic ODE modelling, parameter estimation, and also Boolean networks. Part III is devoted to constraint-based modelling of metabolic networks, including methods to study perturbations. Part IV focusses on advanced topics such as the modelling of interactions between microbes in microbiomes, or between pathogens and their hosts, an introduction to synthetic biology and the design of biological circuits, and, lastly, a discussion on robustness and evolvability that pervade biological systems.

For the most part, the focus is on large-scale models with the goal of capturing *systems-level* effects. The first three parts are relatively independent, and can practically be studied in any order—although there is a logical progression from the static models of Part I to the dynamic models of Part II, and finally the constraint-based models of Part III that somewhat bring the best from both the static and dynamic worlds. The last part draws on nearly all of the methodologies described in the first three, applying it to more complex systems/concepts. While models exist at all levels of biological scale and hierarchy, we will focus more on models at the cellular level—comprising various cellular networks.

Pre-requisites. The only pre-requisite for this text is a familiarity with a high-level programming language aside from an intellectual curiosity towards how mathematical and computational approaches aid in understanding biology.

An engineering mathematics background will be advantageous but not strictly necessary.

TEACHING & LEARNING SYSTEMS BIOLOGY

Much of this textbook has emerged out of a course *Computational Systems Biology*, which I have taught to engineers from many different disciplines at IIT Madras over the last eight years. This book will appeal more to the mathematically inclined, and is intended as a textbook for an advanced interdisciplinary undergraduate course. With the emergence of interdisciplinary graduate programmes in computational and systems biology across the globe, a hands-on approach to various paradigms for modelling biological systems will likely be an attractive reference for many young scientists in these programmes, as well as senior scientists who wish to explore this area. The material in this textbook could be used in such a broad course, or for shorter introductory courses on network biology or metabolic network modelling. Below, I describe brief outlines for these courses:

Computational Systems Biology In contrast to nearly all of the existing textbooks, the present text provides a somewhat more balanced treatment to three key modelling paradigms—static network biology, dynamic modelling, and constraint-based modelling. Thus, depending on the flavour of the course, this book can form the main textbook, supplemented by topic-specific textbooks such as those by Alon [1], Palsson [8], and perhaps other specialised reference books [6, 10, 11]. This book has evolved out of teaching an interdisciplinary class; when used to teach specific majors, it may be supplemented suitably, *e.g.* Kremling [4] for chemical engineers, or Strogatz [5]/Nelson [7]/Sauro [12] for physicists.

Introduction to Network Biology Given the widespread importance of graph theory and network science in modelling biological systems, such a course may be of great interest to an interdisciplinary audience comprising computer scientists, biologists, and engineers. Chapters 2–4 of this textbook provide quite an extensive account of networks, including a detailed account of their applications to biological systems (Chapter 4) and microbial communities (Chapter 11). Newman's wonderful book [2] can be the classic reference for the network science portions, although it could be a bit intimidating to learners from a biology background.

Metabolic Network Modelling A key advance in the field of systems biology was the development of genome-scale metabolic models of microbial, and later, mammalian systems, which has widespread applications from metabolic engineering in the industry to drug target identification. Metabolism is the central function of a vast majority of cells; therefore, modelling metabolic networks can provide deep insights both into the behaviour of individual organisms, as well as how they interact in larger microbial communities. Chapters 8–10, with portions of

Chapter 11, and supplemented by some detailed reference books [8, 10, 11], can make for a very interesting course.

Hands-on training and course projects In all of the above courses, it is necessary to adopt a project-based teaching–learning approach, so that the learners may obtain hands-on experience on the various tools and modelling paradigms. A vital skill to pick up from a course such as this would be the ability to troubleshoot and debug models, as well as the ability to identify the right modelling framework/-paradigms for addressing a given modelling problem. Modelling is inarguably an art; therefore, hands-on experience is indispensable for honing one's skills. The present text emphasises on these by discussing common issues, and how to surmount them, at the end of nearly every chapter. In a typical 15-week semester, I ask the students to spend seven weeks on a project that builds on the topics covered in the course. Students have unanimously suggested that it is a central aspect of learning in my course.

ONLINE RESOURCES

The companion website to this book can be found at:

 https://ramanlab.github.io/
SysBioBook/

and includes the following resources for students and instructors:

Appendices It is desired to make the book as self-contained as possible; therefore, additional material in the form of five online appendices are also included: (A) a concise introduction to biology and biological networks, (B) detailed accounts for the reconstruction of biological networks (an important step that precedes modelling and simulation), compendia of (C) software tools and (D) databases important in systems biology research, and, lastly, (E) a primer to MATLAB. Appendices are located online only at: https:/www.routledge.com/9781138597327.

Companion Code Repository The companion code repository to this book includes some key snippets of code that will enable readers to understand some of the models/methodologies described in this book.

 https://github.com/RamanLab/
SysBioBook-Codes

The idea is to expand this repository over a period of time. If you think there is a piece of code that will help you better understand a concept in the book, please contribute an issue at https://github.com/RamanLab/SysBioBook-Codes/issues, and I will be more than happy to include one or more code snippets to address it.

Figure slides   Editable slides in both Powerpoint and PDF (one figure per page) formats are available freely for instructors wishing to use these in their course.

Material for further reading   While the book contains over 700 references including special pointers to material for further reading, the website offers an annotated reading list separated chapter-wise. This will be helpful in digging deeper into the concepts discussed in the text.

Online Course   A Massive Open Online Course on Computational Systems Biology, which is in part, a basis for this book, is also available through the Indian educational platforms, NPTEL (National Programme on Technology Enhanced Learning) (https://nptel.ac.in/) and SWAYAM (https://swayam.gov.in/). The lectures (including hands-on lab sessions), totalling over 30 hours of content, are also available on YouTube:

https://www.youtube.com/playlist?list=
PLHkR7OTZy5OPhDKvFJ_Xc-PuQFw4-oCZ4

Errata & Feedback   Despite my best efforts and those of my colleagues and students who have carefully gone through the book, it is likely that there may be a few typographical/grammatical errors. Do send your feedback via the website—critical feedback will be much appreciated and duly acknowledged!

COLOPHON

This book has been typeset using LaTeX (thank you, Leslie Lamport!) using the packages `fbb` (Cardo/Bembo for main text), `AlegreyaSans` (sans-serif fonts) and `mathpazo` (Palatino for math/Greek and tabular lining figures (in math)). Nearly all line drawings have been prepared using TikZ (and a few using Inkscape). The bibliography was typeset using `biblatex`. A special word of mention to Shashi Kumar from CRC/KGL for promptly fixing many arcane `hyperref` bugs.

I must also acknowledge the underlying TeX (and of course the venerable Donald Knuth!) and wonderful packages such as TikZ. A special mention for the super-efficient SumatraPDF, and Sublime Text, for being such a powerful editor—it saved me an immense amount of time during the writing of this book. I should also thank the rich online community of TeX gurus who have freely shared their expertise on various fora such as StackExchange, from where I learnt most of my TeX/TikZ tricks to prepare the beautiful illustrations that adorn this book.

ACKNOWLEDGEMENTS

My first thanks go out to Nagasuma Chandra (IISc Bengaluru) and Andreas Wagner (University of Zürich), my PhD and post-doctoral mentors, respectively, who first helped me foray into this exciting field of systems biology and mathematical modelling. I am also grateful to Saraswathi Vishveshwara, my PhD co-advisor, for introducing me to various aspects of graph theory. Since then, I consider myself lucky to have had a wonderful array of colleagues, collaborators, and students, who have enabled me to further my interests in this field.

Sincere thanks to my wonderful colleagues, Vignesh, Himanshu, Raghu, Ravi, Athi, Smita, and Guhan, for being constant sounding boards and sources of support and encouragement. A special mention to Ashok Venkitaraman (University of Cambridge) and K. V. Venkatesh (IIT Bombay), who have been wonderful collaborators as well as mentors over the last few years. It really helped to have a thriving group of interdisciplinary researchers at the Initiative for Biological Systems Engineering (IBSE) and the Robert Bosch Centre for Data Science and Artificial Intelligence (RBCDSAI), providing the ideal environment for authoring a book such as this. My heartfelt thanks to Mani for critically reading multiple chapters and giving elaborate suggestions for improving their content. Thanks are also due to my other colleagues and collaborators—Rukmini, Athi, Swagatika, Biplab, and Riddhiman—for reading through one or more sections and chapters, and sharing critical feedback that helped me in improving the book.

My students took great interest in reading through early versions of this book and shared several useful suggestions, including ideas for some exercises; thanks to Aarthi, Gayathri, Aparajitha, Sahana, Maziya, Sowmya, Sankalpa, Sathvik, Lavanya, Dinesh, and Rachita for making this book better! Further, I would be remiss were I not to acknowledge every author whose work I have read and every student I have taught—they have had a telling influence on my thinking and writing. I also thank Bharathi for encouraging me to create an NPTEL MOOC—the experience helped in many ways in putting together this book.

I am also grateful to Sunil Nair, CRC Press, for pushing me to write this book I had been planning for quite a while. Thanks also to all the other CRC folks, including Talitha, Elliott, Shikha, and Callum for their support and patience during the entire book-writing process.

Special thanks to my research group at IIT Madras, as well as my family, for enduring many delays and my excessive attention to book-writing, particularly during the last few months.

This book is dedicated to my parents and teachers, who have shaped me at every stage of my life, and *Bhagavān Nārāyaṇa*, without whose grace, nothing is possible.

Chennai, India   KARTHIK RAMAN
March 2021

FURTHER READING

[1] U. Alon, *An introduction to systems biology: design principles of biological circuits* (Chapman & Hall/CRC, 2007).
A classic treatise on biological networks, emphasising the relation between network structure and function.

[2] M. E. J. Newman, *Networks: an introduction* (Oxford Univ. Press, 2011).
The most authoritative textbook on network science and applications, more so for the mathematically inclined.

[3] E. O. Voit, *A first course in systems biology*, 1st ed. (Garland Science, 2012).
A solid textbook covering all aspects of systems biology.

[4] A. Kremling, *Systems biology: mathematical modeling and model analysis* (Chapman & Hall/CRC, 2013).
Covers multiple modelling paradigms through a chemical engineer's viewpoint, with a strong emphasis on ODE-based modelling, and control theory.

[5] S. Strogatz, *Nonlinear dynamics and chaos: with applications to physics, biology, chemistry, and engineering*, 2nd ed. (Westview Press, 2014).
The standard textbook for non-linear dynamics.

[6] J. DiStefano III, *Dynamic systems biology modeling and simulation*, 1st ed. (Academic Press, 2015).
A rigorous overview of dynamic modelling with dedicated chapters to system identification, parameter estimation as well as experimental design.

[7] P. Nelson, *Physical models of living systems* (W.H. Freeman & Company, New York, NY, 2015).
An engaging textbook on models of biological phenomena.

[8] B. Ø. Palsson, *Systems biology: constraint-based reconstruction and analysis* (Cambridge University Press, 2015).
The most authoritative text on reconstruction and constraint-based modelling of metabolic networks.

[9] E. Klipp, W. Liebermeister, C. Wierling, A. Kowald, H. Lehrach, and R. Herwig, *Systems biology*, 2nd ed. (Wiley-VCH, Germany, 2016).
An excellent self-contained "handbook" for systems biology.

[10] C. D. Maranas and A. R. Zomorrodi, *Optimization methods in metabolic networks*, 1st ed. (Wiley, 2016).
A detailed account of various constraint-based optimisation techniques.

[11] M. Covert, *Fundamentals of systems biology: from synthetic circuits to whole-cell models* (CRC Press, Taylor & Francis Group, Boca Raton, 2017).
An interesting textbook with a strong focus on dynamic modelling. Also presents a nice account of constraint-based modelling.

[12] H. M. Sauro, *Systems biology: introduction to pathway modeling* (Ambrosius Publishing, 2020).
Another excellent textbook on dynamic modelling, including a solid account of stochastic models.

Introduction to modelling

CONTENTS

L IVING ORGANISMS display remarkable abilities, to adapt and thrive in a variety of environments, fight disease, and sustain life. Underlying every living cell, from the 'simplest' prokaryote to human beings and other complex organisms, is an intricate network of biological molecules[1] performing a very wide variety of functions. This network includes complex macromolecules such as DNA, RNA, and proteins, as well as smaller molecules that drive metabolism, such as ATP or even water. In this chapter, we will see why mathematical modelling provides a valuable framework for us to begin unravelling the seemingly interminable complexity of biological systems.

Rapid advancements in sequencing and other high-throughput technologies over the last two decades have generated vast amounts of biological 'big data'. Spurred by this explosion in data, many computational methods have been developed for analysing such data. Notably, many tools and techniques strive to infer from these data, the complex networks that span metabolism, regulation, and signalling. Simulatable computational models of biological systems and processes, which provide a handle to understand and manipulate the underlying complex networks, form the cornerstone of *systems biology*.

1.1 WHAT IS MODELLING?

Many definitions exist for modelling, but beyond definitions, it is essential to understand the process of modelling, from one key perspective: *what is the overarching goal of the modelling process?* What are the important questions we seek to answer through the modelling exercise? Many a time, the clarity of this question decides

---

[1]See the cover image of the book!

the success of the modelling exercise. For example, are we interested in the dynamics of the system under consideration? Do we only need to know if certain interactions exist? Or, if some interactions are more important than others (in a qualitative sense)? Do we need to know the exact values of specific parameters in the system? And so on.

The questions also dictate the *resolution* of the model. As the famous adage goes, "*Don't model bulldozers with quarks*" [1]. Biological systems abound in time-scales and length-scales; therefore, a careful choice of the scale and resolution of the model is critical. If the scales chosen for modelling are incongruent with the questions being asked, the effect(s) we wish to observe may well be lost.

1.1.1 What are models?

Models essentially capture some physical system of interest, say a space shuttle, or a bacterial cell, or a human population where a communicable disease is spreading. Mathematical models capture these systems typically by means of mathematical equations. Models are *abstractions* of real-world systems; in fact, they are abstractions of (key) parts of a real-world system, capturing features deemed to be essential by the modeller. Modelling is thus a very subjective process, driven by the need to answer specific questions about the real-world system. Models can comprise a bunch of mathematical objects or equations, or even a computer program. Every model is characterised by assumptions on the real-world system, as well as approximations. These assumptions cover:

- variables (things that change),
- parameters (things that do not change, or are assumed to not change), and
- functional forms (that connect the variables and parameters).

Bender [2] defines a model as an "*abstract, simplified, mathematical construct related to a part of reality and created for a particular purpose*". This succinct definition harps on several key points:

- Models are abstractions
- Models are simplifications
- Models are incomplete
- Models are *question-specific*

The last point is very important—it is not unusual to find models that are built or known to be valid only under certain conditions, but are unwittingly applied to other conditions as well. For example, there are models that predict the initial rate of an enzyme-catalysed reaction when no product is present. Often, the same model is used to also predict/fit observations of measurements made at later time-points. There are certain assumptions that have gone into the model building process—if we lose sight of them at some point in time, we are very likely to commit major errors in predictions. Robert May outlines some of these points nicely in his essay titled "*Uses and Abuses of Mathematics in Biology*" [3].

Models can also be thought to divide the world into three sets [2]:

1. things whose effects are deliberately neglected (things beyond the system *boundary*),
2. things that are known to affect the model but which the model is not designed to study (things beyond the *scope* of the model), and
3. things/effects the model is *actually* designed to study.

In any modelling exercise, we must deliberately choose to neglect several things. Otherwise, we must build a 'whole-universe model', for any system to be studied! Therefore, we carefully choose a system boundary and consciously neglect various effects outside of this boundary. For example, while studying bacterial *chemotaxis*[2] in an open vessel, we may choose to neglect the effect of wind in the room or the temperature of the environment.

A particular model of chemotaxis may choose to ignore the temperature of the cell suspension, fully knowing that it can have an effect on chemotaxis. That is, these effects are considered to be beyond the scope of the model. The things or effects the model is actually designed to study are the *variables* of interest. In the chemotaxis example, this would be the concentration of an 'attractant', such as glucose. What we choose to neglect—in terms of both the system boundary and model scope—obviously affects both the complexity of a model, and its accuracy/predictive power. As Einstein purportedly remarked [3], *"models should be as simple as possible, but not more so"*—a practical translation of Occam's razor (see §6.4.1) for modellers.

1.2 WHY BUILD MODELS?

Invariably, a key reason why we strive to model any system is the desire to predict its behaviour, typically at another point in time/space, or under different conditions. Modelling drives conceptual clarification and helps reveal knowledge gaps. We create a model by committing to an understanding of a given system, and when the model predictions are off from reality, they point towards assumptions about the system that do not hold.

Modelling can also help prioritise experiments that are most informative and shed light on system function. Further, many experiments are infeasible or difficult to perform, expensive, and may also not be ethical (*e.g.* while dealing with human/animal subjects).

A key aspect of modelling biological systems, and even biology in general, is the study of systems under perturbation: modelling can potentially answer a number of 'what-if' questions, to understand the possible behaviour of the system under a variety of conditions. These could include the removal of a protein, the addition of a drug, or a change in the environment of a cell. Indeed, many models are used to predict the outcomes of the independent removal of every

---

[2]The movement of a motile organism, such as a bacterium, in the direction corresponding to the concentration gradient of a particular substance, *e.g.* glucose.

single component from a network, or even combinations of such removals, to clearly understand the key role played by each component in the network.

"All biology is computational biology!"

This box takes its title from the provocative article written by Florian Markowetz in 2017 [4]. The article essentially emphasises the central importance of computation and modelling in modern biology. Nevertheless, there is always a palpable tension between experimentalists and modellers. The late Nobel laureate Sydney Brenner decried systems biology as, *"low-input, high-throughput, no-output biology"* [5]! Admittedly, Dr. Brenner was criticising blind "data generation" and "factory science", bereft of solid theory/hypotheses. This is also a cue as to where to start in the *systems biology cycle* (see Figure 1.1).

David Eisenberg, the pre-eminent structural biologist, remarked in ISMB–ECCB 2013, *"Computational biology is the leader in telling us what experiments to do"*. Modern biology has reached a point where modelling and computation are indispensable; yet, all the exciting insights remain to be unravelled via the good old painstaking wet-lab experiments. Nevertheless, a tight collaboration between experimentalists and modellers is central to uncovering key biological principles, especially by taking a systems-level approach to studying and understanding biological complexity. The thoughts of these scientists reiterate the importance of the systems biology cycle, which can be best executed only by marrying modelling with wet-lab experiments.

Finally, it is a very demanding task to understand large complex systems, which are typically highly non–linear—modelling enables the systematic characterisation and interpretation of such systems. A lofty goal of modelling is the ability to understand design principles, or generalise the behaviour of a system, based on the insights generated from the model. Going even further, the culmination of modelling is the creation of 'digital twins', which have become extremely popular in various fields of science and engineering.

Digital twins are essentially *in silico* replicas of their real counterparts—an aircraft engine, the human heart, or even a complete virtual human being. Such models are not built to address specific questions; rather, their purpose is to faithfully replicate the behaviours of the original system to the maximum extent possible. Digital twins are rapidly gaining importance in manufacturing [6], as well as health care [7]. Systems biologists have long toyed with the idea of building virtual hearts [8] and virtual patients [9]; as early as 1997, the Physiome Project (http://physiomeproject.org/) was initiated by the International Union of Physiological Sciences (IUPS) to apply multi–scale engineering modelling approaches in physiology. Remarkable progress has been made since; yet, the long and winding road to a true digital twin of a human being is fraught with challenges and of course, many exciting discoveries. The Epilogue (p. 313) discusses a major step in building such detailed models, with the example of a whole-cell model of *Mycoplasma genitalium*.

1.2.1 Why model biological systems?

Mathematical modelling has been around for over a century, but the systematic modelling of biological systems at large scales has caught on only in the last few decades. This has been primarily catalysed by the explosion of quantitative data on biological systems, arising from various high-throughput measurements. An increasing number of genomes have been sequenced. Transcriptomes[3] have been uncovered, first using microarrays, and more recently, using RNAseq. Further, data on proteomics and metabolomics have burgeoned over the last few years. As a result, we have a very detailed view of the cell—practically an information overload, with terabytes of data! How do we make sense of all this information? One way to organise, integrate, and assimilate these multiple datasets is by building models and networks. Models are useful to ask and answer various questions of complex systems:

- What are the different possible behaviours the system can exhibit?
- How do these behaviours change based on conditions?
- How will the system react to various perturbations, genetic or environmental? That is, how robust is the system?
- What are the most important components/interactions in the system?
- Will the system reach a steady-state, or oscillate?

Modelling has led to several useful and actionable insights across biology, generating new and novel hypotheses and guiding wet-lab experiments. Many useful models exist to understand protein folding and protein–ligand interactions. The field of molecular dynamics relies on applying the laws of physics to predict and understand molecular motion and interactions. Many *in silico*[4] simulations can actually predict the outcome of real-life experiments. As the citation of the Nobel Prize in Chemistry 2013, to Martin Karplus, Michael Levitt, and Arieh Warshel read, "*Today the computer is just as important a tool for chemists as the test tube. Simulations are so realistic that they predict the outcome of traditional experiments.*"[5] While the citation obviously refers to chemists, computation has become an indispensable tool for biological scientists as well. Indeed, modern science is itself becoming increasingly driven by computer simulations; computational models are slowly but surely superseding mathematical models, as we foray into modelling more and more complex systems.

---

[3]The suffix -ome in molecular biology is used to denote a 'complete' set—*e.g.* of RNA transcripts in a cell (transcriptome), proteins (proteome), metabolites (metabolome), reactions (reactome), lipids (lipidome) and so on. The suffix -omics denotes the study and analysis of the various -omes.

[4]Meaning in silicon (computers), as an addition to the standard biological terms: *in vivo*, in living (cells), and *in vitro*, in glass, *i.e.* outside a living cell.

[5]https://www.nobelprize.org/prizes/chemistry/2013/press-release/

Mathematical vs. Computational Models

Essentially, mathematical models are represented by mathematical equations, and computational models are 'translations' of those mathematical equations, and could even be computer programs. In other words, mathematical models can (possibly) be solved analytically, while computational models generally demand 'numerical' solutions, and are decidedly more approximate. Nevertheless, as the systems we model become larger and more complex, we easily go beyond the realm of analytically solving equations. It is nigh impossible to analytically solve a system of tens of non-linear ordinary differential equations (ODEs), as one might encounter in modelling even a simple signalling network. Several powerful numerical algorithms have been developed, which can handle increasingly large systems of ODEs or other mathematical equations.

1.2.2 Why systems biology?

Many terms and phrases have been used to describe systems biology, *e.g.* postgenomic biology, quantitative biology, mathematical modelling of biological processes, multidisciplinary biology, molecular physiology, and the convergence of biology and computer science [10]. Ruedi Aebersold [10] however suggested a crisper definition, as "*the search for the syntax of biological information*"—the study of the dynamic networks of interacting biological elements. Indeed, as we discussed above, the search for design principles or a syntax is indeed an ultimate goal of modelling.

Beyond definitions, it is again essential to appreciate the key philosophies of systems biology, which ushered in the era of building large-scale holistic 'systems-level' models of biological systems, heralding the departure from the simpler *reductionist* studies and models of single molecules, proteins or pathways. As Anderson very succinctly put it, "*More is different*" (see the box on the following page). That is, complex systems exhibit various emergent behaviours, which cannot be entirely predicted or understood, even from a comprehensive understanding of all the constituent *parts*. Jacques Monod said as early as 1954, that "*Anything found to be true of E. coli must also be true of elephants.*" While his words[6] probably stemmed from the then idea of "unity in biochemistry" [11], the post-genomic era has, if anything, underlined the surprising similarity between genomes (*i.e.* components, or parts) of different species, across kingdoms of life. However, despite the striking similarity of the parts, their 'wiring' (*i.e.* interactions) varies markedly across these organisms, further emphasising the importance of systems-level investigations of the emergent complexities.

---

[6]As also those of the Dutch microbiologist Albert Jan Kluyver, "*From the elephant to butyric acid bacterium—it is all the same!*"

As a consequence of such complexity, every perturbation to such complex systems results in cascading effects that pervade various parts of the system. Thus, a systems perspective is necessary to appreciate, understand, and disentangle the complex networks underlying every living cell. This need for a holistic perspective is also emphasised in the classic Indian parable, often quoted by logicians, about *"Blind men and the elephant"*[7], where a number of blind men make drastically wrong conclusions about the 'system' they are examining depending on whether they felt the tusk, tail, trunk, or other parts. Extending the analogy, we are bound to make erroneous inferences about a system by looking at only some parts of it, especially, when the system itself is a result of complex interactions between various components distributed across various parts and subsystems.

"More is different"

This box takes its title from an extraordinary essay by Anderson in 1972 [12], where he wrote *"The ability to reduce everything to simple fundamental laws does not imply the ability to start from those laws and reconstruct the universe. The constructionist hypothesis breaks down when confronted with the twin difficulties of scale and complexity. At each level of complexity entirely new properties appear. Psychology is not applied biology, nor is biology applied chemistry. We can now see that the whole becomes not merely more, but very different from the sum of its parts."* Anderson was awarded the Nobel Prize in Physics in 1977. One could say that Anderson was very prescient, referring to systems biology and emergent biological phenomena, years before the formal arrival of 'systems biology'!

Although catalysed by the advances in high-throughput technologies, the rise of systems biology is due in no small measure to the concurrent advances in computing, mathematical techniques, and algorithms for analysing biological systems. Systems biology also marks the departure from the *'spherical cow'*[8], attempting to capture the complexity of biological systems in truer detail. Systems biology also advocates an iterative cycle of modelling, experimental design, and model recalibration based on the differences between predictions and reality, *i.e.* experimental observations. This *systems biology cycle* of iterant model-guided experimentation and experiment-aided model refinement is illustrated in Figure 1.1.

1.3   CHALLENGES IN MODELLING BIOLOGICAL SYSTEMS

We have convinced ourselves that it is very important to model biological systems, but this means that we are in for several challenges. Unlike most human-made systems or physical phenomena that have been the subjects of mathematical

---

[7]For an image and further discussion, see the textbook by Voit EO, under "Further Reading".
[8]A meme directed at the severe approximations that often characterise modelling.

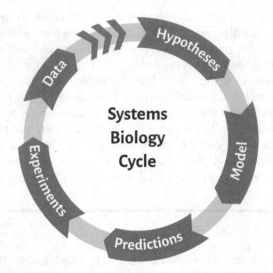

Figure 1.1 **The systems biology cycle.** Beginning with hypotheses, data, or the experiments that generate them, one builds a model. Predictions from this model can guide the design of (new) experiments and generate more informative/insightful data, which can be used to figure out knowledge gaps in the original hypotheses. This parallels the classic *"Design-Build-Test-Learn"* cycle that pervades engineering.

modelling for decades now, biological systems are unique in several respects and consequently, present a number of challenges for modelling.

Firstly, biological systems are extremely complex—a single cell of the simple *E. coli* bacterium contains thousands of different proteins and over a thousand metabolites embedded in highly complex networks. While there are thousands of interacting species, there are trillions of molecules present, including $> 2 \times 10^{10}$ molecules of water[9]. These numbers are staggering when we think of the size of the cell, which is barely a couple of microns in length (with a smaller diameter of $\approx 1\mu m$), and a volume of $\approx 1fL$. Nevertheless, human-made systems such as modern aircrafts do have millions of components. But the difficulty in modelling biological systems vis-à-vis engineered systems emerges from the sheer *complexity*, in terms of the number of interactions between the various components. As discussed in the box on the previous page, many new properties emerge in highly complex systems, with the whole being very different from merely the sum of its parts.

Biological systems are also remarkably *heterogeneous*—even cells in a clonal population will naturally exhibit different behaviours, in terms of their gene expression, and even their growth rate. This heterogeneity arises from the inherent

---

[9]BioNumbers database, BNID 100123 (accessed Nov 4, 2019) [13].

noise in various cellular processes, which brings along other challenges. We usually try to get around these issues by studying population averages for growth, or population averages for expression of genes and so on. The remarkable cellular heterogeneity has spawned an entire discipline of single-cell analysis [14]! The study of transcriptomics or metabolomics at the single-cell level seeks to uncover mechanisms or behaviours that may not be apparent in bulk populations.

Biological systems are also grossly incompletely characterised. Even in well-studied organisms such as *E. coli*, there remain a number of proteins of unknown function, as well as many interactions and metabolic reactions that are yet to be identified and characterised. This makes it extremely challenging to build accurate models of such systems. Interestingly, modelling becomes a very useful tool to identify and plug such knowledge gaps.

Another challenge in biological systems is the existence of multiple scales, both length-scales and more importantly, time-scales. A very wide range of time-scales are commonly observed in living cells: metabolism happens over a few seconds or few minutes, while signalling can happen over a few microseconds. There are processes such as growth and development that happen over months and years. Some processes may be triggered at a particular point, but the process itself may have evolved over a very slow time-scale. Evolution itself happens across several generations, spanning thousands of years. Such a large variation in time-scales in biology presents challenges for modelling: while we may not need to *simultaneously* model evolution and signalling, which are at opposite ends of the time spectrum, we may often need to reconcile and model processes that are at least a few orders apart in their time-scales, *e.g.* metabolism and signalling.

Although evolution is a slow process, it can be observed in organisms such as bacteria in even a few generations. A cell adapts in response to its environment, continuously picking up new mutations, and consequently, new behaviours and abilities. This is another challenge that makes the modelling of biological systems a rather demanding but nonetheless rewarding exercise.

1.4 THE PRACTICE OF MODELLING

We will now study the basic precepts of modelling. How does one go about the modelling process in general? Every modelling exercise must begin with the asking of several questions by the modeller. These questions set the ground for the model, notably with regard to its scope—both in terms of what the model seeks to capture and what the model will strive to predict. Ultimately, we need to figure out what type of model to build, to answer the desired questions, with the data presently available. Modelling, thus, can be a very subjective exercise, although it can be well guided by posing the right questions during model building.

It is instructive to keep a simple reference model in mind for the rest of the discussions in this chapter. Let us consider the popular Michaelis–Menten model of enzyme kinetics. While we will dissect this model more closely in Chapter 5, for the purposes of this discussion, we will merely concern ourselves with a broad picture of the model. The model captures the *initial* rate, v_0, of an enzyme-catalysed reaction as a function f of substrate concentration S:

$$v_0 = f(S) = \frac{k_1 S}{k_2 + S} \tag{1.1}$$

where k_1 and k_2 are some parameters that capture properties of the enzyme and substrate.

Note that the equation specifically mentions v_0, the *initial* rate of the reaction. This is in turn because of various assumptions that have gone into building the model, which we need not discuss at this time. It suffices for us to understand that the scope of the model is restricted to predicting the initial rate of a single substrate enzyme-catalysed reaction, in the absence of inhibitors. Tacitly, we have assumed constant temperature (or even pressure).

Thus, as discussed earlier, we choose for the model to take care of certain essential features or variables, ignoring (tacitly and deliberately) several others. Modelling is thus highly selective, and also subjective.

1.4.1 Scope of the model

Perhaps the most critical question of all is what the scope of a given model is. What is the model going to capture, *i.e.* what part of reality are we going to concern ourselves with? What are all the (input) 'predictor' variables that are going to be accounted for, which will have an effect on the variable(s) we are seeking to predict? Which interactions are going to be ignored? In an enzyme kinetics model, we (practically) always ignore the effect of other enzymes present and focus only on the enzyme of interest. Given how many enzymes show promiscuous activity with multiple substrates, this is a simplifying assumption, but a tenable one.

Next, what kinds of predictions are desirable? Do we seek to merely answer a 'yes–no' question, like will the enzyme be saturated at a substrate concentration of 10 mM, or, we would like to know more about the dynamics, or how the rates of reactions vary with substrate concentrations. Depending on the question, the choice of modelling paradigm would vary. Let us say we want to understand how the initial rate of an enzyme-catalysed reaction changes with substrate concentration, S. This would require a dynamic model.

The next important thing is that we need to have the right kind of data to facilitate this modelling exercise; obviously, we need enzyme and substrate concentrations, or other measurements from which these values can be reasonably inferred. Without any measurements of concentration, it would be impossible to embark on such a modelling exercise. While this sounds trivially obvious, in more complex scenarios, the clarity of whether the available data can enable the creation of the desired model may get lost!

The scope of the model also induces limitations. For instance, in the Michaelis–Menten model above, we assume that there is only a single substrate, which binds to the enzyme at a single site, and the reaction happens in a single step with a single transition state. We also assume that there are no enzyme inhibitors present. The model derived with the above scope cannot be blindly applied to scenarios where there are multiple substrates, or if there is an enzyme inhibitor, although, the model can be systematically extended, as we will see in §5.3.1.

1.4.2 Making assumptions

Every model makes a long list of assumptions about the system, the effect of various variables on the system, and the interactions present within the system. While deciding the scope of the model, some assumptions have already been taken into the fold. We next make some explicit assumptions. For instance, we will assume that the rate of an enzyme-catalysed reaction is directly proportional to, say, the enzyme concentration. Further, we may also make other assumptions about the relative rates of the intermediate reactions, or the relative ratios of different concentrations and so on, which (a) enable the model building, and (b) simplify the model to make it more tractable. Three assumptions/approximations are common in Michaelis–Menten kinetics: the steady-state approximation, the free ligand approximation, and the rapid equilibrium approximation (later obviated by Briggs–Haldane).

1.4.3 Modelling paradigms

Strongly influenced by the scope of the model and the nature of assumptions that we have already made is the choice of modelling paradigm. Many different modelling paradigms exist, each with its own set of advantages and disadvantages. The following box presents an interesting view of models, and can be extended to modelling paradigms.

"All models are wrong, but ..."

This box takes its title from the very famous quote of the pre-eminent statistician, George Box, who said [15] *"Essentially, all models are wrong, but some are useful."* Models are beset with approximations, but that need not necessarily detract from their usefulness. Despite blatant approximations, models provide many insights, illuminating complex systems such as those we will study in this book. Modelling is certainly an art, not in the least because Pablo Picasso once remarked something on the lines of *"Art is the lie that reveals truth"*!

It is conceivable that no one model paradigm can answer all questions—even if it did, it would be far too complex! Therefore, it is important to choose a modelling paradigm that is consistent with our model scope and prediction aspirations. We will briefly overview some broad modelling paradigms here:

Static vs. Dynamic Static models may explain only the steady-state behaviour of a system, or essentially, a single snapshot of the system at some time-point. Dynamic models, on the other hand, seek to explain transient behaviours of a system, and responses of a system to varying inputs/perturbations. In this book, Part I deals with static network models while Part II discusses dynamic models. Part III deals with models that typically model the steady-states of dynamic metabolic networks.

Continuous-time vs. Discrete-time Some dynamic models consider time to be continuous, while other dynamic models consider time in discrete time-steps, essentially a lower *resolution* for time. As the time-step or Δt becomes smaller, the distinction between discrete-time and continuous-time models blurs. In Part II, we will see examples of both continuous-time and discrete-time dynamic models.

Mechanistic vs. Empirical We may have models that are derived from first principles, or a knowledge of mechanisms, and are therefore *explanatory*. Other models may be purely empirical, and data-centric, just attempting to explain the data based on correlations. These include machine learning models, or regression models, which seek to explain data and make predictions. Most of the models described in this book are mechanistic, seeking to explain various observations based on various biological principles.

Deterministic vs. Stochastic Some models may be deterministic, in that the model always produces the same prediction/result once the inputs and conditions are specified. In stochastic models, one or more variables are random, which affect the outcome observed. Consequently, there will not be a single outcome, but a *distribution* of possible outcomes/model predictions. All models discussed in this book are deterministic. Stochastic models are also important in biology, and have been widely simulated using Gillespie's algorithm [16] but only in processes where molecule numbers are very low, *e.g.* where polymerases are involved. For systems-level models, we will neglect stochasticity.

Multi-scale vs. Single-scale Models may span a single or multiple temporal, spatial, and organisational scales. Biological systems are replete in multiple scales, be it time (fast vs. slow cellular processes), space (small molecules to massive macromolecular assemblies), or organisation (cells to tissues to organs and beyond). Nearly all models in this book are single-scale.

Further, these modelling paradigms are applicable to one or more levels of biological hierarchy, ranging from genes and proteins to whole organisms. The paradigms discussed in this textbook focus mostly at the level of cells, or their component molecular networks, although we will also discuss modelling of cell-cell interactions briefly (Chapter 11).

Immaterial of the modelling paradigm we choose, we must be ready to embrace approximations. Models are always approximate (wrong yet useful!). We

Correlation vs. causation

It is always very essential to carefully understand the prediction from a model. An empirical model may indicate correlations in the data, but not causation; it would be perilous to extrapolate correlations to causations, without duly performing targeted experiments to unravel the relationships between various variables in a model. Mechanistic models may point toward causation, but again, interpretations have to be made carefully, considering all alternative realities that the model may represent.

Tyler Vigen has put together a popular website (https://www.tylervigen.com/) and also a satirical book, "Spurious correlations" [17] illustrating how disregard for rigorous statistics can produce many a spurious correlation, between highly unrelated variables, such as the number of math doctorates awarded in the US and the amounts of Uranium stored in US power plants!

should also remember that the intention of modelling is always to answer specific questions. It is unreasonable to expect to answer every possible question using the same model. A model is, therefore, crafted very carefully to answer a specific set of questions, and may nonetheless be tweaked and re-worked to handle other scenarios. An exception to this general rule is the creation of a digital twin (see §1.2), where the goal is to capture the real system as accurately as possible, rather than answer specific questions. Figure 1.2 shows a few models popular in their respective fields, and lists their paradigms/properties.

1.4.4   Building the model

Once we have firmed up the scope of the model and the assumptions, and made a choice of which modelling paradigms work best for our purpose, we must actually *build* the model. Two key aspects exist for any model—*topology* (*i.e.* model structure) and *parameters*. The model structure or topology captures how the different physical components that the model abstracts are wired together. The model topology comprises different *variables*, which represent various physical components, as well as interactions and processes, which connect the different variables. Another important aspect of any model is the set of parameters that characterise these connections. These parameters may be well-known constants such as π or Avogadro number, or enzyme affinities and rate 'constants', or association/dissociation 'constants'. Together, all of these comprise the mathematical equations that make up the model.

In the context of the Michaelis–Menten model, the topology is the set of elementary chemical reactions that the model describes, involving the conversion of substrate to product through a single transition state intermediate. The topology is encoded into carefully chosen mathematical equations, in order to capture the real system to the extent possible. The parameters in these equations are k_1 and k_2 (Equation 1.1), which have more significant meanings as we will discuss

Michaelis–Menten model

A model (rate 'law') for predicting the rate of an enzyme-catalysed reaction.

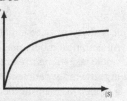

Dynamic • Continuous-time • Mechanistic • Deterministic

Turing patterns

A model of pattern formation in various systems.

Dynamic • Continuous-time • Mechanistic • Deterministic

Basic SIR model for the spread of infectious disease

Models the spread of infectious diseases in a population.

Dynamic • Continuous-time • Mechanistic • Deterministic

Kepler's laws of planetary motion

Three laws that capture the motion of planets in the solar system around the Sun.

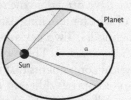

Dynamic • Continuous-time • Empirical • Deterministic

Graph model of protein interactions

A network that captures various kinds of interactions between proteins.

Static • Deterministic

Figure 1.2 **Examples of models.**

Zipf's law

A power law that relates frequency of words in a sample, to its rank in the frequency table.

Static • Deterministic • Empirical

Moore's law

An observation on the rate of increase in the number of transistors in an integrated circuit.

Static • Deterministic • Empirical

Newton's law of cooling

Models the rate of change of the temperature of an object.

$$\frac{dT}{dt} = -k(T - T_a)$$

Dynamic • Continuous-time • Empirical • Deterministic

Minimal model of insulin–glucose dynamics

A simplified model of glucose and insulin levels in the human body.

$$\dot{G}(t) = -(p_1 + X(t))G(t) + p_1 G^*$$

$$\dot{X}(t) = -p_2 X(t) + p_3(I(t) - I^*)$$

Dynamic • Continuous-time • Empirical • Deterministic

Brownian motion (Wiener process)

A model that explains many phenomena, such as the diffusion of suspended particles in a fluid.

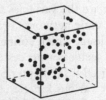

Dynamic • Continuous-time • Empirical • Stochastic

Figure 1.2 **Examples of models. (continued)**

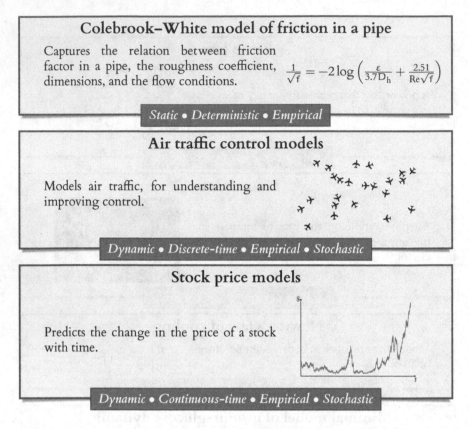

Colebrook–White model of friction in a pipe

Captures the relation between friction factor in a pipe, the roughness coefficient, dimensions, and the flow conditions.

$$\frac{1}{\sqrt{f}} = -2\log\left(\frac{\varepsilon}{3.7 D_h} + \frac{2.51}{\mathrm{Re}\sqrt{f}}\right)$$

Static • Deterministic • Empirical

Air traffic control models

Models air traffic, for understanding and improving control.

Dynamic • Discrete-time • Empirical • Stochastic

Stock price models

Predicts the change in the price of a stock with time.

Dynamic • Continuous-time • Empirical • Stochastic

Figure 1.2 **Examples of models. (continued)**

in §5.3.1. Elucidating the model topology in itself is a major challenge, both in terms of identifying all the interacting components/processes, and the nature of their interactions. This is the process of network *reconstruction*, which is a crucial aspect of modelling in systems biology. Network reconstruction is detailed in Appendix B. Another critical task is parameter estimation, dealt with in detail in Chapter 6.

What are the variables and parameters for the models we will discuss in this book? In static networks (Part I), there really aren't variables or parameters. The network itself, which connects the various components, which in turn could represent molecules or processes in a real system, *is* the model. In a dynamic model involving ODEs (Chapter 5), the variables could be species concentrations, connected by various kinetic equations, which contain a number of parameters. In discrete Boolean models (Chapter 7), there are only variables, which are connected by Boolean equations but no parameters. In the constraint-based models that form the main subject of Part III, the variables are the reaction fluxes, connected by mass-balance equations, with not many parameters save for substrate uptake rates or ATP maintenances (although these are rarely estimated).

1.4.5 Model analysis, debugging and (in)validation

Once the model is set up and the parameters estimated, the model must be simulated and validated. How does one go about simulating the model? If we have experimental data, how the model describes experimental data is one way of assessing how good the model is. Notably, we should use data that has not gone into building/fitting the model itself. The data at hand may or may not falsify the model—while a good fit would increase our confidence in the model, a poor fit would help us question our assumptions and point towards possible knowledge gaps.

But even before we get to formally simulating the model, we must carry out some *sanity checks*, to assess if the model predictions are *reasonable*. For instance, is the model predicting sensible steady-states? Does the model make correct predictions for *boundary cases*, like very low values or very high values or values around which certain behaviours are triggered? In the case of the Michaelis–Menten model, v_0 must be zero for $S = 0$ and v_0 must increase with S and saturate at some point, as one would expect in an enzyme system. In the case of a predator-prey model, if there are no predators, the prey population should go on increasing (till the carrying capacity of the environment). If there are no prey, the predator population should dwindle away and die. Typically, most systems in biology exhibit some sort of saturation (or even oscillations), rather than unbounded (continuously increasing) responses.

Depending on the nature of the model, *i.e.* the modelling paradigms used, appropriate model-specific diagnostics can be run. However, there are standard diagnostics that we might want to run on every model, such as assessing the stability, sensitivity, and robustness of the model and its predictions. These are best studied in the context of dynamic models, as we will see in Chapter 5. For example, how does the system respond to (small) perturbations in the initial conditions or model parameters? Does the model *fail* under certain conditions or make absurd predictions?

Further, is the model able to make good predictions on *unseen data, i.e.* data that has not already informed the model building exercise? If not, the model may have *overfit* the input data, learning it *too well*, measurement/experimental noise included, without being able to generalise for newer inputs/scenarios. A common strategy to avoid over-fitting is *cross-validation*, where some part of the data is kept aside during model building and later brought in to test the predictive ability of the model. In many biological modelling scenarios, it is typical to conduct a *holdout* study, where the oldest data are used to build the model (the newer data being 'held out'), and an attempt is made to predict the newer data using the model. Suppose we have a model that predicts interactions between proteins. This model can be built by using data from experiments/databases till, say, a year ago, and then one can examine what fraction of the new interactions (accumulated over the last year) in databases/experiments are correctly predicted.

An important outcome of model analysis and debugging is the identification of flaws in the underlying hypotheses, or even fundamental limitations of

---

"With four parameters, I can fit an elephant ...

..., and with five, I can make him wiggle his trunk." This statement is famously attributed to the famous Hungarian–American mathematician and physicist (not to mention computer scientist) John von Neumann [18]. His quote is a reminder to the importance of parsimony in models, and also of the dangers of over-fitting.

On a lighter note, some physicists actually drew an elephant using four (complex) parameters and wiggled its trunk using a fifth [19]. This is indeed a useful maxim to keep in mind while modelling! Also, elephants seem to attract the interest of biologists such as Monod, logicians that discuss the blind men parable, as well as physicists concerned about over-fitting—the term *elephant* is, therefore, duly indexed in this book!

---

the model. As a consequence, new experiments may be necessary, to corroborate unusual model predictions. This exercise may also shed further light on knowledge gaps; hence, an iterative cycle of model refinement and experimentation can be very fruitful (see Figure 1.1). Model (in)validation is thus a crucial step in the modelling process.

1.4.6 Simulating the model

Once all the hard work of model building is done, and we are reasonably sure of the model, following various diagnostics, we can now focus on the ultimate aim of the model—to predict various things about a system or enable a better understanding of a system, via *simulations*.

What is a simulation? The model is an abstraction of a real system. Simulation is essentially the process of executing this abstraction—or solving the model equations—to *simulate, i.e. imitate* the behaviour of the real system. A simulation, via the model, tries to capture as many elements as possible of the real system behaviour. Further, a simulation can also compress time (and space). The Indian Mars Orbiter Mission, called *Mangalyaan*, had a 298-day transit from Earth to Mars. The simulations took much lesser time! The simulations could reveal the exact times for the orbital manoeuvres, taking into account the positions of other planetary objects. Simulations thus compress time (or space) to enable understanding of effects that are separated in time (or space).

Why simulate models? Simulations can lend insights into system dynamics. Simulations can also substitute experiments, as mentioned in the Nobel citation, earlier in this chapter (§1.2.1). Simulations can help evaluate our hypotheses, and enable model refinement and the generation of new theories. Simulations can also provide insights into a number of '*what-if*' scenarios, notably perturbations to various aspects of the model, including various parameters or the topology itself.

Simulations are also wonderful pedagogical tools to understand various systems and processes, especially complex ones. For instance, Vax (https://vax.

herokuapp.com/) is an interactive learning environment that visualises the spreading of an epidemic across a network. It also hosts a *gamified* simulation, where a player can choose which healthy people to quarantine in a network, and which sick people to isolate, to prevent an outbreak from transitioning to an epidemic. Essentially, it performs a discrete-time dynamic simulation of the same process that the 'SIR' model in §1.5.2 simulates.

A particularly desirable outcome of the modelling and simulation exercise would be the discovery of design principles underlying the system. This may be possible because modelling and simulation enable a somewhat deeper characterisation of the system, through the study of multiple what-if scenarios. For example, a dose–response model for a drug may shed light on the relative importance of different mechanisms underlying how the drug interacts with the human body.

1.5 EXAMPLES OF MODELS

In this section, we will briefly discuss a few example models. The central purpose here is to understand the scope and objectives of the model, and to get a feel for the *modelling exercise* in general. Further, models such as the Lotka–Volterra predator–prey model or the SIR model also make for lively classroom discussions, as they are easy to understand and relate to. The rest of this book, of course, discusses several models and modelling paradigms in much greater detail.

1.5.1 Lotka–Volterra predator–prey model

The Lotka–Volterra Predator–Prey Model is a classic, deterministic, mechanistic, continuous-time, dynamic, single-scale model that captures the dynamics of two interacting species—a predator and a prey—in an environment. A detailed treatment of such models is available in the classic textbook by J. D. Murray [20]. In the context of such a predator–prey model, one may ask the following questions:

- What is it that we wish to predict?

  - will either population become extinct?
  - how do the population dynamics vary with time?

- What are the factors affecting the birth and death rate of the populations?
- Is there a carrying capacity of the environment for either population?

  - *i.e.* are we going to consider it in the present model?

Depending upon the answers to the above questions, a suitable model can be formulated. The classic predator–prey model seeks to study the population dynamics of both species with respect to time. Consider a predator, lion (L) and a prey, deer (D). The rate of growth of the deer population, in the absence of predation, can be modelled as:

$$\frac{dD}{dt} = bD \tag{1.2}$$

where b is the birth rate of the deer population. We can imagine that this b also subsumes the availability of grass/food for the deer population to survive and reproduce. For the lion population, the growth will be proportional to both the deer and the lion populations:

$$\frac{dL}{dt} = rDL \qquad (1.3)$$

where rD denotes the net rate of growth of the lion population, proportional to the availability of prey (deer). However, the above equations miss out on the death of lion population due to natural causes, and the effect of predation on the deer population. When we include these, the above equations change to:

$$\frac{dD}{dt} = bD - pLD \qquad (1.4)$$

$$\frac{dL}{dt} = rDL - dL \qquad (1.5)$$

This is the form of the original Lotka–Volterra predator–prey model; it is easy to extend this model to account for various other aspects, such as, say, the death of deer due to other causes, the existence of more than one prey/predator, and limitations on the availability of grass (carrying capacity of the environment). A more generalised version of this model is discussed in §11.1.2.1.

1.5.2 SIR model: A classic example

The 'SIR model' is a classic example of a simple yet useful mathematical model to capture the spread of infectious disease in a population. In the context of an infectious disease model, one may ask the following questions:

- What is it that we wish to predict?

  - the number of infected persons at any point of time?
  - whether a particular individual will fall sick or not?
  - whether the infection will blow up into an epidemic, or die down?

- For the spread of infection, is the population reasonably *mixed*? Or, is there a definite structure to the interactions?
- Is the population homogeneous? Or, do different people have different susceptibilities (*e.g.* children, older people)?
- Do people recovering from the disease develop immunity, or are susceptible to relapses?

In any modelling exercise, it is a good idea to begin with a simple model, incrementally adding other features or eliminating simplistic assumptions. In this case, we will look at the simplest SIR model, which was first formulated by Lowell Reed and Wade Hampton Frost in the 1920s (but never published!) [21]. The

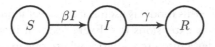

Figure 1.3 **A simple SIR model.** The figure depicts the three states S, I, and R, and the rates at which these epidemic states transit from one to another.

SIR model is also known as the Kermack–McKendrick Model [22, 23]. Many extensions have been proposed to the model since [24]. The model assumes that the entire population is susceptible (S), from which individuals transition to an infected state (I) and later recover, with immunity to the infection (R). The population is assumed to be *well-mixed* and homogeneous; we also ignore births and deaths (due to infection or otherwise), emigration, immigration, etc. An illustration of this basic SIR model is shown in Figure 1.3.

For very large populations, these dynamics can be captured by ODEs as follows:

$$\frac{dS}{dt} = -\beta SI \tag{1.6}$$

$$\frac{dI}{dt} = \beta SI - \gamma I \tag{1.7}$$

$$\frac{dR}{dt} = \gamma I \tag{1.8}$$

where S, I, and R represent the fractions of populations of susceptible, infected, and recovered people. Susceptible people become infected based on coming into contact with the infected people, as captured in the βSI term, β being the infection rate of susceptible individuals. Infected people recover from the infection, with γ being the recovery rate for infected individuals. Note that one equation above is redundant, as $S + I + R = 1$, at any point of time.

A very useful quantity from the model is the value $\mathcal{R}_0$, known as basic reproduction number, or basic reproductive ratio[10], which represents the number of infections that contact with a single infected individual generates, before the individual recovers (or dies):

$$\mathcal{R}_0 = \frac{\beta S}{\gamma} \tag{1.9}$$

When $\mathcal{R}_0 < 1$, the outbreak dies out since $dI/dt < 0$; on the other hand, when $\mathcal{R}_0 > 1$, the outbreak balloons into an epidemic ($dI/dt > 0$). $\mathcal{R}_0$, therefore, is a helpful parameter to quantify the danger presented by an outbreak. Of course, the parameters that contribute to $\mathcal{R}_0$ will vary based on the model equations, for more complex SIR models.

This simple SIR model best captures fast-spreading diseases; those diseases that have very slow dynamics (that are comparable to population dynamics time-scales) and diseases where the immunity is only temporary are not captured by

---

[10]Note however, that $\mathcal{R}_0$ has no relation to R in Equation 1.8 above!

this model. Thus, while this model is useful, it has several obvious limitations and provides great scope for extensions (see Exercises). An interesting extension of this model is to consider a discrete scenario, where one examines the spread of a disease on a social contact network (see §4.4.2). Such analyses can answer very different questions, such as who should be isolated first, or which people should be quarantined first, to minimise the spread of infection. Excellent reviews are available in literature, discussing various aspects and variations of the SIR model [24], and epidemic spreads on networks [25, 26]. §4.4 briefly discusses some of the studies on the spread of diseases on networks.

The SIR model also found its way into popular culture, when an episode of the TV series "Numb3rs"[11] had the lead mathematician in the show using it for tracking the spread of a deadly virus in a city.

1.6   TROUBLESHOOTING

1.6.1   Clarity of scope and objectives

The biggest challenge in modelling comes with the subjective choice of how to model a given system, keeping in mind the scope and the objectives. Clarity on these, therefore, is most critical to the success of any modelling exercise. Depending on the kind of questions we seek to answer, the most appropriate modelling paradigm can be chosen.

Another major challenge is the availability of *suitable* data, both for building the model and for validating it. A model is no good if we are not able to perform any validation on it; although, the predictions the model throws up can be interesting in terms of generating novel hypotheses and of course, designing new experiments. Time and again, we see that the systems biology cycle becomes very important, particularly the synergy between modelling and experimentation.

1.6.2   The breakdown of assumptions

Every model, as we have seen above, has its own set of assumptions, arising partly from the intended scope and application of the model. As a model becomes popular and becomes pervasively used, it is common to lose sight of the original scope/applications of the model, and models begin to be employed in scenarios where the assumptions do not hold. Also, models that have become very popular are taken for granted and are often thought of as *laws*—the law of mass-action, or even the Michaelis–Menten law—somehow conferring an aura of inviolability to a model.

Another classic example is the Higuchi equation—again, an elegant equation that describes the kinetics of drug release from an ointment as a function of the square root of time. However, beyond its extensive use, the model has also been misinterpreted and misused in many scenarios, which are not intended applications of the model, as detailed elsewhere [27].

---

[11]Season 1, Episode 3: "Vector" (https://www.imdb.com/title/tt0663234/).

1.6.3 Is my model fit for purpose?

As an extension of the discussion on scope, it is important to ensure that the model is indeed fit for the intended purpose. Building a steady-state model of a metabolic system, to understand and elucidate the effect of feedback inhibition, would be the modelling equivalent of *"bringing a knife to a gunfight"*. While the inadequacy of steady-state modelling for studying inhibition is relatively obvious, many a time, in complex systems, such errors may not be apparent. Thus, due diligence during both model building and debugging is essential.

1.6.4 Handling uncertainties

Uncertainties are the only certainty in biology. Owing to the very nature of biological systems, many sources of uncertainty exist, which confound the modelling process. As discussed above, we are uncertain about the components themselves, as well as their possible interactions (topology). From the perspective of modelling, there are two broad classes of uncertainty [28]: *aleatoric*, arising from inherent noise/randomness/variability, and *epistemic*, arising from incomplete knowledge of the system. Then, there are uncertainties that arise from experiments—both measurements and conditions. It is also not uncommon to find conflicting hypotheses in literature, about a given system or process. Thus, it becomes challenging to develop a model, which is able to budget for the various uncertainties and variabilities, and make careful predictions. Of course, these predictions cannot be precise, but they should predict a *region of behaviour* for the system, which is consistent with reality. We will discuss this in later sections, notably in §5.6.2 and §6.2.2.2. Paradigms such as probabilistic graphical models [29] are also powerful abstractions to quantify and model uncertainties in a variety of systems.

EXERCISES

1.1 Consider one of the many models mentioned in Figure 1.2. Can you identify the scope of the model, its key assumptions, and the kind of data needed to build the model?

1.2 Consider a system involving two proteins. Each of the proteins can activate or repress (or have no effect) another protein (including itself). What is the total number of *possible* topologies?[12] Can you see that this number becomes even larger, if we include another protein?

1.3 List three improvements you would make to the simple SIR model. Do any of these improvements also require changing the paradigm of the model? What additional questions can the improved model now answer?

1.4 Consider the Lotka–Volterra predator–prey model discussed in §1.5.1. In the basic model, the prey population can explode, when the predator

---

[12]This problem alludes to the work of Ma *et al* (2009), which we will discuss in Chapter 12.

population is very low. However, in practice, the environment (*e.g.* grassland) will not support this, and has a finite *carrying capacity*. How will you model this?

1.5 Identify any model that you have come across at some point in time. It could be a weather model, a model to predict sports scores, or even a mouse model to study a disease. Can you list out its key features, such as its scope, any assumptions, the goal of the model, as well as the various paradigms that pertain to it?

1.6 Consider the SIR model for infectious diseases. The Severe acute respiratory syndrome coronavirus 2 (SARS-CoV-2), more familiarly known as the coronavirus, caused a major pandemic, COVID-19 beginning late 2019, infecting millions of individuals across the globe. In response to this pandemic, several countries announced 'lockdowns' to prevent the rapid spread of the virus. For example, in India, mobility between the 28 states and eight union territories was completely cut off for a few weeks. To model the spread of the infection under such a scenario, what kind of alterations would you make to the basic SIR model? Also see reference [30].

1.7 Another peculiarity of SARS-CoV-2 is how long an individual remains asymptomatic, while still spreading the infection to others. Indeed, this was one of the major reasons that necessitated strong social distancing norms and extended lockdowns in multiple countries. To account for this, would you have to incorporate additional states to the SIR model? Or, is a simple model sufficient? How would you decide?

[LAB EXERCISES]

1.8 In 2009, Munz *et al* [31] published an interesting paper titled *"When zombies attack!: Mathematical modelling of an outbreak of zombie infection"*, which also received much interest from the popular science press[13]. Built on the SIR framework, this S'Z'R model presents interesting conclusions, contrary to popular movie expectations. See if you can simulate their model and reproduce their results.

REFERENCES

[1] N. Goldenfeld and L. P. Kadanoff, "Simple lessons from complexity", *Science* 284:87–89 (1999).

[2] E. A. Bender, *An introduction to mathematical modeling* (Dover Publications, 2000).

[3] R. M. May, "Uses and abuses of mathematics in biology", *Science* 303:790–793 (2004).

---

[13]See, for example, https://www.wired.com/2009/08/zombies/

[4]   F. Markowetz, "All biology is computational biology", *PLoS Biology* **15**:1–4 (2017).

[5]   E. C. Friedberg, "Sydney Brenner", *Nature Reviews Molecular Cell Biology* **9**:8–9 (2008).

[6]   M. Grieves and J. Vickers, "Digital twin: mitigating unpredictable, undesirable emergent behavior in complex systems", in *Transdisciplinary Perspectives on Complex Systems: New Findings and Approaches* (Cham, 2017), pp. 85–113.

[7]   K. Bruynseels, F. Santoni de Sio, and J. van den Hoven, "Digital twins in health care: ethical implications of an emerging engineering paradigm", *Frontiers in Genetics* **9**:31 (2018).

[8]   D. Noble, "Modeling the heart–from genes to cells to the whole organ", *Science* **295**:1678–1682 (2002).

[9]   P. Kohl and D. Noble, "Systems biology and the virtual physiological human", *Molecular Systems Biology* **5**:292 (2009).

[10] R. Aebersold, "Molecular Systems Biology: a new journal for a new biology?", *Molecular Systems Biology* **1**:2005.0005 (2005).

[11] H. C. Friedmann, "From *"Butyribacterium"* to *"E. coli"*: an essay on unity in biochemistry", *Perspectives in Biology and Medicine* **47**:47–66 (2004).

[12] P. W. Anderson, "More is different", *Science* **177**:393–396 (1972).

[13] R. Milo, P. Jorgensen, U. Moran, G. Weber, and M. Springer, "Bionumbers—the database of key numbers in molecular and cell biology", *Nucleic Acids Research* **38**:750–753 (2010).

[14] D. Wang and S. Bodovitz, "Single cell analysis: the new frontier in 'omics'", *Trends in Biotechnology* **28**:281–290 (2010).

[15] G. E. P. Box and N. R. Draper, *Empirical model-building and response surfaces* (Wiley, USA, 1987).

[16] D. T. Gillespie, "Exact stochastic simulation of coupled chemical reactions", *The Journal of Physical Chemistry* **81**:2340–2361 (1977).

[17] T. Vigen, *Spurious correlations* (Hachette Books, New York, 2015).

[18] F. Dyson, "A meeting with Enrico Fermi", *Nature* **427**:297 (2004).

[19] J. Mayer, K. Khairy, and J. Howard, "Drawing an elephant with four complex parameters", *American Journal of Physics* **78**:648 (2010).

[20] J. D. Murray, *Mathematical biology: I. An introduction (interdisciplinary applied mathematics) (pt. 1)* (Springer, New York, USA, 2007).

[21] M. E. J. Newman, "Spread of epidemic disease on networks", *Physical Review E* **66**:016128+ (2002).

[22] W. O. Kermack, A. G. McKendrick, and G. T. Walker, "A contribution to the mathematical theory of epidemics", *Proceedings of the Royal Society of London: Series A, Containing Papers of a Mathematical and Physical Character* **115**:700–721 (1927).

[23] R. M. Anderson and R. M. May, "Population biology of infectious diseases: Part I", *Nature* **280**:361–367 (1979).

[24] H. W. Hethcote, "The mathematics of infectious diseases", *SIAM Review* **42**:599–653 (2000).

[25] M. J. Keeling and K. T. D. Eames, "Networks and epidemic models", *Journal of the Royal Society Interface* **2**:295–307 (2005).

[26] J. C. Miller and I. Z. Kiss, "Epidemic spread in networks: Existing methods and current challenges", *Mathematical Modelling of Natural Phenomena* **9**:4–42 (2014).

[27] J. Siepmann and N. A. Peppas, "Higuchi equation: derivation, applications, use and misuse", *International Journal of Pharmaceutics* **418**:6–12 (2011).

[28] H.-M. Kaltenbach, S. Dimopoulos, and J. Stelling, "Systems analysis of cellular networks under uncertainty", *FEBS Letters* **583**:3923–3930 (2009).

[29] D. Koller and N. Friedman, *Probabilistic graphical models: principles and techniques* (MIT Press, 2009).

[30] W. C. Roda, M. B. Varughese, D. Han, and M. Y. Li, "Why is it difficult to accurately predict the COVID-19 epidemic?", *Infectious Disease Modelling* **5**:271–281 (2020).

[31] P. Munz, I. Hudea, J. Imad, and R. J. Smith, "When zombies attack!: mathematical modelling of an outbreak of zombie infection", in *Infectious Disease Modelling Research Progress* (2009), pp. 133–150.

FURTHER READING

Y. Lazebnik, "Can a biologist fix a radio?–Or, what I learned while studying apoptosis", *Cancer Cell* **2**:179–182 (2002).

L. H. Hartwell, J. J. Hopfield, S. Leibler, and A. W. Murray, "From molecular to modular cell biology", *Nature* **402**:C47–C52 (1999).

M. E. Csete and J. C. Doyle, "Reverse engineering of biological complexity", *Science* **295**:1664–1669 (2002).

M. Eckhardt, J. F. Hultquist, R. M. Kaake, R. Hüttenhain, and N. J. Krogan, "A systems approach to infectious disease", *Nature Reviews Genetics* **21**:339–354 (2020).

E. O. Voit, *A first course in systems biology*, 1st ed. (Garland Science, 2012).

I

Static Modelling

Introduction to graph theory

CONTENTS

G RAPHS are a very important theme in mathematics and computer science, and provide an important way to understand very complex systems such as the networks that pervade biology, in terms of the various entities that exist and their relationships. *Graphs* are essentially the mathematical formalisms that are used to study *networks*, but the two terms are invariably used interchangeably.

2.1 BASICS

Formally, a graph G is denoted as $G(V, E)$, where V represents the set of possible entities (typically called *nodes*, or *vertices*) and E, their relationships (typically referred to as *links*, or *edges*). Every edge connects a pair of vertices from the set V; therefore, $E \subseteq V \times V$. Every pair of vertices in E are *adjacent* to one another. Some simple illustrative graphs are shown in Figure 2.1.

2.1.1 History of graph theory

Graph theory has a very interesting historical origin, and like many other mathematical concepts, is a tribute to the intellectual genius of Leonhard Euler. In 1735, Euler solved a problem, now often referred to as the *Seven Bridges of Königsberg*, a classic problem in recreational mathematics. The residents of the city of Königsberg (present-day Kaliningrad, Russia) had an interesting thought, as to whether they could set out from their homes, cross each of the seven bridges on the river

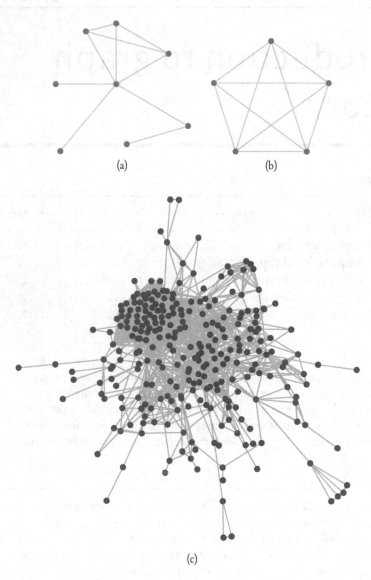

(a)

(b)

(c)

Figure 2.1 **Some simple example graphs.** (a) A simple graph with eight nodes and ten edges. (b) A complete graph of five nodes and all ten possible edges. (c) A snapshot of a real protein–protein interaction network (275 nodes, 2715 edges). Every node represents a protein in the bacterium *Mycoplasma genitalium*, from the STRING database, version 10.5 [1]. Only functional associations and interactions with a confidence of 900 and above, from the single largest component of the graph are shown.

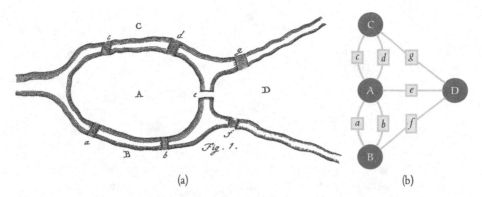

(a)   (b)

Figure 2.2 **Euler's solution to the Königsberg problem**, as described in his 1735 paper entitled "*Solutio problematis ad geometriam situs pertinentis*", or "The solution of a problem relating to the geometry of position" [3]. (a) Euler's sketch of the map of Königsberg (1735), represented as (b) the first ever graph. Euler denoted the land masses by capital letters A–D, and the seven bridges on the river Pregel, by small letters *a–g*.

exactly once and get back home. Euler's famous assessment of the problem goes as follows, "*This question is so banal, but seemed to me worthy of attention in that [neither] geometry, nor algebra, nor even the art of counting was sufficient to solve it.*", as quoted in [2]. Euler of course solved this problem in the negative, showing that no such walk exists.

Figure 2.2a shows Euler's sketch derived from the map of the city. Figure 2.2b is, essentially, the first graph that was ever created! Note the mapping of nodes and edges, as land masses (or islands) and the corresponding bridges. However, it is not a *simple graph*, given that it has parallel edges running between the same pairs of vertices. Also, Euler never drew it in this form, in 1735. Nevertheless, it is easy to paraphrase Euler's 1735 solution (see references [2, 3]) in modern graph theory parlance, namely in terms of the number of nodes with odd and even number of edges! Paths that begin and end at the same vertex of a graph, passing through every single edge in the graph exactly once, are now called *Eulerian cycles*, in honour of this solution of Euler. Figure 2.3 shows some simple examples of graphs with and without Eulerian paths. Finding Eulerian paths in graphs has more far-reaching consequences than solving the recreational mathematics problems of Figures 2.2 and 2.3, notably in modern biology, for assembling reads from a next-generation sequencer [4].

2.1.2 Examples of graphs

To fix ideas, let us review a few examples of graphs. Every graph, as discussed above, needs to be described by a set of nodes/vertices and a set of edges. In social networks, the nodes are individual people, and their friendships are edges. In city

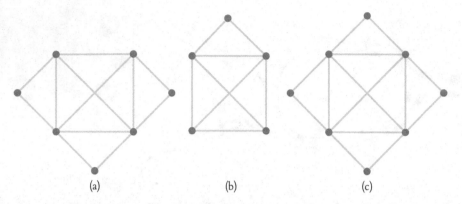

(a) (b) (c)

Figure 2.3 **Eulerian paths on graphs.** (a–b) These graphs have Eulerian paths, which means you can draw them without taking your pencil off paper, passing through all edges exactly once. (c) This graph does not have an Eulerian path.

TABLE 2.1 **Some examples of graphs.**

Network	Nodes	Edges
Facebook	*People*	*Friendships*
Twitter	*People/Businesses*	*'Follows'*
Protein interaction network	*Proteins*	*Interactions*
Gene regulatory network	*Genes*	*Regulatory effects*
Metabolic network	*Metabolites*	*Reactions*
Citation network	*Research articles*	*Citations*
Co-authorship network	*Authors*	*Co-authors*
Food web	*Species*	*Who eats whom*
Protein structure	*Amino acid residues*	*Contact maps*
Molecular graphs	*Atoms*	*Bonds*

road networks, roads form the edges connecting various junctions within the city. In the context of biology, we have protein interaction networks, where proteins are obviously the nodes, and their interactions are captured by means of edges. Table 2.1 lists a few networks, indicating what their nodes and edges typically describe. Some of these examples are more rigorously dealt with in §2.5.

2.2   WHY GRAPHS?

Graphs are useful to answer many interesting questions about different types of systems. For instance, social network analysis, or *sociometry*, as it was originally called, is dated to the early 1930s [5]. In the context of social networks, many interesting questions have been asked of graphs:

Do I know someone who knows someone ... who knows X? (*existence of a path*)

How long is that chain to X^1? (*shortest path problem*)

Is everyone in the world connected to one another? (*identifying connected components*)

Who has the most friends, or who is the most influential person? (*most connected nodes, centrality analyses*)

Can you predict if X and Y are friends ('friend suggestions')? (*link prediction*)

Many graph algorithms have been developed to address questions such as these. Table 2.2 lists some of the popular graph algorithms. Interestingly, very similar questions are often of interest in the context of biological networks, and can be addressed using pretty much the same graph algorithms:

Is there a way to produce metabolite X from A? (*existence of a path*) This is the classic problem of *retrosynthesis*, as we will also see in §4.5.1.

How long is the *optimal* chain from X to A? (*shortest path problem*) How many reactions (or enzymes) do we need to synthesise a molecule? Can we find the most biologically plausible path, if there are multiple paths?

Are all proteins connected to others by a path? (*identifying connected components*) Or similarly, what are all the metabolites that are producible, from a given set of starting molecules? (see §11.1.1.2)

Which is the most influential protein in a network? (*most connected nodes, centrality analyses*) Perhaps, one wants to target the most connected protein in a pathogen's protein network, to try and kill it.

Are proteins X and Y likely to interact? (*link prediction*) Given the interaction profiles (akin to interest or friendship profiles in social networks) of two proteins, link prediction algorithms can predict if they are likely to interact (§4.7).

In this book, we will not particularly focus on graph algorithms; there are many excellent textbooks on this subject [6, 7]. Instead, we will move over to the different types of graphs that exist (§2.3), their computational representations (§2.4), and focus on how graphs can be used to represent and study biological networks (§2.5).

2.3 TYPES OF GRAPHS

Graphs can be classified into different types, based on the nature of the connections, the strengths of the connections and so on. In this section, we will review some of these basic concepts.

---

[1] Preferring the shortest of such chains.

TABLE 2.2 **Popular graph algorithms.** These are the graph algorithm pages on Wikipedia that have had the most page views in the years 2014–2018.

[1]	Dijkstra's [shortest path] algorithm
[2]	A* search algorithm
[3]	PageRank [centrality measure]
[4]	Travelling salesman problem
[5]	Breadth-first search
[6]	Depth-first search
[7]	Tree traversal
[8]	Kruskal's algorithm
[9]	Prim's algorithm
[10]	Topological sorting
[11]	Floyd–Warshall algorithm
[12]	Bellman–Ford algorithm

2.3.1 Simple vs. non-simple graphs

Simple graphs are graphs without *self-loops* or *multi-edges*: a self-loop involves an edge from a node to itself; a multi-edge means that there is more than one edge between a pair of nodes in a graph. Self-loops and multi-edges (or 'parallel edges') necessitate special care while working with graphs and graph algorithms—simple graphs are devoid of such structures. In the following sections and for most of the book, we will concern ourselves only with simple graphs, although self-loops do occur in gene regulatory networks, where a gene (strictly speaking, its product) can inhibit its own synthesis, thus resulting in a vertex connected to itself.

2.3.2 Directed vs. undirected graphs

Perhaps the most important classification of graphs is based on the symmetry of the edges, into *directed* and *undirected* graphs. In an undirected graph, when a node A is connected by an edge to B, it means that the node B is also connected (by the same edge) to A. In other words, A and B are *adjacent* to one another: $(A, B) \in E \Rightarrow (B, A) \in E$. On the other hand, in a directed graph, also called a *digraph*, there is a sense of direction to the edges (Figure 2.4a,b). While A may be connected by an edge to B, there may be no edge connecting B back to A.

A classic example of directed networks is that of city networks, with its one-way roads, which will result in some pairs of junctions being connected only in one direction. In social networks, Facebook friendships are undirected, whereas Twitter *follows* are directed—you may follow a celebrity, but they may not follow you back. In the context of biology, protein interaction networks are typically undirected, while, gene regulatory networks are directed—a gene A may activate a gene B, but gene B may have no effect on the activation/repression of gene A. Can you think about the directionality of all the graphs mentioned in Table 2.1?

Note that edges are not counted twice in undirected graphs; there is just no notion of directionality in the edge. An edge from A to B is equivalent to an edge from B to A. An undirected graph with all possible edges will therefore have $\binom{|V|}{2} = \frac{|V|(|V|-1)}{2}$ edges, while a directed graph with all possible edges will have twice as many.

Connections, edges, paths, ...

A number of terms exist, to describe how nodes are 'connected' to one another in a graph. This box seeks to disambiguate and clarify the use of these terms. When we say that two nodes A and B are connected, it is a little vague—it is best to say that there is an edge (or in case of a directed graph, a directed edge) from A to B. One can also say that A and B are adjacent to each other. Simply saying A and B are connected could also mean that they are merely reachable, or connected by a path. In other words, the two nodes belong in same *connected component* (see §3.1.1).

2.3.3 Weighted vs. unweighted graphs

The edges between nodes in a graph may be associated with *weights*, which typically represent some sort of costs (Figure 2.4c,d). In an airline map, for example, the edges between airports may capture distances or flight fares. Google Maps often computes two different routes, optimising two different costs, such as time (fastest route) or distance (shortest route). In a majority of computer science applications, edge weights in graphs denote costs. In contrast, in most biological applications, edge weights denote the confidence in the interaction, in the case of protein interaction networks, or similarity, in the case of co–expression networks (see Appendix B).

2.3.4 Other graph types

Graphs can be classified in many more ways; we will here only visit a few of these, which are relevant to our study of networks in biology.

Sparse vs. Dense. Sparse graphs have relatively fewer edges per node on average, compared to dense graphs (Figure 2.4e,f). Density of a graph is defined as the fraction of possible edges that actually exist in the graph; see Equation 3.5. Many graphs are naturally sparse, owing to the maximum number of edges that emanate from a given node, *e.g.* rarely do we have more than four roads from junctions in road networks. Most biological networks are also quite sparse.

Cyclic vs. Acyclic. Graphs can be classified as cyclic or acyclic depending on the existence of one or more *cycles*. A cycle in a graph is a path from a node to itself.

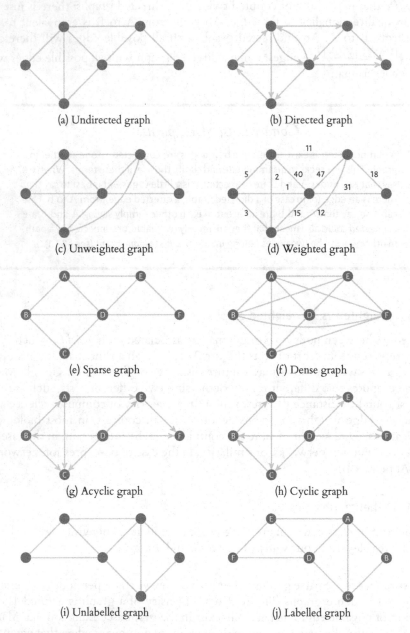

Figure 2.4 **Different types of graphs.** (a–b) Note the edge directions as shown by arrows, in the directed graph. (c–d) Edges are labelled by weights. (e–f) The sparse graph has far fewer edges. (g–h) Direction of edge BC is changed from (g) to (h), resulting in the introduction of a cycle ACBA. (i–j) Nodes may or may not be associated with labels.

In the graph of Figure 2.4h, there is a cycle involving nodes A, C, and B. On the other hand, there is no such cycle in Figure 2.4g. Trees in computer science are examples of undirected acyclic graphs. Directed Acyclic Graphs, also known as DAGs, naturally arise in many scenarios, such as in scheduling problems.

Labelled vs. Unlabelled. Nodes in a graph may or may not have a label. In a labelled graph, every vertex is assigned a unique label, distinguishing nodes from one another. Figure 2.4i is an unlabelled graph, while Figure 2.4j is labelled. Real-world graphs are naturally labelled, *e.g.* using protein names or metabolite names. In some applications, the labels need not be unique, like while representing molecules as graphs, with nodes being labelled by atoms. Labels are typically ignored while testing for graph isomorphism, to determine if two graphs are topologically identical if we ignore node labels.

Isomorphic vs. Non-isomorphic. Two graphs G_1 and G_2 are said to be *isomorphic*, if there exists a one-to-one function that maps every vertex in G_1 to every vertex in G_2, such that two vertices in G_1 are connected by an edge if and only if their corresponding (mapped) vertices in G_2 are connected by an edge[2]. For instance, the graphs in Figures 2.4a,i are isomorphic, as are those in Figures 2.4i,j. However, despite having the same number of edges and nodes, the graphs in Figures 2.4e,j are not isomorphic. Graph isomorphisms have interesting implications in chemoinformatics, for the study of metabolic networks, as we will see in §4.5.1.

Bipartite Graphs. A bipartite graph is a special type of graph, where there are two disjoint independent[3] sets of nodes, such that no two nodes within the same set have edges between them. Bipartite graphs arise naturally in scenarios where there are two types of objects (*e.g.* flights and airports, or actors and movies), or two sets of people (*e.g.* men and women). Film actor networks are bipartite, where the two sets of nodes represent actors and movies, and edges connect actors to the movies they starred in. Bipartite graphs are *two-colourable*, in that every node can be assigned a colour such that no two adjacent nodes in the entire graph have the same colour. It is also helpful to think of this in terms of shapes—a graph of squares and ellipses, where no ellipse is adjacent to another ellipse, and no square is connected to another square (as we will see in Figure 2.10). Bipartite graphs are also very useful to represent a wide range of biological networks, capturing diverse kinds of information, ranging from disease–gene networks or drug–target networks (§4.2), to metabolic networks (§2.5.5.3). It is possible to make projections of bipartite graphs, to compress the information. Naturally, two different projections are possible, for each node type (colour/shape)—the edges in the projection connect every pair of nodes that were separated by two edges in the bipartite graph. We will discuss examples in §4.4.2.

---

[2]Therefore, they must obviously have the same number of vertices and edges.
[3]A set of vertices such that no two vertices in the set are adjacent to one another.

Implicit Graphs. These are graphs that are not explicitly constructed and stored for any analysis, owing to their prohibitive size. Rather, they can be studied, and algorithms can be run on them, based on definitions of how nodes connect to one another. A simple example of an implicit graph would be a *hypercube graph*. A hypercube graph in d dimensions has 2^d nodes, with two nodes connected to one another if they vary *exactly* in one dimension. For d = 3, we have a cube, with eight nodes and 12 edges. For very large values of d, the graph will be too large to construct and store; yet, it is possible to compute shortest paths or run other algorithms efficiently on these graphs, owing to the knowledge of the rules that define connectivity between the nodes. Motivating use cases for implicit graphs are the large genotype spaces that we will study in §13.2.

Regular Graphs. These are graphs where every node has the same *degree, i.e.* number of neighbours, or adjacent nodes.

Complete Graphs. These are graphs where all possible edges exist, as in Figure 2.1b. All complete graphs are regular. A *clique* is a part of a larger graph, *i.e. sub-graph*, which is complete. In fact, a clique is an *induced sub-graph* that is complete. An induced sub-graph (or more precisely, a *vertex-induced* sub-graph) of a graph G is a subset of vertices of G, and every edge of G whose both end-points are in this subset. A sub-graph, on the other hand, is any subset of nodes and edges from the graph.

2.3.5 Hypergraphs

Graphs typically encode only *binary* relationships. Often, there may be n-way relationships among certain sets of entities. These can be better captured in hypergraphs, where the set of *hyper-edges* is not just a subset of $V \times V$, but rather 2^V, *i.e.* the power set of vertices (excluding, of course, the null set φ). In other words, any subset of vertices can be designated as a hyper-edge.

For instance, in a protein network, protein interactions can be represented by edges. An edge between proteins A and B indicates that A complexes with B. However, if a particular protein complex contains multiple sub-units, as in the case of haemoglobin, representing them in a simple graph is harder. G-protein coupled receptors, or GPCRs, interact with hetero-trimeric G proteins, comprising G_α, G_β, and G_γ sub-units[4]. G_β, and G_γ bind to one another, and this complex binds to G_α. It is not possible to accurately represent this information in a simple graph—at best, we would have edges between all pairs of proteins, G_α, G_β, and G_γ. On the other hand, the corresponding hypergraph will have $V = \{G_\alpha, G_\beta, G_\gamma\}$ and $E = \{\{G_\alpha, G_\beta, G_\gamma\}, \{G_\beta, G_\gamma\}\}$. This clearly informs that there is no binary interaction (complexation) between G_α and G_β or G_α and G_γ.

---

[4]Appendix A gives a gentle introduction to biology, including biomacromolecules such as proteins.

Metabolic or signalling networks may also be represented as hypergraphs. In fact, many different types of biological networks can be represented as hypergraphs, since they better capture n-way relations, vis-à-vis normal graphs, which typically capture only binary interactions/relationships [8, 9]. Outside of biology, hypergraphs have been used to represent networks in various fields [10], including co-authorship networks [11] or even railway networks [12].

2.4 COMPUTATIONAL REPRESENTATIONS OF GRAPHS

Now that we have a graph, we need to represent it mathematically, or computationally, in terms of a data structure. In this section, we will review some basic concepts of computational representations of graphs.

2.4.1 Data structures

A number of data structures, ranging from simple to efficient, are available for handling graphs.

Edge List. This is the simplest representation of a graph, as a simple list of all edges. Although inefficient for computations, it happens to be the easiest way to store graphs in a file and exchange them in a human–readable format. Figure 2.5c illustrates the edge list for the directed graph in Figure 2.5b. Formats such as PSI-MI TAB (see Appendix C) are edge lists at their core. An edge list obviously contains $|E|$ rows, two columns representing the interacting entities (node 'IDs') and a possible third column representing edge weights, if any. Additional columns may capture other attributes of the edges.

Adjacency List. The adjacency list maintains a list of all the nodes adjacent to a given node in the graph. More efficient representations of the adjacency list can be achieved by using adjacency maps. The adjacency list has $|V|$ rows, with each row having d_i entries (listed against node i), where d_i is the number of neighbours of the ith node. Figure 2.5d illustrates the adjacency list for the directed graph in Figure 2.5b.

2.4.2 Adjacency matrix

A very straightforward representation of graphs is the matrix representation, known as an *adjacency matrix*. A graph with $|V|$ nodes and $|E|$ edges can be represented by a $|V| \times |V|$ adjacency matrix containing $|E|$ non-zero entries. An entry a_{ij} in the matrix is unity (or the edge weight, if any) if there is an edge from i to j, and 0 otherwise. Notably, the adjacency matrix corresponding to an undirected graph is symmetric, as can also be observed from Figure 2.5a, while that for a directed graph need not be symmetric (Figure 2.5b). So, to save space in practice, adjacency matrix of undirected graphs can be stored as upper triangular

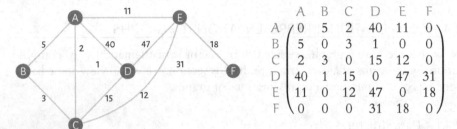

	A	B	C	D	E	F
A	0	5	2	40	11	0
B	5	0	3	1	0	0
C	2	3	0	15	12	0
D	40	1	15	0	47	31
E	11	0	12	47	0	18
F	0	0	0	31	18	0

(a) A weighted undirected graph and its corresponding adjacency matrix

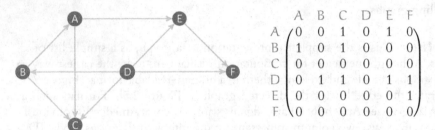

	A	B	C	D	E	F
A	0	0	1	0	1	0
B	1	0	1	0	0	0
C	0	0	0	0	0	0
D	0	1	1	0	1	1
E	0	0	0	0	0	1
F	0	0	0	0	0	0

(b) An unweighted directed graph and its corresponding adjacency matrix

Source	Target
A	C
A	E
B	A
B	C
D	B
D	C
D	E
D	F
E	F

Source	List of adjacent nodes
A	C, E
B	A, C
C	—
D	B, C, E, F
E	F
F	—

(c) Edge list data structure   (d) Adjacency list data structure

Figure 2.5 **Data structures for graph representation.**

matrices. The diagonal of the adjacency matrix is typically zero, unless there are 'self-loops', which are uncommon in most (simple) graphs.

The adjacency matrix also has interesting mathematical properties. A_{ij} in the adjacency matrix can be thought of counting the number of paths of length unity in the graph between nodes i and j (*i.e.* the existence of an edge). A_{ij}^2 measures the number of paths of length two between i and j, and so on!

Sparse matrices and their representation.   Most biological networks are very sparse, and consequently, their adjacency matrices have relatively fewer numbers of non-zero entries. A full matrix requires $\sim |V|^2$ space for storing all the entries. If the number of non-zero entries, $|E|$, is much smaller, one naive way to store the matrix would be to just store i, j and a_{ij}, for every entry in the matrix. This would require only $\sim 3|E|$ storage space, which will result in substantial savings when $|E| \ll |V|^2$. Sparse matrices are also briefly discussed in Appendix E. Sparse matrices are very important in the context of large networks, since the *full* matrices may not even fit in the computer memory, preventing one from working freely with such networks. A number of algorithms have been designed to work efficiently with sparse matrices.

Sparse Matrix Formats

There are more efficient ways of storing sparse matrices in computer memory, such as the *compressed sparse row* format or *compressed sparse column* format. A nice place to learn more about such formats is the manual of the SciPy library (https://docs.scipy.org/doc/scipy/reference/sparse.html).

2.4.3   The Laplacian matrix

The Laplacian matrix, L, of an undirected unweighted graph is defined as

$$L = D - A \tag{2.1}$$

where D is a diagonal *degree* matrix, such that d_{ii} is the number of neighbours of i, and A is the adjacency matrix. If we were to ignore the weights of the graph in Figure 2.5a, its Laplacian would be

$$L = \begin{bmatrix} 4 & 0 & 0 & 0 & 0 & 0 \\ 0 & 3 & 0 & 0 & 0 & 0 \\ 0 & 0 & 4 & 0 & 0 & 0 \\ 0 & 0 & 0 & 5 & 0 & 0 \\ 0 & 0 & 0 & 0 & 4 & 0 \\ 0 & 0 & 0 & 0 & 0 & 2 \end{bmatrix} - \begin{bmatrix} 0 & 1 & 1 & 1 & 1 & 0 \\ 1 & 0 & 1 & 1 & 0 & 0 \\ 1 & 1 & 0 & 1 & 1 & 0 \\ 1 & 1 & 1 & 0 & 1 & 1 \\ 1 & 0 & 1 & 1 & 0 & 1 \\ 0 & 0 & 0 & 1 & 1 & 0 \end{bmatrix} = \begin{bmatrix} 4 & -1 & -1 & -1 & -1 & 0 \\ -1 & 3 & -1 & -1 & 0 & 0 \\ -1 & -1 & 4 & -1 & -1 & 0 \\ -1 & -1 & -1 & 5 & -1 & -1 \\ -1 & 0 & -1 & -1 & 4 & -1 \\ 0 & 0 & 0 & -1 & -1 & 2 \end{bmatrix} \tag{2.2}$$

The Laplacian has many interesting properties and is very useful for graph clustering. An entire field of study, *spectral graph theory*, is dedicated to the study of eigenvalues and eigenvectors of the Laplacian (or even the adjacency) matrix [13].

File formats and standards

Beyond data structures, it is also important to learn of the various file formats and standards for the representations of networks and models. SBML (Systems Biology Markup Language) is the *de facto* standard for the representation of models in systems biology. It is a representation based on XML (eXtensible Markup Language), which supports a modular architecture for the representation of systems biology models of various types, as discussed in Appendix D. SBGN (Systems Biology Graphical Notation) is another important standard, specialised for representing biological models and networks (again, see Appendix D).

The Cytoscape manual lists a number of compatible network file formats that Cytoscape can read: http://manual.cytoscape.org/en/stable/Supported_Network_File_Formats.html. Salient formats include the SIF ('Simple Interaction File'), which basically resembles an edge list or an adjacency list (both formats are acceptable as SIF), and the GML (Graph Modelling Language), and its XML-ised version, XGMML. Other formats include PSI-MI (Proteomics Standards Initiative Molecular Interactions; http://www.psidev.info/mif; see Appendix C), GraphML, `Cytoscape.js` JSON (JavaScript Object Notation) and of course, SBML. Such a wide array of file formats can be intimidating; however, here is a simple principle for guiding the choice for your applications: the file format should be human-readable, machine-readable and a standard supported by various existing tools.

2.5   GRAPH REPRESENTATIONS OF BIOLOGICAL NETWORKS

Many biological networks are naturally represented as graphs, which also makes them amenable to several interesting analyses. In this section, we will briefly overview some classic biological networks and how they are typically represented. For those without a strong background biology, it would be helpful to go through Appendix A, to particularly understand the *flow of information* in biology via the key macromolecules, DNA, RNA, and proteins, ultimately leading to biological function. Appendix B outlines the reconstruction of the various networks discussed here in greater detail.

2.5.1   Networks of protein interactions and functional associations

Amongst the most widely studied biological networks are protein interaction networks, as well as networks of *functional associations*[5]. Their popularity is

---

[5]Functional associations connect proteins that both jointly contribute to a biological function—they may or may not physically interact [1].

exemplified by the existence of several databases (such as those listed in Appendix C), which catalogue these interactions and functional associations. In all these cases, proteins are the nodes of the network, with edges representing various types of associations, ranging from physical interactions/complexation to co-occurrence in PubMed abstracts as identified via text-mining algorithms, or co-expression in transcriptomes. These networks are generally undirected. While databases such as the STRING (see Figure 2.6) host functional association networks (physical interaction networks are essentially a subset of these), others such as the DIP (Database of Interacting Proteins) [14] host only experimentally determined protein–protein interactions. It is important to note that these are distinct types of networks, as elaborated in Appendix B.

Databases such as the STRING host experimentally and computationally derived networks of such protein interactions and associations (see Appendix C). In STRING, the edges can denote molecular action (such as inhibition, catalysis, or activation), evidence (separately marking edges arising from different axes such as co-expression, co-occurrence or experimental evidence), or confidence (a score that quantifies how likely an interaction is). Note that some of these networks are directed, such as those that capture molecular action (*e.g.* inhibition, as in Figure 2.6a), while others that capture evidence in terms of co-expression or co-occurrence are undirected (Figures 2.6b,c).

Try it out …

The STRING database [1] is accessible from https://string-db.org/. Punch in a simple query such as *glyA*, and you will notice the network you obtain is quite similar to the one shown in Figure 2.6. Try changing settings to show different types of edges, such as 'evidence', 'confidence' and 'molecular action'.

There are also other databases, such as the STITCH [15], which augment protein interaction data, with interactions of small molecules such as drugs and metabolites. In such networks, the nodes are easily defined, as proteins/small molecules; however, the edges can be defined in many ways, and can exist in many *layers*. An example network involving both proteins and small molecules is also illustrated, in Figure 2.6d.

2.5.2 Signalling networks

Signalling networks, as described in Appendix B, comprise different kinds of cellular components, from receptors, small molecules such as chemical messengers, to intracellular signalling proteins. They can also involve the genes encoding for transcription factors, kinases, etc. The formal reconstruction of signalling networks is discussed in Appendix B. Once such networks are reconstructed, they can be subject to various types of network analyses. The STITCH network

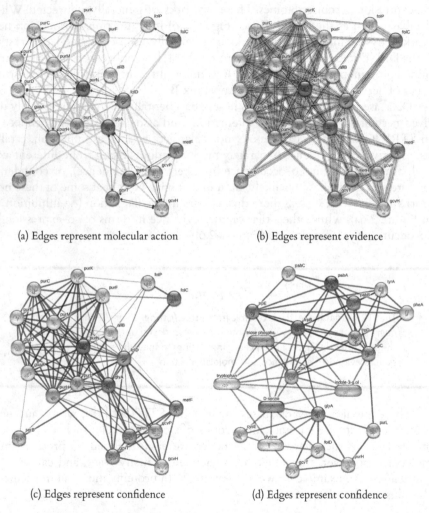

(a) Edges represent molecular action

(b) Edges represent evidence

(c) Edges represent confidence

(d) Edges represent confidence

Figure 2.6 **Illustration of protein networks from STRING/STITCH.** In panels (a–c) the nodes are the same, centred around the protein GlyA from *E. coli*. (d) includes small molecules such as amino acids, distinguished by elongated nodes. The edges carry different meanings as indicated.

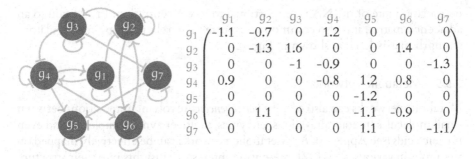

$$
\begin{array}{c}
\quad\; g_1 \quad\; g_2 \quad\; g_3 \quad\; g_4 \quad\; g_5 \quad\; g_6 \quad\; g_7 \\
\begin{array}{c} g_1 \\ g_2 \\ g_3 \\ g_4 \\ g_5 \\ g_6 \\ g_7 \end{array}
\left(
\begin{array}{ccccccc}
-1.1 & -0.7 & 0 & 1.2 & 0 & 0 & 0 \\
0 & -1.3 & 1.6 & 0 & 0 & 1.4 & 0 \\
0 & 0 & -1 & -0.9 & 0 & 0 & -1.3 \\
0.9 & 0 & 0 & -0.8 & 1.2 & 0.8 & 0 \\
0 & 0 & 0 & 0 & -1.2 & 0 & 0 \\
0 & 1.1 & 0 & 0 & -1.1 & -0.9 & 0 \\
0 & 0 & 0 & 0 & 1.1 & 0 & -1.1
\end{array}
\right)
\end{array}
$$

Figure 2.7 **A hypothetical gene regulatory network.** The normal arrowheads indicate activation, and the bar-headed arrows indicate inhibition/repression. Note that all nodes have (negative) self-loops. The corresponding matrix of regulatory strengths is also indicated.

illustrated in Figure 2.6d closely resembles a signalling network, with a number of proteins and small molecules. Signalling networks can be undirected, with a mere 'connectivity reconstruction', or directed, if a 'causal reconstruction' is performed.

2.5.3 Protein structure networks

At a very different resolution, it is also possible to represent a single protein as a network/graph. Many interesting studies have proposed alternate useful representations of protein structures as graphs, uncovering many interesting properties [16–18]. For instance, nodes can be the amino acid residues in the protein structure, and edges can connect nodes (residues) that have a pairwise interaction energy lower than a particular threshold. Alternatively, distance cut-offs can also be used. The database NAPS (Network Analysis of Protein Structures) [19] provides five different ways to construct protein networks out of structures from the Protein Data Bank (PDB). A nice review of the applications of network science in understanding protein structure, energy, and conformation is available elsewhere [20]. These applications are also discussed in §4.6 and §4.6.1.

2.5.4 Gene regulatory networks

The reconstruction of gene regulatory networks is an interesting computational challenge, as detailed in Appendix B. Gene regulatory networks are typically directed, illustrating directional interactions, such as inhibition, repression, and activation between various genes. Gene networks essentially map all interactions between genes and other genes (or gene products, *i.e.* proteins) on a directed graph. The edge weights can take both positive and negative real values, depending on activation or repression, respectively, as illustrated in Figure 2.7. Most of the genes will have a negative edge self-loop in these networks since the degradation rate of mRNA is proportional to its concentration. In addition to an adjacency matrix, it is also common to capture these relationships in a regulatory strength matrix [21], as shown in Figure 2.7.

2.5.5 Metabolic networks

Metabolic networks can also be very *heterogeneous*, involving interactions between small molecules (metabolites), ions, enzymes, multi-enzyme complexes, and even nucleic acids (see Appendix A). Metabolic networks can be effectively mapped on to graphs in various ways [22], presenting interesting insights into their structure and function [23, 24], as we discuss below.

2.5.5.1 Substrate graphs

In this representation, the nodes are metabolites, and the edges represent reactions. Every metabolite on the left-hand side of a reaction is connected to every metabolite on the right-hand side, via an edge. Consider the well-known pathway of glycolysis, which happens in ten steps (Table 2.3). The nodes are the metabolites, of which there are eighteen. The graph is directed, with many edges going both ways, owing to reversible reactions. The first reaction gives rise to four nodes and $2 \times 2 = 4$ edges, while the sixth reaction gives rise to $3 \times 3 = 9$ edges. Figure 2.8 illustrates substrate graphs for glycolysis. Figure 2.8a is visibly crowded, making it difficult to discern the key reactions taking place; more importantly, if one were to employ this graph for path-finding, it is possible to come up with erroneous conclusions such as a two-step glycolysis, from Glucose → ADP → Pyruvate! However, if one eliminates *current/currency* metabolites, such as co-factors and ATP/ADP, etc. [25], the graph of Figure 2.8b appears less cluttered and is more informative, even for path-finding. Despite this, the substrate graph representation fails to capture the fact that more than one metabolite may be necessary for a reaction to happen, necessitating the development of other representations such as hypergraphs [26] or bipartite graphs [27, 28].

2.5.5.2 Reaction or enzyme graphs

In this representation, every reaction is represented by a node; edges connect reaction R_i to R_j if any product of R_i is a reactant in R_j. R1 and R2 share G6P, while R3 and R10 share ATP. Ideally, we would want to connect only reactions that can successively take place; *spurious* connections are introduced by the presence of currency metabolites such as ATP or NADH. The reaction graph is seldom useful as is, but a variant of it, employing reaction rules instead of just reactions, has been used to predict pathways (as we will see in §4.5.1). Figure 2.9 illustrates the reaction graph for glycolysis, which comprises ten nodes corresponding to the ten reactions shown in Table 2.3. Reversible reactions are typically represented by a single node as shown in the figure; alternatively, they can also be represented by two separate nodes, for the forward and backwards reactions, respectively.

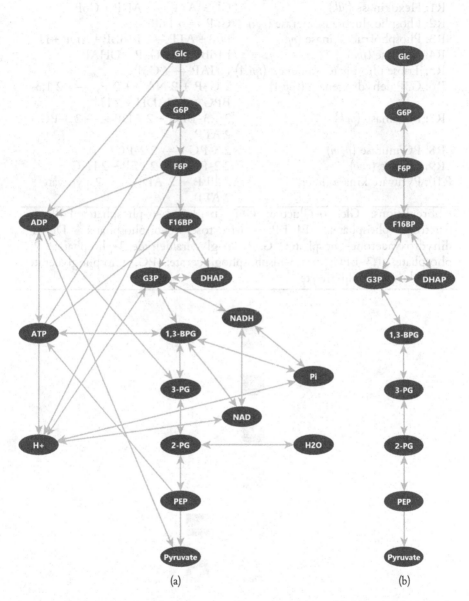

Figure 2.8 **Substrate graph illustration: glycolysis.** (a) all metabolites are shown. (b) currency metabolites removed.

TABLE 2.3   **Reactions of the glycolysis pathway.**

R1: Hexokinase (*glk*)	Glc + ATP $\longrightarrow$ ADP + G6P
R2: Phosphoglucose isomerase (*pgi*)	G6P $\longleftrightarrow$ F6P
R3: Phosphofructokinase (*pfk*)	F6P + ATP $\longrightarrow$ F16BP + ADP + H$^+$
R4: Aldolase (*fba*)	F16BP $\longleftrightarrow$ G3P + DHAP
R5: Triose phosphate isomerase (*tpiA*)	DHAP $\longleftrightarrow$ G3P
R6: G3P dehydrogenase (*gapA*)	2 G3P + 2 NAD + 2 P$_i$ $\longleftrightarrow$ 2 1,3-BPG + 2 NADH + 2 H$^+$
R7: PG kinase (*pgk*)	2 1,3-BPG + 2 ADP $\longleftrightarrow$ 2 3-PG + 2 ATP
R8: PG mutase (*pgm*)	2 3-PG $\longleftrightarrow$ 2 2-PG
R9: Enolase (*eno*)	2 2-PG $\longleftrightarrow$ 2 PEP + 2 H$_2$O
R10: Pyruvate kinase (*pyk*)	2 PEP + 2 ADP $\longrightarrow$ 2 pyruvate + 2 ATP

Abbreviations: Glc, D–Glucose; G6P, D–glucose-6-phosphate; F6P, D–fructose-6-phosphate; F16BP, D–fructose-1,6-bisphosphate; DHAP, dihydroxyacetone-phosphate; G3P, D–glyceraldehyde-3-phosphate; P$_i$, phosphate; 1,3-BPG, D–1,3-bisphosphoglycerate; PG, phosphoglycerate; PEP, phosphoenolpyruvate

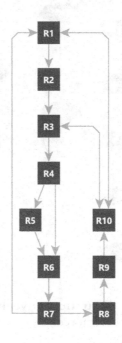

Figure 2.9 **Reaction graph illustration: glycolysis.** Every node represents one of the reactions from glycolysis, as listed in Table 2.3.

2.5.5.3 Bipartite graphs

As described earlier (§2.3.4), bipartite graphs contain two types of nodes, with no connections between nodes of the same type. In the context of metabolic networks, these can be metabolites and reactions. All (reactant) metabolites have outgoing edges to reaction nodes, and all reaction nodes have outgoing edges to (product) metabolite nodes. Reversible reactions must be indicated by two nodes, one each for the forward and backward reactions. For simplicity, these are omitted in Figure 2.10.

This kind of representation is highly informative compared to substrate and reaction graphs; it also clearly depicts that more than one reactant may be necessary for a reaction to occur. The bipartite graph representation has been exploited in many path-finding algorithms [27, 28]. Note that substrate and reaction graphs are nothing but projections of the bipartite graph. *Petri nets* are also directed bipartite graphs, and are useful to model many biological networks (see §7.3.4).

Alternatively, metabolic networks can also be represented as hypergraphs. Hyperedges in these hypergraphs can represent an entire reaction [26], or a set of reactants on the reactant/product side of the reaction [8]. Hypergraph representations are versatile and can be used for effective path-finding and other applications in biological networks. Overall, we have a large armoury of representations, to be employed depending upon the nature of the questions one wants to ask of these networks.

We have thus far seen several different networks in biology, ranging from protein interaction networks, to gene regulatory and signalling networks, metabolic networks, and even protein structure networks. There are many software tools available for visualisation and analyses of these networks, as detailed below in §2.7. For many purposes, visualisation of these networks is not important, or even informative, owing to the very high complexity and size of these graphs. Most times, the graphs are simply represented using simple circular nodes, with edges and arrowheads to indicate direction. However, particularly in the case of signalling networks or gene regulatory networks, it is often necessary to build a *process diagram* systematically, as specified in standards such as SBGN (Systems Biology Graphical Notation; see Appendix D).

2.6 COMMON CHALLENGES & TROUBLESHOOTING

2.6.1 Choosing a representation

A critical aspect of model building, in the context of graphs/networks, is the choice of representation itself. As discussed earlier (§1.4), the choice of representation is often driven by the *question* underlying the modelling exercise itself.

An instructive example is that of a protein network. Amino acid residues or C_α atoms can be picked as nodes for the network, and edges could be derived

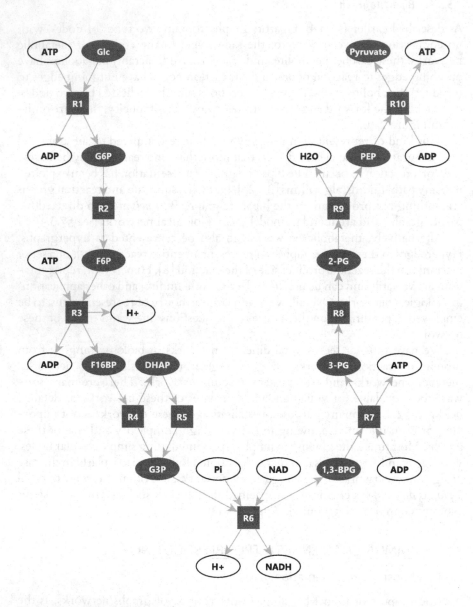

Figure 2.10 **Bipartite graph illustration: glycolysis.** Every reaction is represented by a square-shaped node, while elliptical nodes represent metabolites. For de-cluttering and pedagogy, the currency metabolite nodes have been duplicated (and indicated as plain nodes, without colour). In practice, all nodes with identical labels must be regarded as one.

from contact maps[6], or even interaction energies. Many questions arise: What is the best distance threshold to use? How do we weight the edges? Again, this will depend on whether we want to study the folding dynamics of the protein, or identify key residues that dictate protein function, or possible communication pathways within the protein structure (also see reference [19] and §4.6).

Another classic example is the study of protein–protein interaction/functional association networks, from databases such as the STRING. The STRING, for instance, harbours networks for thousands of organisms. Every network has proteins for nodes, but the edges can be derived in different ways (also see Figure 2.6). How do we pick a network for analysis? Do we use the STRING network as is, in which case, the edges have a range of 'confidence' values? Or, do we pick a suitable threshold, *e.g.* 700 (high confidence), or 400 (medium confidence) or lower (low confidence).

The choice of threshold has practical implications, in terms of the size and density of the resulting network, as well as the reliability of any predictions made. It is also possible to leave out certain evidence channels, such as text-mining or homology, or retain only experimentally observed interactions, depending on the application. A common strategy would be to start with a relatively sparse small network, resulting from a high threshold (*e.g.* 900), working our way towards lower thresholds, and consequently, larger, more complex networks. It is conceivable that as the network changes, many of the observations will change, but it is also possible that the key network parameters (as will be discussed in §3.1) tell a coherent story about the structure of the network and the relative importance of the different nodes.

2.6.2 Loading and creating graphs

Another key practical challenge is the creation of graphs themselves, from representations such as edge lists. Graphs may be represented in a variety of formats, beginning with simple edge lists, adjacency lists or markup languages such as GraphML. Many software tools already provide interfaces to import from and export to multiple such formats. However, it is important to be able to readily load large graphs for analysis into tools like Cytoscape or MATLAB, as discussed in Appendix D.

2.7 SOFTWARE TOOLS

Many tools have emerged over the last few years to enable network analysis, particularly for biological and social networks. A comprehensive analysis and comparison of these tools has been published elsewhere [29, 30].

---

[6]A simplified representation of protein structures in terms of a matrix that captures the distance between every pair of residues, binarised using a specific distance threshold. These distances could be computed between C_α–C_α atoms, or even C_β–C_β atoms, and binarised to discard all distances greater than, say, 6Å.

Cytoscape (http://cytoscape.org) is an open-source software tool (and platform) for the analysis and visualisation of networks [31]. Notably, a lot of *apps* are available for Cytoscape, which extend it for various domains ranging from social network analysis to network biology.

NetworkX (https://networkx.github.io) is an open-source Python package that enables the creation and manipulation of various kinds of networks [32]. It has several built-in algorithms for network analysis.

GraphViz [33] (https://graphviz.org) is an open-source graph visualisation software, with many layout algorithms.

yEd (https://www.yworks.com/products/yed) is a free desktop application for creating various kinds of diagrams, including graphs. Developed by yWorks GmbH, it notably supports various automatic layout algorithms.

Appendix D provides a brief overview of some of the more popular tools and software libraries for analysing large networks.

EXERCISES

2.1 Euler showed that there was no Eulerian circuit/path, in the Königsberg problem. However, this can change if new bridges are constructed, or existing bridges are lost. Can you identify a scenario, where two bridges are lost, and you can indeed have an Eulerian path in the graph?

2.2 As discussed in Appendix B, it is possible to reconstruct protein functional associations based on the genomic context of a set of proteins. Such methods are extensively employed in databases such as the STRING. The figure below represents a hypothetical scenario where six proteins A–F are present in one or more genomes as indicated. In such a scenario, which of the proteins A–F are likely to be functionally associated? On the basis of which approach (again, see Appendix B), do you predict these associations?

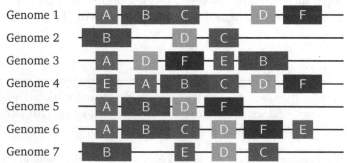

2.3 If you write down the adjacency matrix of a bipartite graph, how does it look like? Do you see some structure? (*Hint: Group the nodes by type, when you build the matrix.*)

2.4 Is a hypercube graph bipartite? Is it regular?

2.5 Pick one of the graph algorithms from Table 2.2 (or elsewhere) and search the literature for its possible applications in biology. Write a short essay on the same, also emphasising what kinds of networks the algorithm finds applications in.

[LAB EXERCISES]

2.6 Download the protein–protein functional association network for *Mycoplasma genitalium* from the STRING database. Filter out the edges to retain only those with a confidence score of 900 or more. Load the edge list into Cytoscape (see Appendix D) and visualise.

2.7 Write a piece of code in MATLAB to read an edge list, such as the file you obtained above, from STRING, and convert it to a sparse adjacency matrix. Read the above network into MATLAB using this piece of code, and visualise the matrix using spy. Would you classify the network as dense or sparse? Find out the amount of memory utilised by MATLAB for storing this matrix. How does this memory requirement reduce if you convert it to a sparse matrix?

2.8 Write a piece of code to convert an adjacency matrix to an adjacency list. As discussed in §2.6.2, it is important to be able to swiftly switch between formats. Can you think about the time and space complexity of these different data structures to store matrices? For example, is it faster to insert an edge into an adjacency matrix or adjacency list? In which data structure is it easier to insert or remove vertices? While working with very large networks, such considerations can become very important. Consult the textbook by Goodrich *et al* [34] for a discussion on the space–time complexity of various graph data structures.

2.9 Create the network corresponding to C_α atoms in the structure of the enzyme NDM-1 (New Delhi metallo-β-lactamase), which is used by bacteria to inactivate a wide range of penicillin-like antibiotics (PDB ID: 4ey1), using the NAPS portal (§2.5.3). How many edges are there in the network? Can you see how the different networks generated based on interaction strength, or C_β atoms differ from one another in their density?

2.10 Obtain the metabolic network of *E. coli* from the ModelSEED database (see Appendix C). Construct a substrate graph. Visualise in Cytoscape, sizing nodes by the number of connections, using VizMapper. What are the top 10 most connected nodes in the graph?

2.11 Write a piece of code to take as input a bipartite graph and node types, and output the two possible projections. The output can be a weighted or unweighted graph. What edge weights would you use in case of a weighted graph?

REFERENCES

[1] D. Szklarczyk, A. L. Gable, D. Lyon, et al., "STRING v11: protein–protein association networks with increased coverage, supporting functional discovery in genome-wide experimental datasets", *Nucleic Acids Research* **47**:D607–D613 (2019).

[2] B. Hopkins and R. J. Wilson, "The truth about Königsberg", *The College Mathematics Journal* **35**:198–207 (2004).

[3] L. Euler, "Solutio problematis ad geometriam situs pertinentis", *Commentarii Academiae Scientiarum Imperialis Petropolitanae* **8**:128–140 (1736).

[4] P. A. Pevzner, H. Tang, and M. S. Waterman, "An Eulerian path approach to DNA fragment assembly", *Proceedings of the National Academy of Sciences USA* **98**:9748–9753 (2001).

[5] L. Freeman, *The development of social network analysis: a study in the sociology of science* (Empirical Press, 2004).

[6] N. Deo, *Graph theory: with applications to engineering and computer science* (Prentice-Hall of India ; Prentice-Hall International, 1994).

[7] S. Even, *Graph algorithms, 2nd ed.* (Cambridge University Press, 2011).

[8] S. Klamt, U.-U. Haus, and F. Theis, "Hypergraphs and cellular networks", *PLoS Computational Biology* **5**:e1000385+ (2009).

[9] A. Ritz, A. N. Tegge, H. Kim, C. L. Poirel, and T. M. Murali, "Signaling hypergraphs", *Trends in Biotechnology* **32**:356–362 (2014).

[10] A. Bretto, "Applications of hypergraph theory: a brief overview", in *Hypergraph theory: an introduction*, Mathematical Engineering (Heidelberg, 2013), pp. 111–116.

[11] R. I. Lung, N. Gaskó, and M. A. Suciu, "A hypergraph model for representing scientific output", *Scientometrics* **117**:1361–1379 (2018).

[12] S. N. Satchidanand, S. K. Jain, A. Maurya, and B. Ravindran, "Studying Indian railways network using hypergraphs", in 2014 Sixth International Conference on Communication Systems and Networks (COMSNETS) (2014), pp. 1–6.

[13] B. Nica, *A brief introduction to spectral graph theory* (European Mathematical Society, Zürich, 2018).

[14] I. Xenarios, L. Salwinski, X. J. J. Duan, P. Higney, S.-M. M. Kim, and D. Eisenberg, "DIP, the database of interacting proteins: a research tool for studying cellular networks of protein interactions", *Nucleic Acids Research* **30**:303–305 (2002).

[15] D. Szklarczyk, A. Santos, C. von Mering, L. J. Jensen, P. Bork, and M. Kuhn, "STITCH 5: augmenting protein-chemical interaction networks with tissue and affinity data", *Nucleic Acids Research* **44**:380–384 (2016).

[16] K. V. Brinda and S. Vishveshwara, "A network representation of protein structures: implications for protein stability", *Biophysical Journal* **89**:4159–4170 (2005).

[17] M. Vendruscolo, N. V. Dokholyan, E. Paci, and M. Karplus, "Small-world view of the amino acids that play a key role in protein folding", *Physical Review E* **65**:061910–061910 (2002).

[18] L. H. Greene and V. A. Higman, "Uncovering network systems within protein structures", *Journal of Molecular Biology* **334**:781–791 (2003).

[19] B. Chakrabarty and N. Parekh, "NAPS: network analysis of protein structures", *Nucleic Acids Research* **44**:375–382 (2016).

[20] C. Böde, I. A. Kovács, M. S. Szalay, R. Palotai, T. Korcsmáros, and P. Csermely, "Network analysis of protein dynamics", *FEBS Letters* **581**:2776–2782 (2007).

[21] A. de la Fuente, P. Brazhnik, and P. Mendes, "Linking the genes: inferring quantitative gene networks from microarray data", *Trends in Genetics* **18**:395–398 (2002).

[22] J. van Helden, L. Wernisch, D. Gilbert, and S. J. Wodak, "Graph-based analysis of metabolic networks", in *Bioinformatics and Genome Analysis*, Vol. 38, Ernst Schering Research Foundation Workshop (2002), pp. 245–274.

[23] H. Jeong, B. Tombor, R. Albert, Z. N. Oltvai, and A.-L. Barabási, "The large-scale organization of metabolic networks", *Nature* **407**:651–654 (2000).

[24] A. Wagner and D. A. Fell, "The small world inside large metabolic networks", *Proceedings of the Royal Society B: Biological Sciences* **268**:1803–1810 (2001).

[25] H. Ma and A. P. Zeng, "Reconstruction of metabolic networks from genome data and analysis of their global structure for various organisms", *Bioinformatics* **19**:270–277 (2003).

[26] A. Mithani, G. M. Preston, and J. Hein, "Rahnuma: hypergraph-based tool for metabolic pathway prediction and network comparison", *Bioinformatics* **25**:1831–1832 (2009).

[27] D. Croes, F. Couche, S. J. Wodak, and J. van Helden, "Inferring meaningful pathways in weighted metabolic networks", *Journal of Molecular Biology* **356**:222–236 (2006).

[28] A. Ravikrishnan, M. Nasre, and K. Raman, "Enumerating all possible biosynthetic pathways in metabolic networks", *Scientific Reports* **8**:9932 (2018).

[29] M. Suderman and M. Hallett, "Tools for visually exploring biological networks", *Bioinformatics* **23**:2651–2659 (2007).

[30]   G. Agapito, P. H. H. Guzzi, and M. Cannataro, "Visualization of protein interaction networks: problems and solutions", *BMC Bioinformatics* **14 Suppl 1**:S1 (2013).

[31]   P. Shannon, "Cytoscape: a software environment for integrated models of biomolecular interaction networks", *Genome Research* **13**:2498–2504 (2003).

[32]   A. A. Hagberg, D. A. Schult, and P. J. Swart, "Exploring network structure, dynamics, and function using NetworkX", in Proceedings of the 7th Python in Science Conference (SciPy2008) (2008), pp. 11–15.

[33]   J. Ellson, E. R. Gansner, E. Koutsofios, S. C. North, and G. Woodhull, "Graphviz and Dynagraph — static and dynamic graph drawing tools", in *Graph Drawing Software*, Mathematics and Visualization (Berlin, Heidelberg, 2004), pp. 127–148.

[34]   M. T. Goodrich, R. Tamassia, and M. H. Goldwasser, *Data structures and algorithms in Python*, 1st edition (Wiley, Hoboken, NJ, 2013).

FURTHER READING

B. Bollobás, *Modern graph theory*, 1st ed., Graduate Texts in Mathematics 184 (Springer-Verlag New York, 1998).

A.-L. Barabási, *Network Science*, Illustrated Edition (Cambridge University Press, Cambridge, United Kingdom, 2016).

L. H. Greene, "Protein structure networks", *Briefings in Functional Genomics* **11**:469–478 (2012).

B. A. Shoemaker and A. R. Panchenko, "Deciphering protein–protein interactions. Part II. Computational methods to predict protein and domain interaction partners", *PLoS Computational Biology* **3**:e43+ (2007).

Structure of networks

CONTENTS

Now that we have familiarised ourselves with the basic concepts of graph theory, as well as how various kinds of biological networks can be modelled as graphs, we turn our attention to how these networks can be studied, characterised, classified, and further analysed. In particular, we will explore an array of network parameters that provide more insights into a network's structure and organisation. Following that, we will survey some key canonical network models. We will then study how to identify clusters, or communities, in networks, followed by a discussion on motifs. Lastly, we will study perturbations to various networks.

3.1   NETWORK PARAMETERS

A number of parameters can be computed for networks, their constituent nodes and edges, each shedding a different insight into network structure. These parameters can provide answers to various questions about a network: What is the average number of neighbours to any given node? What is the average distance between two nodes on the network? Are all nodes reachable from one another? Which is the most influential node in a given network? And so on. Recall that we asked some of these questions earlier, while comparing social networks with biological networks (§2.2). It is common to represent the number of nodes in a network by n (the same as $|V|$), and the number of edges in a network by m ($= |E|$). Figure 3.1 shows two example networks, which we will use to demonstrate the various parameters described below.

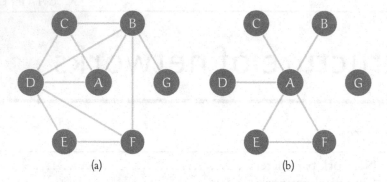

Figure 3.1 **Example networks, for computing various parameters.**

3.1.1 Fundamental parameters

Degree. The most elementary property of a node, *degree* is the number of neighbours a node has in a given network. For directed networks, the degree k is made up of two numbers, the *in-degree* k^{in} and the *out-degree* k^{out}, which count the number of incoming and outgoing edges, respectively. Recall the definition of the adjacency matrix (§2.4.2), where every entry a_{ij} in the matrix encodes the presence or absence of an edge between nodes i and j. For an unweighted, undirected graph, described by a (binary) adjacency matrix A,

$$k_i = \sum_{j=1}^{n} a_{ij} \tag{3.1}$$

For a directed graph,

$$k_i^{in} = \sum_{j=1}^{n} a_{ij} \tag{3.2}$$

$$k_j^{out} = \sum_{i=1}^{n} a_{ij} \tag{3.3}$$

The degree of A in Figure 3.1a is 3; the nodes B through G have degrees of 5, 3, 5, 2, 3, and 1, respectively.

Average degree. The average degree, or mean degree of a node in an undirected graph (also denoted as c), is given as

$$\langle k \rangle = \frac{1}{n} \sum_{i=1}^{n} k_i = \frac{2m}{n} \tag{3.4}$$

Can you see why $\sum_{i=1}^{n} k_i = 2m$? The average degree of the network in Figure 3.1a, $\langle k \rangle = 22/7 \approx 3.14$.

Degree distribution. As the name suggests, this is the *distribution* of the node degrees in a graph. It is usually characterised as $N(k)$ or its normalised version, $P(k)$, which respectively measure the number and probability of nodes in the network with degree exactly equal to k. The degree distribution is particularly useful in classifying network topologies, as we will see in §3.2.

Density. The density of a network is defined as the fraction of all possible edges the network contains. For an undirected network,

$$\rho = \frac{2|E|}{|V|(|V| - 1)} \tag{3.5}$$

For the network in Figure 3.1a, the density is $\frac{2 \times 11}{7 \times 6} \approx 0.52$. For the network in Figure 3.1b, the density is $\frac{2 \times 6}{7 \times 6} \approx 0.29$.

Clustering coefficient. Defined by Watts and Strogatz in 1998 [1], C can be thought to measure the *cliquishness*[1] of a node's neighbourhood, *i.e.* what fraction of (possible) edges in a node's neighbourhood do actually exist? Mathematically,

$$C_i = \frac{\textit{number of edges between the neighbours of } i}{\textit{total number of possible edges between the neighbours of } i} \tag{3.6}$$

$$= \frac{\textit{number of triangles connected to } i}{\textit{number of triples centred on } i} = \frac{\Delta(i)}{\tau(i)} \tag{3.7}$$

If the denominator of the expression is zero, so is the numerator (this happens for nodes with degree zero or one), and C_i is defined to be 0. Note that if a node has k, neighbours, the total possible edges between those neighbours is $\binom{k}{2}$.

The mean clustering coefficient of a network can be defined as

$$\langle C \rangle = \frac{1}{n} \sum_{i=1}^{n} C_i \tag{3.8}$$

For the network in Figure 3.1a, the clustering coefficients, for nodes A through G are

$$C = \left[\frac{3}{\binom{3}{2}}, \quad \frac{4}{\binom{5}{2}}, \quad \frac{3}{\binom{3}{2}}, \quad \frac{5}{\binom{5}{2}}, \quad \frac{1}{\binom{2}{2}}, \quad \frac{2}{\binom{3}{2}}, \quad 0 \right]$$

$$\approx \quad [1.0, \qquad 0.4, \qquad 1.0, \qquad 0.5, \qquad 1.0, \qquad 0.67, \qquad 0]$$

$$\langle C \rangle \approx \quad 4.56/7 \approx \quad 0.65$$

---

[1]A clique is a subset of vertices in an undirected graph, where every pair of vertices in the subset are adjacent to one another.

For the network in Figure 3.1b, the clustering coefficient of A is $\frac{1}{\binom{5}{2}} = 0.1$, since most of its neighbours do not have edges between them. Nodes E and F both have only two neighbours each, connected by a single edge, resulting in $C = 1$. Nodes B, C, D, and G have $C = 0$.

Another measure, known as the *"fraction of transitive triples"*, attributed to Barrat and Weigt by Newman [2] is defined as

$$\mathcal{C} = \frac{3 \times \text{number of triangles in the network}}{\text{number of 'connected triples' in the network}} = \frac{3n(\vee)}{n(\triangledown)} \quad (3.9)$$

$$= \frac{6 \times \text{number of triangles in the network}}{\text{number of paths of length two in the network}} \quad (3.10)$$

Here, 'connected triples' refer to a single node with edges running to an unordered pair of other nodes. $\mathcal{C}$ computes the fraction of triples that have their third edge filled in completing the triangle. The factor of three in the numerator can be understood in terms of each triangle contributing to three triples; this also ensures that $0 \leqslant C \leqslant 1$. In other words, all these definitions try to measure (in slightly different ways) the probability of your "friend's friend also being your friend", *i.e.* a triangle. For the network in Figure 3.1b, there are twelve triples, viz. {A(BC), A(BD), A(BE), A(BF), A(CD), A(CE), A(CF), A(DE), A(DF), A(EF), F(AE), E(FA)}, with only a single triangle (AEF), resulting in $\mathcal{C} = \frac{3 \times 1}{12} = 0.25$. Note that the value of $\langle C \rangle$ is distinct, and equal to $(0.1 + 0 + 0 + 0 + 1 + 1 + 0)/7 = 0.3$.

Geodesic path. Geodesic, or shortest path represents the path with the fewest number of edges between a given pair of vertices (in the case of an unweighted network), denoted d_{ij}. There is no shorter path between i and j with length $< d_{ij}$. In 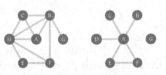 weighted networks, the shortest path will be the one with least total edge weight (edges invariably represent costs, as discussed in §2.3). Of course, it is possible to have multiple shortest paths (of the same length/distance) between a given pair of nodes. An extension of this concept is that of k-shortest paths, essentially the first (top) k shortest paths between a pair of nodes in a network. The matrices of shortest paths, showing d_{ij} for every pair of nodes, for the networks in Figure 3.1 can be computed as:

	A	B	C	D	E	F	G			A	B	C	D	E	F	G
	0	1	1	1	2	2	2		A	0	1	1	1	1	1	∞
	1	0	1	1	2	1	1		B	1	0	2	2	2	2	∞
	1	1	0	1	2	2	2		C	1	2	0	2	2	2	∞
	1	1	1	0	1	1	2		D	1	2	2	0	2	2	∞
	2	2	2	1	0	1	3		E	1	2	2	2	0	1	∞
	2	1	2	1	1	0	2		F	1	2	2	2	1	0	∞
	2	1	2	2	3	2	0		G	∞	∞	∞	∞	∞	∞	0

Further, for the network in Figure 3.1a, the shortest paths (excluding the edges) can be enumerated as: ADE, ABF/ADF, ABG, BDE/BFE, CDE, CBF/CDF, CBG, DBG, EDBG/EFBG, FBG. The number of pairs of nodes connected by shortest paths in a connected network is $\binom{n}{2}$, including the m edges.

Erdős number

A fun application of network theory to scientific collaboration networks is in terms of the popular Erdős number, which measures the distance between two mathematicians on a math co-authorship network. Erdős is defined to have an Erdős number of zero. His immediate co-authors have an Erdős number of one and their co-authors have an Erdős number of two, and so on. The famous Indian mathematician, Srinivasa Ramanujan, has an Erdős number of three, while Albert Einstein has an Erdős number of two. You can explore more about Erdős numbers at *The Erdős Number Project* of Oakland University at https://oakland.edu/enp/. We will encounter Erdős soon, when we study the theory of Erdős–Rényi random networks (§3.2.1).

A more frivolous application of the same concept is the 'Bacon number', measuring the distance from movie star Kevin Bacon, on a film actor network. You can read more about it at https://oracleofbacon.org/.

Connected component. This represents sets of nodes, where every pair of nodes have a path (not an edge) between them. A graph may have more than one connected component. In the case of directed graphs, there is the equivalent notion of a *strongly connected component*, where every pair of nodes are connected to each other in both directions. That is, for every pair of vertices {A, B} in a strongly connected component, there is a path from A to B, as well as B to A. The graph is *weakly connected* if the underlying undirected graph is connected. Connected components are *maximal*, *i.e.* they denote the *maximal connected sub-graph* of a graph or the *maximal strongly connected sub-graph* of a graph.

The graph in Figure 3.1a has a single connected component, while the one in Figure 3.1b has two connected components, one of which is just the *orphan* node, G. In the social network space, LinkedIn[2] does inform you if someone is "outside of your network"—it means that the two of you do not belong to the same connected component.

The largest connected component in a network is at times (incorrectly) referred to as the "giant component". Newman [3] alerts that this usage is sloppy, and distinguishes the use of "largest component" vs. "giant component" carefully: technically, only a network component whose size grows in proportion to the number of nodes in the network is called a giant component. Ma and Zeng [4]

---

[2]An undirected network.

have studied the connectivity structure and the size of the giant strong component in metabolic networks, although they seem to use the definition loosely.

Diameter. This is the length of the longest geodesic (longest shortest path!) in the graph. For a graph with more than one component, it refers to the largest of the diameters of the individual components [5]. Diameter measures the maximum separation between two nodes in a network (that do have a path between them).

$$d = \max_{i,j} d_{ij} \tag{3.11}$$

The networks in Figure 3.1a,b have diameters of 3 ($= d_{EG}$) and 2, respectively.

Characteristic path length. This measures the average separation between two nodes in a network. Again, this measure is computable only for a connected component:

$$\ell = \frac{1}{\binom{n}{2}} \sum_{i<j} d_{ij} \tag{3.12}$$

A low ℓ would indicate that a network is tightly connected; it is easy to reach nodes from one another through a few edges. For the network in Figure 3.1a, $\ell = \frac{32}{21}$. For the network in Figure 3.1b, we consider the largest connected component, with ℓ being $\frac{24}{15}$. This measure is more reflective of network navigability and behaviour, compared to diameter. Diameter captures only the distance between a single pair of vertices—that are farthest from one another in the graph.

3.1.2   Measures of centrality

The parameters discussed above quantify different characteristics of networks as well as nodes in these networks. Beyond these fundamental parameters, a large set of parameters have been developed to quantify the *centrality*, or (relative) importance, of a node in a network. There are obviously many ways in which a node's importance can be assessed, and these give rise to various centrality measures.

Degree centrality. Again, the simplest centrality measure is degree itself. In social networks, nodes with higher degrees (more friends) are likely to be more influential. Similar arguments can also be made for protein networks.

Closeness centrality. Closeness centrality [6] tries to quantify the centrality of a node, based on its proximity to all other nodes in the network. The mean geodesic distance from a given node to all other nodes in a network is given as

$$L_i = \frac{1}{n-1} \sum_{j(\neq i)} d_{ij} \tag{3.13}$$

The reciprocal of this measure is known as closeness centrality:

$$C_c(i) = \frac{1}{L_i} = \frac{n-1}{\sum_{j(\neq i)} d_{ij}} \tag{3.14}$$

Of course, this measure can be defined only for the nodes connected to a given node, *i.e.* nodes in a given component; otherwise, the ∞'s will clobber the centrality values.

Newman's closeness centrality. An alternative suggestion by Newman [3] is to re-define closeness as the harmonic mean of the distances to the remaining $n - 1$ nodes. Such a measure is not affected by the ∞'s, and naturally gives more weightage to nodes that are closer to i than to those farther away.

$$C_{NC}(i) = \frac{1}{n-1} \sum_{j(\neq i)} \frac{1}{d_{ij}} \tag{3.15}$$

Eigenvector centrality. This measure can be thought of as an extension of degree centrality. Instead of a uniform contribution of 1 from every neighbour in the case of degree centrality, it possibly makes more sense to value contributions from more influential neighbours (*i.e.* neighbours with higher centralities) highly, and vice-versa. Bonacich (1987) [7, 8] defined it as

$$e_i = \alpha \sum_j a_{ij} e_j \tag{3.16}$$

where e_i denotes the eigenvector centrality of node i, α is some constant and a_{ij} denotes the elements of the adjacency matrix. If we replaced α with $1/\lambda$, and **e** with **x**, we end up with the familiar matrix equation

$$\mathbf{Ax} = \lambda\mathbf{x} \tag{3.17}$$

clearly justifying the name of eigenvector centrality. Here, λ is the largest eigenvalue of the adjacency matrix. Other popular centrality measures derived from eigenvector centrality include Katz centrality [9] and Google PageRank [10]. Named after Google co-founder Lawrence Page, PageRank ranks websites in Google's search results, based on the number and quality of links to a given page.

Betweenness centrality. Originally defined by Anthonisse (1971) [11], and also attributed to Freeman (1977) [12], betweenness centrality captures a very different aspect of node centrality, in terms of how often a node is situated on paths *between* other nodes. That is, these nodes are more important in communication in the network. Formally,

$$C_B(k) = \sum_{i \neq k \neq j \in V} \frac{\sigma_{ij}(k)}{\sigma_{ij}} \tag{3.18}$$

where σ_{ij} is the number of shortest paths from i to j, and $\sigma_{ij}(k)$ is the number of shortest paths from i to j that pass through the node k.

To illustrate, the betweenness centrality of node B in Figure 3.1a can be computed by first listing out all the shortest paths in the network. Recall the shortest paths enumerated on page 61: ADE, ABF/ADF, ABG, BDE/BFE, CDE, CBF/CDF, CBG, DBG, EDBG/EFBG, FBG. The number of pairs of nodes connected by shortest paths (excluding edges) in a connected network is $\binom{n}{2} - m$, in this case 10.

$C_B(B)$ is obtained by summing the contribution of B to each of these 10 sets of shortest paths, eliminating paths that begin or end in B ($i \neq k \neq j \in V$ in Equation 3.18).

$$C_B(B) = \frac{\sigma_{AE}(B)}{\sigma_{AE}} + \frac{\sigma_{AF}(B)}{\sigma_{AF}} + \frac{\sigma_{AG}(B)}{\sigma_{AG}} + \frac{\sigma_{CE}(B)}{\sigma_{CE}} + \frac{\sigma_{CF}(B)}{\sigma_{CF}}$$
$$+ \frac{\sigma_{CG}(B)}{\sigma_{CG}} + \frac{\sigma_{DG}(B)}{\sigma_{DG}} + \frac{\sigma_{EG}(B)}{\sigma_{EG}} + \frac{\sigma_{FG}(B)}{\sigma_{FG}}$$
$$= \frac{0}{1} + \frac{1}{2} + \frac{1}{1} + \frac{0}{1} + \frac{1}{2} + \frac{1}{1} + \frac{1}{1} + \frac{2}{2} + \frac{1}{1} = 6$$

Node A does not occur in the above list of paths at all, except as a terminus. Therefore, $C_B(A) = 0$. Likewise, for nodes C, E, and G. Can you see that $C_B(D) = 1 + 0.5 + 0.5 + 1 + 0.5 + 0.5 = 4$ and $C_B(F) = 0.5 + 0.5 = 1$? Figure 3.2 illustrates simple toy networks, where the nodes vary in their importance, based on different centrality measures.

A similar definition to the above, for "edge betweenness" was given by Girvan and Newman [13]. Centrality definitions have also been extended for arbitrary pairs of nodes or groups of nodes [14, 15].

Random walk centralities. A *random walk* on a graph involves taking a series of random steps, visiting one node after another, traversing the corresponding connecting edges. Starting at a given node in a graph, a random walker will explore all outgoing edges with equal probability. Some interesting centrality measures have been defined based on random walks, such as "random walk betweenness centrality" [16] or "random walk closeness centrality". As a random walker explores a graph, the (stationary) probability of visiting different nodes is proportional to their accessibility (or importance). For instance, a highly connected node is much more likely to be visited than a node with few edges.

A potent tool to study the local structure of graphs is the concept of random walks with restart (RWR) [17]. RWR builds on the idea of a random walker on a graph, but resets the walk at any step with a fixed probability α, so that the exploration is reset. $\alpha = 0.25$ would mean that the exploration is reset once in four steps, on average. Similar concepts have also been used in the classic paper on PageRank [10], where a random web surfer keeps clicking links, and resets to a random page. In the case of RWR, the restart takes the random walker back to the original starting node. The parameter α affects the extent of local exploration.

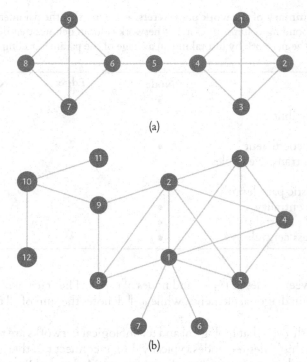

(a)

(b)

Figure 3.2 **Toy networks to illustrate centralities.** In (a), nodes 4 and 6 have the highest degrees (4), but node 5 has the highest betweenness centrality. In (b), node 1 evidently has the highest degree, but the other centralities are not that obvious: node 9 has the highest betweenness centrality, while node 2 has the highest closeness centrality.

A more elaborate treatment of these and other centrality measures is available in the excellent textbook by Newman [3].

3.1.3   Mixing patterns: Assortativity

Assortativity, or assortative mixing is the preference for a network's nodes to connect to others that are similar in properties (*e.g.* degree, protein essentiality). Newman [18] defined a measure to quantify the discrete assortative mixing of networks as follows:

$$r = \frac{\text{Tr } \mathbf{e} - \|\mathbf{e}^2\|}{1 - \|\mathbf{e}^2\|} \tag{3.19}$$

where $\mathbf{e}$ is a $n \times n$ matrix indicating the fraction of edges between nodes of n different types (*e.g.* boys and girls, essential and non-essential proteins, enzymes of six different classes[3]). e_{ij} in the matrix represents the fraction of edges in the

---

[3]See Exercise 3.12.

TABLE 3.1 **Summary of network parameters.** • denotes that the parameter is defined for the corresponding column. A $\langle\rangle$ in the network column denotes that the parameter is defined for the network by just taking an average of the parameter computed for nodes.

Parameter	Node	Edge	Network
Degree	•		$\langle\rangle$
Degree distribution			•
Density			•
Clustering coefficient	•		$\langle\rangle$
Fraction of transitive triples			•
Diameter			•
Characteristic path length			•
Closeness centrality	•		
Eigenvector centrality	•		
Betweenness centrality	•	•	

network between nodes of type i and nodes of type j. The trace of a matrix is the sum of the main diagonal elements, while $\| \cdot \|$ denotes the sum of all the elements in a matrix.

Newman showed that biological and technological networks are typically disassortative, *i.e.* high-degree nodes typically do not connect to other high-degree nodes; they tend to connect to low-degree nodes. Maslov and Sneppen [19] had already illustrated this tendency in both protein interaction networks and regulatory networks of yeast. On the other hand, social networks typically show assortative mixing, with high-degree nodes more often connecting to high-degree nodes. Protein structure networks, particularly the sub-networks comprising hydrophobic amino acids, have also been shown to be assortative, in that hubs preferentially associate with other hubs, and that too, in a hierarchical fashion [20].

3.2   CANONICAL NETWORK MODELS

Given a network, we would like to understand its various characteristics, such as the network parameters outlined in the above sections. Beyond that, it is natural to classify a network or bin our observations on a real network into certain canonical networks. We also want to create models that mimic real networks, but based on some theory/assumption. For example, if we suppose that people connect to one another in X fashion, and build a network that happens to match a real social network in many properties, our assumption X of how people connect to one another likely signifies a correct underlying principle of how people connect to one another in the real world.

As we have discussed previously, a lofty goal of modelling is to actually glean some *design principles*. In the following sections, we will discuss some important network models that have been developed to understand real-world networks. We begin with a model of random networks, which of course does not correspond to real-world networks, but will help develop our intuitions on random networks, before moving on to more realistic models of networks.

3.2.1 Erdős–Rényi (ER) network model

The ER network model for random networks was proposed by Erdős and Rényi in 1959 [21]. A random network should contain nodes that are not very different from one another—*i.e.* there should be no special nodes, and all connections should be placed *randomly*. How do we create such a random network, with n nodes and m edges? The network has $\binom{n}{2}$ possible node-pairs, out of which we must connect m distinct node-pairs with edges at random—therefore, we must choose these m node-pairs uniformly randomly from the set of all possible pairs. This model is often referred to as $G(n, m)$.

Alternatively, a simpler model, which is very similar, involves connecting nodes in the graph based on a defined probability p. If p is defined as

$$p = \frac{m}{\binom{n}{2}} \tag{3.20}$$

and we connect every pair of nodes with this probability, the *expected* number of edges in such a graph will be m. This is essentially the $G(n, p)$ model of Erdős and Rényi.

The key defining feature of any network model is its degree distribution. How does the degree distribution of an ER random network model look like? The degree distribution is nothing but a count of the number of neighbours for every node. The probability of having an edge to each of the other $n - 1$ nodes is p (Equation 3.20). We now need to compute the distribution of the number of edges, which is the same as the distribution of the number of successes in a sequence of independent Bernoulli trials with a probability p of success—a binomial distribution:

$$P(k) = \binom{n - 1}{k} p^k (1 - p)^{n-1-k} \tag{3.21}$$

Note that as $n \to \infty$, this turns out to be a Poisson distribution (see [3] for a lucid derivation):

$$P(k) = \frac{\lambda^k}{k!} e^{-\lambda}, \quad \text{where } \lambda = np \tag{3.22}$$

For this reason, the ER model is often referred to as a *Poisson random graph* or even a *Bernoulli random graph*.

Figure 3.3a shows the degree distribution for an ER network generated with $n = 10,000$ and $p = 0.001$, and how the binomial distribution overlaps very closely. The expected number of edges is $\binom{10000}{2} \times 0.001 = 49995$. Figure 3.3b illustrates the same for a network where $n = 10,000$ but p is even lower, at 0.0001.

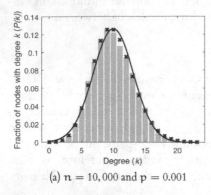

(a) $n = 10,000$ and $p = 0.001$

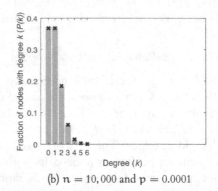

(b) $n = 10,000$ and $p = 0.0001$

Figure 3.3 Degree distribution of ER random networks. The crosses represent corresponding binomial probability distributions. Note that in (a), it is even possible to fit a normal distribution $\approx N(9.97, 3.16)$ (shown by the continuous black curve), but clearly impossible to do so in (b). Of course, Poisson distributions with $\lambda = 10$ and $\lambda = 1$ will overlap very closely with the distributions in (a) and (b), respectively.

Erdős and Rényi showed a number of interesting properties of the graph, such as the size of the largest connected component, for different values of p [22]. Drawing from Equations 3.4 and 3.20, the mean degree of such a random network will be

$$\frac{2 \times \binom{n}{2} \times p}{n} = (n-1)p \tag{3.23}$$

Try it out ...

Generate ER random networks for different values of p, for a given value of n, say $10,000$. Note how the various properties of the network change, notably the number of connected components, as you increase the value of p. For $p = 1$, you will have a complete graph.

What is the clustering coefficient in the ER random network? Recall that the clustering coefficient quantifies the probability that two nodes that have a common neighbour are also neighbours themselves. In a random network, this probability is essentially constant, independent of how the nodes are connected. Therefore, C is the same as p! The characteristic path length of an ER network $\ell \approx \dfrac{\ln N}{\ln \langle k \rangle}$. The diameter is also of a similar order, and grows slowly with N, as $\ln N$ [3].

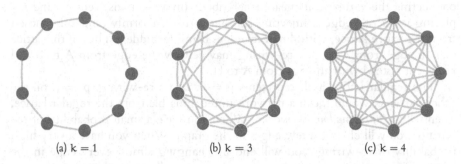

(a) k = 1   (b) k = 3   (c) k = 4

Figure 3.4 **Regular lattices.** Every lattice has $n = 9$ nodes, each node connected to $2k$ of its nearest neighbours. All three graphs are regular; (c) is *complete* as well.

Although ER random networks capture a significant feature common to many real networks, namely short path lengths/diameter, they do not capture another all-important feature of real networks, namely clustering. Therefore, while ER models are useful as a canonical random network, they often cannot effectively be used to model real-world networks. Real networks also differ markedly in their degree distribution, as we will particularly note in §3.2.3. ER random networks also do not capture any of the mixing patterns that exist in real networks.

Note however that there are other possible models of random networks, that need not follow the Poisson degree distribution, but can in fact follow any arbitrary degree distribution [23]. The *configuration model* is the most widely studied generalised random graph model. We will discuss this further in §3.4.1.

3.2.2 Small-world networks

In 1998, Watts and Strogatz proposed a small-world network model [1], in an attempt to model the short path lengths of real networks, as well as the clustering. So, what kind of a network did Watts and Strogatz build? They started with a *regular lattice* and re-wired it, observing many interesting properties.

3.2.2.1 Regular lattices and their re-wiring

A regular lattice, as the name suggests, is a regular graph, with the property that every node in the graph is connected to $2k$ of its nearest neighbours. Regular lattices with nine nodes and values of $k = 1, 3, 4$ are shown in Figure 3.4. You can also think of this as a social network of people seated in a round-table, where everyone knows their immediate (closest) $2k$ neighbours. Watts and Strogatz began re-wiring edges in these networks, at random, with a defined probability, p. Re-wiring an edge involved picking a vertex at random, severing the edge connecting the vertex to its nearest neighbour (in a clockwise sense)[4], and replacing it with an edge connecting another vertex uniformly randomly chosen from the entire lattice, with duplicate edges being forbidden. Thus, if the lattice had an edge from A to B, a re-wiring may destroy the edge from A to B and replace it instead with an edge from A to H.

The regular lattice itself represents $p = 0$, *i.e.* no re-wiring. $p = 1$, on the other extreme, would mean a network unrecognisable from the regular lattice, essentially a *'fully' random* network[5]. When you have a small probability of re-wiring, you will change a few edges in the graph. When you have a very high probability of re-wiring, you will end up changing almost every edge in the graph. In a sense, with increasing p, you incrementally make the network more and more *disordered*. What Watts and Strogatz observed was that, for a broad range of values of p, between the extremes of $p = 0$ and $p = 1$, there are networks that display some very interesting properties, which they called the *small-world properties*.

Networks Resources

Index of Complex Networks (ICON) ICON (https://icon.colorado.edu/) is a project based at the University of Colorado Boulder, and coordinated by Aaron Clauset. Note that ICON is not a *repository*, but an *index*. ICON indexes hundreds of networks scattered across the web and literature. ICON provides basic metadata about the network, in terms of its domain, size, bibliographic reference, as well as a link to the dataset.

The KONECT Project KONECT (http://konect.cc) stands for Koblenz Network Collection. Developed by the lab of Jérôme Kunegis [24], the KONECT project had 1,326 network datasets in 24 categories at the time of this writing. Source code for KONECT is available from GitHub, and includes a network analysis toolbox for GNU Octave, a network extraction library, as well as code to generate the KONECT web pages, including all statistics and plots.

Network Repository Network Repository [25] (http://networkrepository.com/) hosts a comprehensive and representative set of many popular and frequently used datasets in academia/industry. At the time of this writing, the repository provided 6,657 networks across a variety of domains, ranging from biological networks, to chemoinformatics, social networks, collaboration networks, and retweet networks, to name a few.

3.2.2.2   Small-world properties

What are the small-world properties? The re-wired networks discussed above, for some intermediate values of $0 < p < 1$, show very high clustering (similar

---

[4]Strictly speaking, this is not necessary, but we are here recapitulating the exact details of the experiment by Watts and Strogatz [1].

[5]Also see Exercise 3.9.

TABLE 3.2 **Understanding small-worldness.** Shown are the key properties of three different networks discussed in the classic paper by Watts and Strogatz [1]. Watts and Strogatz considered all these networks to be undirected, for the purposes of these analyses. Adapted by permission from Springer Nature: D. J. Watts and S. H. Strogatz, "Collective dynamics of 'small-world' networks", *Nature* **393**:440–442 (1998). © McMillan Publishers Ltd. (1998).

Property	Film actors	Power grid	C. elegans
N	225,226	4,941	282
$\langle k \rangle$	61	2.67	14
ℓ_{real}	3.65	18.7	2.65
ℓ_{rand}	2.99	12.4	2.25
C_{real}	0.79	0.080	0.28
C_{rand}	0.00027	0.005	0.05
$\ln N / \ln k$	2.998	8.661	2.138
$p = \langle k \rangle / N - 1$	0.00027	0.00054	0.050

to regular lattices and real-world networks), yet low characteristic path lengths (like Erdős–Rényi random networks, and also many real networks). The regular lattices have high clustering but also high characteristic path lengths. Table 3.2 shows the original three networks analysed by Watts and Strogatz, and their various properties. Note that they all satisfy the key small-world properties:

$$\ell_{real} \gtrsim \ell_{rand} \tag{3.24}$$

$$C_{real} \gg C_{rand} \tag{3.25}$$

Protein structure networks and protein conformational networks (see §2.5.3 and §4.6) are typically small-world networks [20]. This can be understood in terms of the compactness of protein structure, and its exceptionally high dynamism, respectively.

Despite the key properties explained by the small-world model, it still notably fails to explain the existence of very high degree nodes, or *hubs*, commonly found in real-world networks. Again, the degree distribution of many real-world networks is markedly different from those of small-world networks.

3.2.3 Scale-free networks

So, how does the degree distribution of real-world networks look like? Figure 3.5 shows the degree distribution of some real-world networks. Notably, all these networks harbour a sizeable number of nodes with very large degrees—in the *tail* of the degree distribution, corresponding to high-degree nodes, or *hubs*—and a very large number of nodes with low degrees. Barabási and Albert, in their classic paper of 1999 [26] showed that many such real-world networks display a *power-law* degree distribution.

3.2.3.1 Power-law degree distribution

The degree distribution in such networks is given by

$$P(k) \propto k^{-\gamma} \tag{3.26}$$

If you scale the degree, to say, αk, the distribution at that *scale* becomes

$$P(\alpha k) = C(\alpha k)^{-\gamma} \tag{3.27}$$
$$= C'k^{-\gamma} \tag{3.28}$$
$$\propto k^{-\gamma} \tag{3.29}$$

essentially the same. This is a reason that power-laws are used synonymously with scale-free, in the context of networks.

Distributions such as Equation 3.26 (or Equation 3.29) are *power laws*, where γ is the power-law exponent. Very often, the power-law distribution may not be followed over the entire range of degrees, but many networks demonstrate this behaviour over a (sufficiently) broad range of degrees. Many real-world networks, including biological networks, have $2 < \gamma < 3$, with a characteristic path length of $\ell \sim \log \log N$, thus significantly smaller than (ER) random networks [27]. Most interestingly, these networks can be generated by a *preferential attachment model*, as we will see below.

3.2.3.2 Preferential attachment model

A key aspect of many real-world networks is that they *grow*. New nodes and links are continually added, and certainly not at random. It is easy to imagine the growth of a number of networks, be it a scientific citation network, where new papers are published citing existing papers, or a yeast protein interaction network, evolving via gene duplications and mutations, to the World Wide Web (WWW), where new pages are added every day, linking to existing pages in different ways. Barabási and Albert showed that many such networks could be generated by a *rich get richer* model, or technically, a *preferential attachment* model. The key underlying principle of this model is that incoming (new) nodes do not connect at random to the existing networks, but attach to existing nodes in a preferential fashion, based on degree.

Starting with a network containing m_0 vertices, at every time-point, a new node is added to the network, establishing $m \leqslant m_0$ links to existing nodes. The m links to the existing nodes are independently established, each with a probability proportional to the current degree of the nodes, *i.e.*

$$p_i = \frac{k_i}{\sum_j k_j} \tag{3.30}$$

Here, p_i is the probability that an incoming node connects to node i, and k_i is the degree of node i in the current network, with the denominator representing the sum of degrees of all the nodes currently in the network.

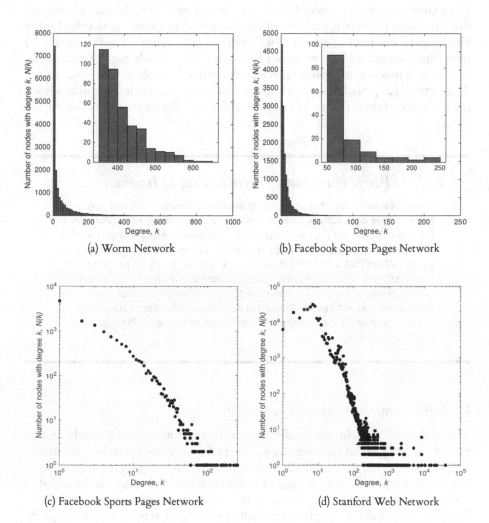

(a) Worm Network

(b) Facebook Sports Pages Network

(c) Facebook Sports Pages Network

(d) Stanford Web Network

Figure 3.5 **Degree distribution of some real networks.** In panels (a) and (b), which correspond to the Worm Network and the network of Facebook Sports Pages, the power-law distribution seems apparent; the insets further show that there are a number of nodes with high degrees. Panel (c) shows a log-log plot of the same network in (b). Panel (d) shows another network, the Stanford Web network, a directed graph, for which the degree distribution shown includes both the in- and out-degrees for every node. All these networks were obtained from https://networkrepository.com/.

Such a process of network *evolution*, or growth, leads to power-law networks characterised by $\gamma = 3$, with a sizeable number of hub nodes with high degrees, and a large number of *peripheral* nodes with low degree. It is also easy to imagine this process of preferential attachment in the context of many networks: highly cited papers will tend to attract more citations and highly connected (popular) web pages are more likely to attract links from newer web pages. Gene duplication, a common evolutionary mechanism, can also result in scale-free degree distributions [27]. Such a network structure also has interesting implications for the resilience of these networks and their ability to tolerate attacks, as we will see in §3.5.

Power laws: From Pareto to Zipf to Barabási

It is interesting to note that power law distributions abound in the real world, not just in degree distributions. Newman, in a paper interestingly titled *"Power laws, Pareto distributions and Zipf's law"* [28] nicely catalogues several distributions, from word frequency (often called Zipf's law [29]), citations of scientific papers (observed by Price [30], who called it *cumulative advantage*, as against preferential attachment), web hits, wealth and population distributions! The Pareto distribution is often colloquially referred to as the "80-20 rule", deriving from an observation by the Italian economist Vilfredo Pareto that 80% of a society's wealth is held by 20% of the population. This is similar to the distribution of degrees in Barabási-Albert's power law networks.

3.2.4 Other models of network generation

The preferential attachment model of Barabási and Albert is very widely used to model and understand the properties of several real-world networks. Nevertheless, there are other models of network generation, which propose alterations to the original approach of §3.2.3.2. For instance, nodes and edges can be added to the network at different points in time, not just during the arrival of a new node. Further, there could be non-linear variations of the preferential attachment probability of Equation 3.30. Bianconi and Barabási [31] proposed a variation of the preferential attachment model, where every node has an associated *fitness* (unrelated to degree), which influences the evolution of the network, in addition to degree-based preferential attachment.

3.2.4.1 *Vertex copying models*

Kleinberg and co-workers proposed an interesting model based on vertex copying. Essentially, instead of new nodes *arriving*, new nodes are created by *copying* existing nodes, and their links. While this was originally devised to explain the growth of the "Web graph", the method makes even more sense in the context of biological networks, where gene duplication is a common mechanism. Essentially, in this process, a node u is duplicated, to form v, copying over (most of) node u's existing links. This process also results in a power-law degree distribution. A formal description of this model is available in [3, 32].

3.2.4.2 Graphlet arrival models

Graphlet arrival models have also been proposed, to mimic the arrival of linked nodes, or *graphlets*. Filkov *et al* [33] showed that a simple graphlet arrival model, modelling the connection of a peripheral node and central node from a three-node graphlet (V)—with different probabilities—can capture interesting properties of real networks, notably assortativity. In a sense, both vertex copying models and graphlet arrival models discussed above are essentially perturbations to a network. A more common perturbation is the loss of nodes (and edges) from a network, as we will see later (§3.5).

3.3   COMMUNITY DETECTION

A very important aspect of studying network structure, is to search for and uncover *community structure* or clusters[6]. Many networks naturally contain clusters or communities—essentially groups of nodes that are more closely connected to one another, than to the rest of the network. These communities are also often known as *modules*; modularity is an important characteristic of many engineered and biological systems (also see §12.4.1). Many functions in any living cell are carried out in a modular fashion.

Numerous modules are present in biological and technological networks, and can be uncovered through an analysis of community structure. For instance, in a protein interaction network, community detection could uncover proteins that work together in a pathway. In Wikipedia, it could identify groups of related articles, and so on. Splitting a network into communities also makes it more tractable for simplified visualisation or other analyses. In the following sections, we will pose the problem mathematically, and overview a few algorithms that can uncover community structure.

Zachary's Karate Club Dataset

This is a social network, capturing friendships among 34 students at a university karate club. The dataset was published by Wayne Zachary in 1977 [34], as part of an anthropological study. During the course of the study, the club split in two, owing to a dispute—these two factions form the 'ground truth' communities for this

---

[6]The 'clustering' referred to in this section, though conceptually related, is strictly distinct from its use in the term clustering coefficient (§3.1.1). The sub-graphs corresponding to tightly-knit clusters will have higher average clustering coefficients. Also, it is common to use the terms clusters and communities interchangeably.

dataset. It turned out that many community detection algorithms are able to accurately predict the two factions (communities), reproducibly. This dataset has since become a classic 'gold standard' dataset for community detection.

Newman, in his textbook on Networks, notes that *"There also exists a tongue-in-cheek society, the 'Karate Club Club', membership in which is bestowed upon the first speaker to use the karate club network as an example in their presentation during any conference attended by the previous recipient of the same honour. A small plastic trophy is passed from hand to hand to commemorate the occasion."* See http://networkkarate.tumblr.com.

3.3.1 Modularity maximisation

If we can quantify the modularity of a network, then, a good community structure would be one that maximises this property, resulting in an optimisation problem. Thus, we need to translate our community definition above—"groups of nodes that are more closely connected to one another, than to the rest of the network"—into a mathematical form. Essentially, we need to compute a quantity that tells how non-random the edges are, between nodes belonging to the same community. Newman [35] defined the modularity, Q, of a network as:

$$Q = \frac{1}{2m} \sum_{ij} \left(A_{ij} - \frac{k_i k_j}{2m} \right) \delta_{g_i g_j} \tag{3.31}$$

where g_i and g_j denote the groups that the ith and j^{th} nodes belong to, respectively, and δ_{xy} denotes the Kronecker delta function. The rest of the symbols carry their usual meanings—m is the number of edges, A is the adjacency matrix, k_i and k_j are the degrees of the ith and j^{th} nodes, respectively. Finding a group assignment that maximises Q becomes a discrete optimisation problem—*modularity maximisation*—and is an effective way to identify communities in many networks. Many heuristics and algorithms have been proposed to solve such optimisation problems, as detailed in an excellent survey on community detection by Fortunato [36].

Hierarchical clustering

Hierarchical clustering is a class of methods in machine learning/unsupervised learning. Hierarchical clustering can be agglomerative or divisive—depending on whether you build the clusters starting from a single pair of data points ('bottom-up') or by decomposing the complete network ('top-down'). Agglomerative clustering is more widely used; the similarity between any two items or data points is computed using some distance metric (*e.g.* Euclidean distance, Hamming distance, cosine distance, etc.). Following this, data points are incrementally clustered. Every point begins in its own cluster. Subsequently, distances are computed between clusters, and clusters that are closest to one another are coalesced (*agglomerated*).

The distance between two clusters can be computed in different ways, as a function of all pairwise distances between points in two clusters (*e.g.* mean, minimum, maximum, median). At the end of the exercise, the entire set of points becomes a single cluster. Clusters are generally represented as *dendrograms*; a cut-off of distances is picked to divide the points into clusters. Very similar strategies can be used for clustering of nodes in a network into communities.

A simple example is shown below, illustrating the steps of the agglomeration, beginning with the distance matrix for five elements A–E in five clusters, $\{A, B, C, D, E\}$, then $\{AB, C, DE\}$, then going to $\{ABC, DE\}$ and finally ending at $\{ABCDE\}$, alongside the final dendrogram. At each step, pairwise distances are used to merge the closest clusters. The distance between two clusters is taken as the maximum of the pairwise distances between the constituent points ('complete linkage'). A distance cut-off of 4.0 as shown (dotted line) will results in two clusters: $\{ABC, DE\}$.

	A	B	C	D	E
A	0				
B	1	0			
C	3	2	0		
D	7	6	4	0	
E	8	7	5	1	0

$\rightarrow$

	AB	C	DE
AB	0		
C	3	0	
DE	8	5	0

$\rightarrow$

	ABC	DE
ABC	0	
DE	8	0

$\rightarrow$

3.3.2 Similarity-based clustering

The concept of hierarchical clustering discussed above can be readily extended to clustering nodes in networks. Given a network, we need a similarity metric to adjudge the similarity of the nodes. Once we have this, the nodes can be clustered hierarchically using any of the hierarchical clustering algorithms. The key challenge here is to come up with a good measure of similarity, which accounts for the network structure.

Many different such similarity measures have been proposed in literature [13]:

Number of node–independent paths Two paths connecting a pair of nodes i and j are "node-independent" if the paths have no nodes in common (other than i and j, of course).

Number of edge–independent paths Similarly, two paths that share no edges in common are "edge-independent".

Total paths This counts *all* possible paths (of any length) between a pair of nodes; paths of length l are weighted by a factor α^l, with small values of α, to obtain a weighted count of the paths.

Adjacency-based measures The adjacency vector (row of the adjacency matrix) corresponding to the nodes can be used to compute similarities, such as cosine similarity or Hamming distances.

Walhout and co-workers [37] have studied the utility of various similarity metrics to cluster genes based on their interaction profiles. Such clustering approaches are also useful to annotate genes based on "guilt by association"— proteins or genes with a similar function are likely to have similar interaction profiles. They also have developed a web-based tool, GAIN (Guide for Association Index for Networks; http://csbio.cs.umn.edu/similarity_index/) for calculating and comparing interaction-profile similarities and defining modules of genes with similar profiles.

3.3.3 Girvan–Newman algorithm

A very popular algorithm for community detection was given by Girvan and Newman [13]. This is somewhat an inverse of the agglomerative hierarchical clustering methods[7], where instead of identifying the edges most central to communities first up, the algorithm identifies those that are *least central*, or *most between* communities. Recall the concept of edge betweenness discussed earlier (§3.1, p. 64). Using edge betweenness, Girvan and Newman applied the following algorithm:

1. Compute the betweenness for all edges in the network
2. Remove the edge with the highest betweenness
3. Re-compute betweennesses for all edges (affected by the removal)
4. Repeat from *Step 2*, until no edges remain in the network

What will we have at the end? All nodes and no edges. But if we were to play the whole process backwards, we will find that nodes get connected pair by pair. You will first add one edge, and then another, and so on. This is the equivalent of hierarchical clustering in some sense. To intuit this algorithm, think of a pair of cliques, connected by a single *bridge edge*[8]. It is easy to imagine that the bridge edge will have the highest betweenness and will be first removed by the Girvan–Newman algorithm, revealing the two separate communities. Recall a similar structure in Figure 3.2a (page 65), where node 5 has the highest (node) betweenness (and is a cut-vertex), and edges 4–5 and 5–6 have the highest edge betweenness (and are bridge edges in the graph).

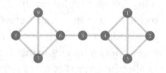

3.3.4 Other methods

There are many other methods for community detection: Louvain [39] is a fast agglomerative algorithm for community detection. CEIL is an algorithm that

---

[7]Indeed, this approach is known as divisive hierarchical clustering, or DIANA (DIvisive ANAlysis).

[8]Indeed, bridge edges in graphs are defined as edges whose removal increases the number of connected components in the graph. Nodes with the same property are not called 'bridge nodes'—they are known as cut-vertices [38].

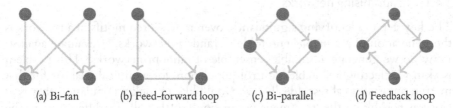

(a) Bi-fan (b) Feed-forward loop (c) Bi-parallel (d) Feedback loop

Figure 3.6 **Examples of motifs in a network.** Many such motifs are present in real networks.

can detect small communities in very large networks. A distributed version of the same, which can notably work on very large graphs is DCEIL [40]. There are also methods that are based on statistical inference and on information theory. Spectral clustering is another class of methods for clustering [41], which draws on the properties of the Laplacian matrix (recall §2.4.3). See [42] for a detailed account of methods, and lucid discussions, including strategies to evaluate algorithm performance.

3.3.5 Community detection in biological networks

Community detection in biological networks is an interesting and very challenging problem. A classic application of community detection to biological networks involves the identification of disease-specific modules from a variety of networks, and is detailed in the following chapter (§4.2). Indeed, this was the subject of a DREAM (Dialogue on Reverse Engineering Assessment and Methods[9]) Challenge in 2018, on Module Identification, which involved the task of identifying disease-relevant modules in six biological networks (see §4.2.2).

3.4 NETWORK MOTIFS

Another key aspect of network structure is the existence of motifs, or repeated patterns in networks. However, not all repeated patterns in a network are motifs—network motifs are only the *statistically significantly over-represented sub-graphs* in a network, with respect to 'randomised' networks [43]. The randomised networks essentially represent a 'suitable' *control*, to study the significance of various motifs in a given 'real' network. Figure 3.6 illustrates a few commonly observed motifs in biological networks, such as the *Escherichia coli* transcriptional regulatory network [43]. These motifs are typically further organised into functional modules.

---

[9]Beginning in 2006, DREAM Challenges have been crowdsourcing challenging problems addressing important questions in biology and medicine. The Index points to other DREAM Challenges discussed in this book.

3.4.1 Randomising networks

The key step in identifying significantly over-represented motifs in a network is the generation of a 'suitable' ensemble of random networks, to evaluate against. How do we generate a 'suitable' ensemble of random networks? This problem is akin to selecting a suitable 'control population' for a clinical trial, or for that matter, any statistical test. Ideally, we would like to pick a population that is as close as possible to the 'treatment population', so that all the differences in the effect can be attributed to the change in treatment, rather than the population itself. So, what properties do we retain in the randomised network, vis-à-vis the original network? Alon and co-workers [43] used a randomisation scheme from [23] to generate random networks with exactly the same number of nodes and edges, and node degrees. That is, not just the degree distribution of the original network was retained, but also the individual node degrees, or degree *sequence*. For instance, if the degrees of nodes in the original network are $\{1, 2, 2, 4, 7, 7\}$, the very same degrees are present in the random network as well. This is in fact the famous generalised random graph model, known as the *configuration model*.

3.4.1.1 *Configuration model*

Beyond the Erdős–Rényi random graphs of §3.2.1, it is possible to generate random graphs of arbitrary degree distributions [23]. A more specific case of this is to generate a random graph with an arbitrary *sequence* of degrees. Thus, the degree of each node is specified as $k_1, k_2, \ldots, k_n$. Now, each node i can be given k_i 'stubs' of edges or half-edges. This satisfies the degree requirement of each node. Following this, we need to tie the stubs together, so that we have the requisite number of edges. This is done at *random, i.e.* we uniformly sample a pair of unconnected stubs from the network, and connect them, repeating the process until no more stubs remain. This presents one random realisation of a network that has the desired degree sequence (and therefore, degree distribution). Importantly, any such network, with the specified degree sequence, is equally probable (and networks with any other degree sequence have zero probability). An ensemble of networks can be generated thus, and can be examined for the presence of various motifs.

Some pitfalls exist in the configuration model. While tying the stubs together, we may produce multiple edges between the same pair of nodes, or a self-edge. While such networks could potentially be discarded from the ensemble, that would violate the equal probability of all possible networks, making it harder to analyse such ensembles analytically. In practice, these issues can be neglected, and the configuration model is generally the model of choice for generating random ensembles.

3.4.1.2 *Edge switching model*

Another model to generate random networks, is by *edge switching* [19, 44]. This is done by first selecting a random pair of edges, say AB and CD. These edges are now re-wired to form the edges AD and BC. If either of these edges already exists, we discard this edge switch attempt and select a new pair of edges. Repeated re-wiring of the network in this fashion produces a randomised version of the original network. Repeating the entire process multiple times will yield us an ensemble, which we can use for statistical testing.

A third method to randomise networks, "go with the winners", was proposed by Milo *et al* [45].

3.4.1.3 Identifying significant motifs

For each sub-graph in the network, it is possible to compute how significantly it is over-represented. It is easy to compute a p–value based on how likely we are to observe the given sub-graph in the random ensemble, with a frequency higher than that in the given network. A z–score can also be calculated:

$$z_{S,G} = \frac{N_{S,G} - \mu_S}{\sigma_S} \qquad (3.32)$$

where $z_{S,G}$ represents the z–score for the given sub-graph S in a graph G, $N_{S,G}$ represents the number of times the sub-graph is found in G, μ_S represents the mean number of occurrences of the sub-graph in the random ensemble, and σ_S the standard deviation of the same. Beyond identifying motifs, such ensembles can be used to evaluate the significance of any network parameter, *e.g.* clustering coefficient.

In large graphs, motifs might number in hundreds of millions, and it becomes near impossible to exactly enumerate them. As an alternative, excellent sampling algorithms have been developed, to fairly accurately detect motifs based on sampling and determining sub-graph 'concentrations' [46].

3.5   PERTURBATIONS TO NETWORKS

A key aspect of systems biology, and biology in general, is the study of perturbations to any system or network. In the context of a network, what are the kinds of perturbations we can perform? And, how do we quantify the *effect* of these perturbations?

The classic perturbation study in molecular biology is a 'gene knock–out' study, where a particular gene is knocked out (or silenced), and the phenotype of the organism is studied. The *in silico* equivalent of that would be to remove a node from the network, along with all its edges. Another very interesting perturbation is that of removing specific edges in a network, known as *edgetic* (from edge-specific genetic; see also §4.2.3) perturbations. Biologically, this would mean disrupting specific molecular interactions, *e.g.* protein–protein interactions, without removing the protein itself. This is far more challenging, although a recent study in 2018 has shown how it is indeed possible to selectively target a specific domain in a protein [47], essentially achieving a selective modulation of intracellular signalling. Vidal and co-workers developed a methodology to use 'edgetic' alleles that lacked *single* interactions, while retaining all others (see §4.2.3).

The study of such perturbations is interesting from various perspectives. For instance, it can shed light on the nodes/edges most crucial to network function/-connectivity. This may be particularly interesting in the context of a pathogenic organism, where one would like to *bring down* the (network of the) organism. It is easy to imagine that the structure of the network has deep implications for its resilience to attacks, as we will discuss in §3.5.2. It is also interesting to think about the impact of network structure and organisation on its function in this context, much like the sequence–structure–function hypothesis that forms the bedrock of sequence analysis in bioinformatics.

3.5.1 Quantifying effects of perturbation

How do we study the impact of any perturbation? Or, how does one quantify the damage to the network, if any, following a perturbation? Indeed, we could study any of the properties discussed in §3.1. Specifically, the diameter of the network or its characteristic path length (ℓ) are particularly useful to study the extent of disruption to the network. Barabási and co-workers [48] also computed the fraction of nodes in the largest component (S), and the average size of the rest of the clusters ($\langle s \rangle$). As nodes (and their corresponding edges) are lost from the network, it is easy to imagine that both the diameter and the ℓ would increase. However, beyond a point, where the diameter peaks, the network disintegrates into multiple components ($S \simeq 0$), and the diameter and ℓ become smaller.

3.5.1.1 Other measures

In addition to the measures proposed above, Wingender and co-workers [49] have proposed a measure that particularly looks at the shortest paths in a network. The measure, called *pairwise disconnectivity index*, quantifies the fraction of paths in a directed network that have been lost due to the perturbation:

$$\mathrm{Dis}(v) = \frac{N_0 - N_{-v}}{N_0} = 1 - \frac{N_{-v}}{N_0} \tag{3.33}$$

where N_0 is the number of ordered pairs of vertices connected by at least one path in the original network; N_{-v} measures the same quantity in the network following the removal of node v and corresponding edges. In other words, this index measures the fraction of (originally connected) node pairs in a network, which become disconnected following the removal of node v.

3.5.2 Network structure and attack strategies

Barabási and co-workers [48] have studied the resilience of real-world networks such as the Internet [50] and the WWW [51], random and power-law networks to both random failures and targeted attacks. Random failures involved removing

TABLE 3.3 **Small-world vs. Power-law networks.** Key features compared between small-world and power-law networks.

Features	Small-world	Power-law
Nodes/Degrees	Egalitarian/homogeneous	Hubs and peripheral nodes
Defining features	Local clusters, distant links	Power-law, high clustering
ℓ	— Small average node-to-node distance in both —	
Growth	No growth	Preferential attachment
Random failure	Deteriorates	Very stable
Targeted attack	No particular targets	Resilient, but weaker

nodes from the network at random. On the other hand, targeted attacks involved removing nodes from the network in the order of highest degree (re-calculated after every removal). The study showed that scale-free networks are largely insensitive to random node removals, showing little changes in diameter and ℓ. It is easy to reason this; most nodes in scale-free networks are *unimportant*, or peripheral, with low degree and lying on very few paths. Their removal, therefore, is unlikely to significantly impact the network structure/performance.

In contrast, under targeted attacks, they deteriorate faster—ℓ increased sharply with the fraction of hubs removed; $\approx 18\%$ of the nodes in a scale-free network needed to be removed before the network disintegrated, forming isolated clusters. Note, however, that this is a sizeable fraction; so scale-free networks such as the WWW can be regarded as resilient, even in the context of targeted attacks [52]. Of course, they are extremely resilient to random failures, with unchanged diameters even after losing 5% of nodes. Random networks, on the other hand, show no substantial difference in their response to random failures vs. targeted attacks, owing to the homogeneity of their network structure.

Another comprehensive study on network resilience was reported by Holme *et al* [53], who examined four different strategies for targeted attack, based on removing nodes from the network in the descending order of degrees, or betweenness centrality, calculated for the initial network or re-calculated after each removal. They observed that strategies removing nodes based on re-calculated centrality values were more disruptive to the network, underlining the importance of the changes in network structure that happen as key nodes (and edges) are removed from the network. Table 3.3 illustrates a comparison of various key features between small-world and power-law networks, including their resilience to failure/attack.

3.6 TROUBLESHOOTING

3.6.1 Is your network *really* scale-free?

Following the publication of the original scale-free network result by Barabási and Albert [26], a number of studies enthusiastically reported power-law distributions in a variety of networks. Many power-laws do not stand up rigorous statistical scrutiny, and either vanish or may not be statistically provable [54].

Newman and co-workers have developed a statistical framework to systematically quantify power-law behaviour in empirical data [54]. Their method illustrates the importance of a robust statistical test to evaluate degree distributions before concluding their scale-freeness. In particular, it is critical to watch out for the common pitfalls—not every distribution that gives a straight (looking) line in a log–log plot has an underlying power law! Similarly, regression on log–log plots is fraught with errors; it is better to use maximum likelihood estimates and perform goodness-of-fit tests.

No matter what the structure of the network is, it is important to see if the processes such as preferential attachment also make sense in the context of your network. For instance, preferential attachment cannot fully explain the structure of metabolic networks [55]. In reality, it is more important to understand the various aspects of a network's structure, rather than to quickly label it as small-world or a power-law. Rather, it is more interesting and fruitful to study various properties of the network and explore their probable causes—such as a high clustering, or a skewed (long-tailed) degree distribution. For instance, the heavy tail might be all that is interesting, implying a large number of high-degree nodes.

Another key issue here is that of false positives and false negatives—many edges, or even nodes, may exist in reality and are missed out in the current reconstruction of the network. Alternatively, there may be spurious nodes and edges in the network. These further confound the computation of properties such as scale-freeness. Thus, it is imperative to understand the errors or incompleteness of a network before we come to conclusions about its properties.

3.7   SOFTWARE TOOLS

The companion code repository contains codes for generating Erdős–Rényi $G(n, p)$ networks, regular lattices and Barabási–Albert networks.

igraph is a package for network analysis that has interfaces for other languages such as R and Python, and can handle large graphs.

Pajek is a freely available network analysis tool for Microsoft Windows [56]. It scales very well for the analysis of very large networks ($> 10^5$ nodes).

Gephi is a free open-source software [57], for visualising and exploring various kinds of graphs.

R/MATLAB/Python also have very good libraries/built-ins for performing network analysis: https://statnet.org (powerful suite for network analysis in R, in addition to igraph), https://mathworks.com/help/matlab/graph-and-network-algorithms.html, and of course https://networkx.github.io.

FANMOD (http://theinf1.informatik.uni-jena.de/motifs/) [58] is a cross-platform tool that can identify motifs in various types of networks. FANMOD identifies motifs in a given network, based on comparisons against a random ensemble, comprising a user-specified number of networks.

EXERCISES

3.1 Show that the mean degree $\langle k \rangle$ of a vertex in an undirected graph is $\frac{2|E|}{|V|}$.

3.2 Compute the clustering coefficient of the network (an octahedral graph) in the figure below:

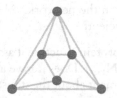

3.3 For the undirected network given below, compute (a) the betweenness centrality of each of the nodes and (b) the characteristic path length.

3.4 Show that the clustering coefficient of a regular lattice where each of the n nodes is connected to its 2k nearest neighbours, is $\frac{3k-3}{4k-2}$.

3.5 Outline a strategy to generate a power-law network with a degree distribution approximately given by $P(k) \sim k^{-2.25}$.

3.6 Dr. B has a hypothesis that PageRank is a very good determiner of a node's importance in a network and that targeting top PageRank nodes will be a successful strategy to quickly disrupt a network. Can you suggest some computational experiments that Dr. B should perform to test the validity of her hypothesis?

[LAB EXERCISES]

3.7 Write a piece of code to re-wire a given regular lattice (specified by n and k), with probability β. Try your code out with $n = 20, k = 4$, and $\beta = 0.4$, and comment.

3.8 How would you generate a very large Erdős–Rényi random network, e.g. $G(10^5, 10^{-5})$? You will not be able to use `erdos_renyi.m` from the Companion Code Repository, as MATLAB will complain that it is out of memory. While one may start out writing codes "that just work", it is important to move on to write code that is memory-efficient and avoids redundant computations.

3.9 Consider the Watts–Strogatz network discussed in §3.2.2. When re-wired with p = 1, these networks become *random*. How would you *test* if they are *different* from Erdős–Rényi random networks created with an identical number of nodes and edges?

3.10 Create a Barabási–Albert network with n = 1,000 nodes, beginning with a single edge (two nodes), as described in §3.2.3. What is the maximum degree that you find in the network? Also, why does the network have a zero clustering coefficient?

3.11 Consider a protein–protein functional association network obtained from STRING for *E. coli* (N_{Eco}). Does it have a *high* clustering coefficient? How will you make a statement about the statistical significance of the clustering coefficient value?

3.12 Consider the same network, N_{Eco}, as the problem above. If you refine the network to include only enzymes (*how will you do this?*) and further consider the six main classes of enzymes (first digit of the Enzyme Commission classification), can you quantify the assortativity of the network? Since there are six discrete classes, you can employ Newman's measure of assortativity, as in Equation 3.19.

3.13 Perturb the largest connected component of the above N_{Eco} network by removing 100 nodes with the highest betweenness centrality, re-calculating the betweenness centrality after each removal, as outlined in §3.5.2. Plot how the diameter of the network changes, as node deletion progresses in the above fashion. Can you also plot the other metrics discussed in §3.5.1?

3.14 Consider the metabolic network corresponding to *E. coli* from the previous chapter lab exercise. Can you examine the degree distribution of the metabolites? What distribution does it follow? Remember the discussion from §3.6.1, and use plfit routines by the authors in [54] to statistically examine the distribution and its fit to a power law.

Do your observations change if you dropped the *currency* metabolites (§2.5.5.1) such as ATP, NADPH, water, oxygen, etc.?

REFERENCES

[1] D. J. Watts and S. H. Strogatz, "Collective dynamics of 'small-world' networks", *Nature* **393**:440–442 (1998).

[2] M. E. J. Newman, "The structure and function of complex networks", *SIAM Review* **45**:167–256 (2003).

[3] M. E. J. Newman, *Networks: an introduction* (Oxford Univ. Press, 2011).

[4] H.-W. Ma and A.-P. Zeng, "The connectivity structure, giant strong component and centrality of metabolic networks", *Bioinformatics* **19**:1423–1430 (2003).

[5]   O. Riordan and N. Wormald, "The diameter of sparse random graphs", *Combinatorics, Probability and Computing* **19**:835–926 (2010).

[6]   G. Sabidussi, "The centrality index of a graph", *Psychometrika* **31**:581–603 (1966).

[7]   P. Bonacich, "Power and centrality: a family of measures", *American Journal of Sociology* **92**:1170–1182 (1987).

[8]   P. Bonacich, "Some unique properties of eigenvector centrality", *Social Networks* **29**:555–564 (2007).

[9]   L. Katz, "A new status index derived from sociometric analysis", *Psychometrika* **18**:39–43 (1953).

[10] S. Brin and L. Page, "The anatomy of a large-scale hypertextual web search engine", *Computer Networks and ISDN Systems* **30**:107–117 (1998).

[11] J. M. Anthonisse, *The rush in a directed graph*, tech. rep. (Amsterdam, 1971).

[12] L. C. Freeman, "A set of measures of centrality based on betweenness", *Sociometry* **40**:35–41 (1977).

[13] M. Girvan and M. E. J. Newman, "Community structure in social and biological networks", *Proceedings of the National Academy of Sciences USA* **99**:7821–7826 (2002).

[14] M. G. Everett and S. P. Borgatti, "The centrality of groups and classes", *The Journal of Mathematical Sociology* **23**:181–201 (1999).

[15] E. D. Kolaczyk, D. B. Chua, and M. Barthélemy, "Group betweenness and co-betweenness: inter-related notions of coalition centrality", *Social Networks* **31**:190–203 (2009).

[16] M. J. Newman, "A measure of betweenness centrality based on random walks", *Social Networks* **27**:39–54 (2005).

[17] H. Tong, C. Faloutsos, and J. Pan, "Fast random walk with restart and its applications", in Sixth International Conference on Data Mining (ICDM '06) (2006), pp. 613–622.

[18] M. E. J. Newman, "Mixing patterns in networks", *Physical Review E* **67**:026126+ (2003).

[19] S. Maslov and K. Sneppen, "Specificity and stability in topology of protein networks", *Science* **296**:910–913 (2002).

[20] C. Böde, I. A. Kovács, M. S. Szalay, R. Palotai, T. Korcsmáros, and P. Csermely, "Network analysis of protein dynamics", *FEBS Letters* **581**:2776–2782 (2007).

[21] P. Erdős and A. Rényi, "On random graphs I", *Publicationes Mathematicae Debrecen* **6**:290 (1959).

[22] P. Erdős and A. Rényi, "On the evolution of random graphs", in Publication of the Mathematical Institute of the Hungarian Academy of Sciences (1960), pp. 17–61.

[23] M. E. J. Newman, S. H. Strogatz, and D. J. Watts, "Random graphs with arbitrary degree distributions and their applications", *Physical Review E* **64**:026118 (2001).

[24] J. Kunegis, "KONECT – The Koblenz Network Collection", in Proceedings of the 22nd International Conference on World Wide Web (2013), pp. 1343–1350.

[25] R. A. Rossi and N. K. Ahmed, "The Network Data Repository with interactive graph analytics and visualization", in Proceedings of the Twenty-Ninth AAAI Conference on Artificial Intelligence, AAAI'15 (2015), pp. 4292–4293.

[26] A.-L. Barabási and R. Albert, "Emergence of scaling in random networks", *Science* **286**:509–512 (1999).

[27] A.-L. Barabási and Z. N. Oltvai, "Network biology: understanding the cell's functional organization", *Nature Reviews Genetics* **5**:101–113 (2004).

[28] M. E. J. Newman, "Power laws, Pareto distributions and Zipf's law", *Contemporary Physics* **46**:323–351 (2005).

[29] G. K. Zipf, *Human behavior and the principle of least effort* (Addison–Wesley, Reading, MA, 1949).

[30] D. D. S. Price, "A general theory of bibliometric and other cumulative advantage processes", *Journal of the American Society for Information Science* **27**:292–306 (1976).

[31] G. Bianconi and A.-L. Barabási, "Competition and multiscaling in evolving networks", *EPL (Europhysics Letters)* **54**:436 (2001).

[32] J. M. Kleinberg, R. Kumar, P. Raghavan, S. Rajagopalan, and A. S. Tomkins, "The web as a graph: measurements, models, and methods", in Computing and Combinatorics (1999), pp. 1–17.

[33] V. Filkov, Z. M. Saul, S. Roy, R. M. D'Souza, and P. T. Devanbu, "Modeling and verifying a broad array of network properties", *EPL (Europhysics Letters)* **86**:28003 (2009).

[34] W. W. Zachary, "An information flow model for conflict and fission in small groups", *Journal of Anthropological Research* **33**:452–473 (1977).

[35] M. E. J. Newman, "Analysis of weighted networks", *Physical Review E* **70**:056131 (2004).

[36] S. Fortunato, "Community detection in graphs", *Physics Reports* **486**:75–174 (2010).

[37] J. I. F. Bass, A. Diallo, J. Nelson, J. M. Soto, C. L. Myers, and A. J. M. Walhout, "Using networks to measure similarity between genes: association index selection", *Nature Methods* **10**:1169–1176 (2013).

[38] B. Bollobás, *Modern graph theory*, 1st ed., Graduate Texts in Mathematics 184 (Springer-Verlag New York, 1998).

[39] V. D. Blondel, J.-L. Guillaume, R. Lambiotte, and E. Lefebvre, "Fast unfolding of communities in large networks", *Journal of Statistical Mechanics: Theory and Experiment* **2008**:P10008 (2008).

[40] A. Jain, R. Nasre, and B. Ravindran, "DCEIL: distributed community detection with the CEIL score", in 2017 IEEE 19th international conference on high performance computing and communications; IEEE 15th international conference on smart city; IEEE 3rd international conference on data science and systems (HPCC/SmartCity/DSS) (2017), pp. 146–153.

[41] U. von Luxburg, "A tutorial on spectral clustering", *Statistics and Computing* **17**:395–416 (2007).

[42] M. Newman, "Community structure", in *Networks*, 2nd ed. (2018), pp. 494–568.

[43] S. S. Shen-Orr, R. Milo, S. Mangan, and U. Alon, "Network motifs in the transcriptional regulation network of *Escherichia coli*", *Nature Genetics* **31**:64–68 (2002).

[44] R. Kannan, P. Tetali, S. Vempala, S. Vempala, and S. Vempala, "Simple Markov-chain algorithms for generating bipartite graphs and tournaments", in Proceedings of the Eighth Annual ACM–SIAM Symposium on Discrete Algorithms, SODA '97 (1997), pp. 193–200.

[45] R. Milo, N. Kashtan, S. Itzkovitz, M. E. J. Newman, and U. Alon, *On the uniform generation of random graphs with prescribed degree sequences*, 2003.

[46] N. Kashtan, S. Itzkovitz, R. Milo, and U. Alon, "Efficient sampling algorithm for estimating subgraph concentrations and detecting network motifs", *Bioinformatics* **20**:1746–1758 (2004).

[47] J. Periasamy, V. Kurdekar, S. Jasti, et al., "Targeting phosphopeptide recognition by the human BRCA1 tandem BRCT domain to interrupt BRCA1-dependent signaling", *Cell Chemical Biology* **25**:677–690.e12 (2018).

[48] R. Albert, H. Jeong, and A.-L. Barabási, "Error and attack tolerance of complex networks", *Nature* **406**:378–382 (2000).

[49] A. P. Potapov, B. Goemann, and E. Wingender, "The pairwise disconnectivity index as a new metric for the topological analysis of regulatory networks", *BMC Bioinformatics* **9**:227 (2008).

[50] R. Albert, H. Jeong, and A.-L. Barabási, "Diameter of the world-wide web", *Nature* **401**:130–131 (1999).

[51] M. Faloutsos, P. Faloutsos, and C. Faloutsos, "On power-law relationships of the internet topology", *ACM SIGCOMM Computer Communication Review* **29**:251–262 (1999).

[52] A. Broder, R. Kumar, F. Maghoul, et al., "Graph structure in the web", *Computer Networks* **33**:309–320 (2000).

[53]   P. Holme, B. J. Kim, C. N. Yoon, and S. K. Han, "Attack vulnerability of complex networks", *Physical Review E* **65**:056109+ (2002).

[54]   A. Clauset, C. R. Shalizi, and M. E. J. Newman, "Power-law distributions in empirical data", *SIAM Review* **51**:661–703 (2009).

[55]   G. Lima-Mendez and J. van Helden, "The powerful law of the power law and other myths in network biology", *Molecular BioSystems* **5**:1482–1493 (2009).

[56]   V. Batagelj and A. Mrvar, "Pajek— Analysis and Visualization of Large Networks", in Graph Drawing, Lecture Notes in Computer Science (2002), pp. 477–478.

[57]   M. Bastian, S. Heymann, and M. Jacomy, "Gephi: an open source software for exploring and manipulating networks", in Third International AAAI Conference on Weblogs and Social Media (2009).

[58]   S. Wernicke and F. Rasche, "FANMOD: a tool for fast network motif detection", *Bioinformatics* **22**:1152–1153 (2006).

FURTHER READING

S. Boccaletti, V. Latora, Y. Moreno, M. Chavez, and D. Hwang, "Complex networks: structure and dynamics", *Physics Reports* **424**:175–308 (2006).

D. J. Watts, *Six degrees: the new science* (Vintage Books USA, 2004).

S. Milgram, "The small-world problem", *Psychology Today* **2**:60–67 (1967).

A.-L. Barabási, *Linked: how everything is connected to everything else and what it means*, Reissue (Plume, 2003).

M. Newman, A.-L. Barabási, and D. J. Watts, *The structure and dynamics of networks: (Princeton studies in complexity)*, 1st ed. (Princeton University Press, 2006).

A. Lancichinetti and S. Fortunato, "Community detection algorithms: a comparative analysis", *Physical Review E* **80**:056117 (2009).

U. Alon, *An introduction to systems biology: design principles of biological circuits* (Chapman & Hall/CRC, 2007).

A. D. Broido and A. Clauset, "Scale-free networks are rare", *Nature Communications* **10**:1017 (2019).

M. Newman, "Power-law distribution", *Significance* **14**:10–11 (2017).

Applications of network biology

CONTENTS

I N THE previous chapters, we have studied various concepts underlying graph theory and network biology. In particular, we have looked at how various biological networks can be modelled as graphs (§2.5), and various network models (§3.2). We also studied key aspects of network structure, such as communities (§3.3) and motifs (§3.4), and reviewed studies on perturbations to networks (§3.5). All of the approaches and models discussed in the previous chapters lend themselves to a variety of applications. Many of these applications are somewhat direct, ranging from the prediction of proteins central to cellular survival to modules of proteins that are important in health and disease, as well as fundamental insights into biological network structure and function. Other applications are rather novel, particularly in terms of casting a challenging problem into the framework of network biology, and then leveraging the various tools to interrogate network structure and function or identify key components of the network.

Every application of network biology typically involves the identification of *key* nodes, edges, paths, or sub-networks (*e.g.* motifs, or communities) from any given network: these could be essential proteins, key interactions, shortest routes for synthesising a metabolite, or disease-relevant protein modules. The major challenge in any of these cases lies in how the network (model) is constructed, to answer relevant biological questions.

4.1 THE CENTRALITY–LETHALITY HYPOTHESIS

A very classic application of network biology is in the identification of functionally important (typically essential) proteins in a given protein–protein interaction network (PPIN). The classic study in this regard was by Jeong *et al* [1], who proposed the famous *centrality–lethality hypothesis*. The hypothesis essentially posits that nodes *central* in a biological network, typically in terms of high degree centrality, are far more likely to be essential for cellular function compared to nodes with lesser centrality.

Specifically, Jeong *et al* analysed the PPIN of *S. cerevisiae*, with 1,870 nodes and 2,240 edges, corresponding to identified direct physical interactions, and integrated experimentally available phenotypic data on single gene deletions. They showed that a large fraction of the proteins in this network (93%) had a degree $\leqslant 5$, and $\approx 21\%$ of them were essential. On the other hand, proteins that were *central*, with degree $\geqslant 15$ comprised only $\approx 0.7\%$, but a much larger fraction of these proteins (62%) were essential for yeast survival. Thus, they concluded that proteins that occupied a more *central* position in the PPIN, *i.e.* proteins with a higher degree, were more likely to be *lethal*, *i.e.* result in a lethal phenotype on knock-out.

The centrality–lethality hypothesis has also prompted the analysis of 'hub' nodes or highly connected nodes in a variety of networks, probing their importance in network function. In 2014, Raman and co-workers [2] showed that the centrality–lethality 'rule' holds good for 20 organisms, by integrating protein networks obtained from the STRING database [3] with essentiality data obtained from the Database of Essential Genes (DEG) [4]. Notably, they showed that in addition to degree centrality, betweenness centrality is also significantly higher for essential genes, compared to the network average. Thus, the centrality–lethality rule is a very useful approach to identify putative essential genes from a network.

Of course, mere degree centrality is not necessarily a robust indicator of lethality; numerous methods have since been developed to predict essential genes with improved accuracies, by leveraging other network parameters and gene/protein features, as we will discuss in the following section. Of course, it is again important to remember that the networks are often incomplete and may also have spurious links, which can confound our predictions (see §3.6.1). Some nice discussions about the essentiality of hubs, and the centrality–lethality rule in general, are presented in references [5] and [6].

4.1.1 Predicting essential genes using network measures

The centrality–lethality rule already emphasises the importance of network properties, such as degree in determining lethality. As a natural extension of this idea, it is also possible that there are various other network features or properties that determine the functional importance of a node. Many studies have predominantly used features based on DNA or protein sequences [7, 8] or physicochemical features combining them with a few basic network features [9]. More recently, in

2018, Azhagesan *et al* [10] predicted essential genes across diverse organisms using *pure* network-based features, even in the absence of other biological knowledge. They leveraged the network structure by extracting hundreds of features that captured the local, global, and neighbourhood properties of the network. Further augmenting the network features with sequence-based features led to a marginal improvement in classification. Their approach can be readily exploited to predict essentiality for any organism with a known interactome, such as those in the STRING database.

4.2 NETWORKS AND MODULES IN DISEASE

4.2.1 Disease networks

Given the extreme complexity of the networks within a cell, there are very few diseases that are caused by mutations in a single gene—most diseases are a consequence of a complex interplay between the various networks within and between cells and tissues in the body [11]. Disease is typically a consequence of dysfunction in these networks; it is therefore vital to study the network of 'interactions' between various genes and diseases. Diseases themselves cannot be easily studied in isolation, as they can be highly interconnected. Indeed, many diseases, such as diabetes and obesity, are known to be "co-morbid", in that patients diagnosed with one disease are often diagnosed with the other.

Barabási, Vidal, and co-workers first developed the concept of disease networks [12]. They represented the disease network as a bipartite graph, where there are two sets of nodes, corresponding to diseases (genetic disorders) and disease genes. Being a bipartite graph (see §2.3.4), edges connect only diseases and genes, with no edges between the diseases or genes themselves, as shown in Figure 4.1. The network, which they called the *diseasome*, contained 1,284 genetic disorders and 1,777 disease genes from the Online Mendelian Inheritance in Man (OMIM) database [13]. Figure 4.1 (centre panel) shows a small subset of these disorder–disease gene associations, with circles and rectangles representing disorders and disease genes, respectively. An edge connects a disorder to a disease gene if mutations in the gene lead to that disorder. The size of each circle is proportional to the number of genes participating in the disorder; the disease nodes themselves are coloured based on the class to which the disorder belongs.

Beginning with this graph, they generated two 'projected networks', namely the human disease network, and the disease gene network. In the human disease network, two diseases are connected by an edge if they share at least one gene in which mutations are associated with either disorder (Figure 4.1, left panel). The width of a link is proportional to the number of genes that are implicated in both diseases. For example, three genes are implicated in both breast cancer and prostate cancer, resulting in a link of weight three between them.

In the disease gene network, two genes are connected by an edge if they are associated with the same disorder (Figure 4.1, right panel). The edge thickness here is proportional to the number of diseases with which the two genes are

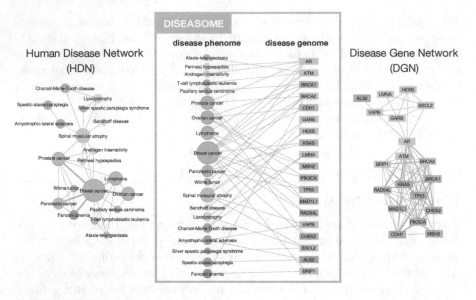

Figure 4.1 **Construction of the diseasome bipartite network.** (Centre) Diseasome network. (Left) The human disease network projection of the diseasome bipartite graph. (Right) The disease gene network projection. Adapted with permission, from K.-I. Goh, M. E. Cusick, D. Valle, B. Childs, M. Vidal, and A.-L. Barabási, "The human disease network", *Proceedings of the National Academy of Sciences USA* **104**:8685–8690 (2007). © (2007) National Academy of Sciences, U.S.A.

commonly associated. They uncovered many interesting patterns in the network; for instance, they found that genes associated with similar disorders were more likely to code for proteins that would interact with one another. Such genes also were more likely to be expressed together, suggesting the existence of disease-specific modules. Recent studies have proposed many novel methods for identifying disease-specific modules [14], as we discuss in the following section.

Huttlin and co-workers [15] developed BioPlex 2.0 (<u>B</u>iophysical <u>I</u>nteractions of <u>ORFEOME</u>-derived com<u>plex</u>es), which employs robust affinity purification-mass spectrometry to experimentally identify a network of human protein-protein interactions. Notably, they also identified 442 communities associated with more than 2,000 disease annotations, suggesting possible disease-relevant modules. BioPlex 2.0 is a massive resource, with > 56,000 interactions and nearly 30,000 previously unknown associations.

Co-morbidity networks have also been constructed in the past, to again identify the diseases that have substantial "genetic overlap" with one another [16]. Another study characterised a symptoms–disease network [17] to unravel the connections between clinical symptoms of a disease and the underlying molecular interactions. They showed that diseases with similar symptoms had a large number of shared genetic associations and protein interactions. Such networks can be interrogated to also shed light on unexpected disease associations and gain insights into disease aetiology.

Drug–target networks have also been constructed, which provide insights towards improving *polypharmacology, i.e.* to design and use molecules that act on multiple targets. Vidal, Barabási, and co-workers [18] constructed a bipartite graph again, comprising drugs (then approved by the US Food and Drug Administration (USFDA)) and their cognate target proteins, linked by drug–target binary associations. Polypharmacology is double-edged—while unintended polypharmacology can lead to adverse drug effects, it can also improve therapeutic efficacy and provide avenues for drug re-purposing [19].

Network approaches are also useful for studying host–pathogen interactions (§11.2.1.1); a very interesting approach has been used to study the coronavirus—human protein interaction networks and predict possible drug re-purposing, as we discuss in the Epilogue (p. 311).

In summary, a number of interesting approaches have been developed to study the network basis of disease. As always, how a network models diseases and genes empowers it to answer various interesting questions about disease associations and aetiology. With the rapid increase in high-throughput data on genetic mutations and phenotypes, this is becoming an exciting, challenging, and also rewarding endeavour.

4.2.2 Identification of disease modules

Biological networks have long known to be highly 'modular' [20], where a group of interacting molecules carry out specific functions. These groups of physically or functionally linked molecules, typically referred to as *modules, clusters,* or *communities,* have functional consequences, such as the orchestration of a signalling network, or in case of dysfunction, the emergence of disease.

The identification of modules in biological networks, therefore, has been a long-standing problem of interest. Ghiassian *et al* [21] observed network properties and patterns corresponding to disease-associated proteins of a comprehensive corpus of 70 complex diseases. Based on the insights obtained, they developed a DIseAse MOdule Detection (DIAMOnD) algorithm drawing on a new metric known as connectivity significance. The metric evaluates if a certain protein has more connections to 'seed' proteins associated with a disease than expected by chance.

More recently, in 2018, the DREAM community put out an interesting challenge, on Module Identification. This challenge involved anonymised datasets—diverse human molecular networks, corresponding to protein interactions, signalling, co-expression, homology, and cancer genetic dependencies derived from loss-of-function screens in cancer cell lines [14]. The anonymisation as done to remove the gene names and network identity, so as to enable a 'blinded' assessment. The study thus strives to benchmark *unsupervised* clustering algorithms, which cannot rely on any prior biological information such as known disease genes.

The challenge task itself was to identify modules of proteins from each of these networks, that are 'enriched' for a given disease/trait. However, in the absence of ground truth 'true' modules in these networks, the evaluation of any algorithm becomes another challenge. The challenge organisers introduced a framework to empirically score any set of modules based on their association with diseases (and complex traits) using 180 genome-wide association studies (GWAS[1]) datasets. Specifically, they used the Pascal tool [22], which aggregates trait-association p-values of single nucleotide polymorphisms (SNPs) at the level of genes and modules. Modules that yielded a significant score for at least one GWAS trait were designated as trait-associated.

The community paper [14] that emerged from the challenge benchmarked as many as 75 community detection methods, for their ability to capture disease-relevant modules. These modules were evaluated for 'disease enrichment', in the context of 180 GWAS datasets. Some of the key recommendations from the challenge consortium were as follows:

1. Methods from diverse categories were found to be useful, viz. kernel clustering, modularity optimisation, random-walk based, local, hybrid, and ensemble-based methods. However, they must be used to identify complementary modules.

2. The modules identified via different methods should be used as is—forming a consensus was seldom found to be helpful (except while integrating > 20 methods).

3. Diverse networks (such as those in the module identification challenge) provide complementary information. However, the module identification methods are better applied to individual networks without merging.

Three top methods from the challenge, based on kernel clustering, modularity optimisation, and random walk, were made available as part of a tool called MONET [23]. In addition to the above, Ravindran and co-workers [24] made a few more interesting observations. First, they proposed a suite of methods to identify smaller 'core' modules from large networks, by effectively decomposing the larger modules routinely identified by classic community detection algorithms. Their approaches improved the otherwise poor performance of classic community detection algorithms, for detecting trait-specific modules in these networks, by 50%.

Interestingly, they also showed that the network topological definition of 'good' modules does not agree with trait-specific modules using measures such as modularity or conductance. Further, they showed that with a knowledge of *seed* nodes, which are known to be involved in diseases (or traits), it is possible to identify far more relevant modules. They also illustrated that overlapping community detection is more appropriate in the context of biological networks, as it

---

[1] Pronounced 'gee-was'

is common that the same gene (node) may be involved in multiple diseases, and therefore, should be a part of multiple disease modules.

4.2.3 Edgetic perturbation models

Perturbations to a network are a common approach to study and understand important nodes in a network, as discussed in §3.5. Indeed, this stems from the premise that phenotypic changes in an organism, especially human diseases, arise as a consequence of such network perturbations [25]. These perturbations, as also discussed earlier, could be the wholesale loss of a particular protein (and its interactions), to the loss of only some select interactions, to a very specific perturbation, leading to the loss of only a single interaction, *i.e.* edge. In the context of network topology, these translate to the loss of a node, or merely the loss of some edges or even a single edge between a node and its neighbours.

Vidal and co-workers have pioneered the study of edgetic perturbations [26, 27]. It is easy to imagine that not all mutations in a gene will have the same effect on the function/interactions of its protein product. Deletions and truncating mutations can be thought of as node removals, as the protein function is likely lost or drastically altered. In contrast, point mutations leading to single amino acids changes and small insertions/deletions are likely more *edgetic* in their effect [26], *i.e.* these are more likely to disrupt only some interactions (edges) of a protein with its neighbours. By experimentally characterising mutant alleles, they showed many evidences pointing towards the validity of their edgetic perturbation models. Their analyses suggested that nearly 50% of the 50,000 Mendelian alleles then available in gene mutation databases can be modelled as edgetic. Further, for genes associated with multiple disorders, they showed that edgetic alleles responsible for these disorders are often located in different interaction domains. Thus, different edgetic perturbations can result in very different phenotypes.

Another study, by Lindquist, Vidal, and co-workers [28] further impressed upon the importance of edgetic studies. Functionally profiling thousands of missense mutations spanning a spectrum of Mendelian disorders, they showed that a majority of disease-associated alleles exhibit wild-type binding profiles/folding/stability. On the other hand, two-thirds of disease-associated alleles perturbed macromolecular interactions—both transcription factor binding and binary protein–protein interactions—with the majority of them being edgetic in nature, affecting only a subset of interactions. In contrast, the natural 'healthy' variants mostly preserved the wild-type interactions. They also showed that different mutations in the same gene, which led to diverse interaction profiles, resulted in distinct disease phenotypes.

4.3 DIFFERENTIAL NETWORK ANALYSIS

A routine task in systems biology is the comparative analysis of different cells or states, *e.g.* a healthy cell vs. a cancerous cell. While performing such analyses, it is common to identify differentially expressed genes or proteins. However, such

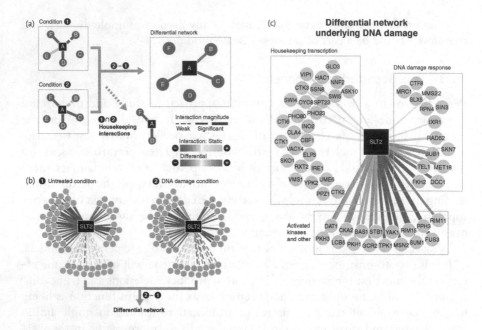

Figure 4.2 **Differential genetic interaction mapping with dE-MAP.** (a) Schematic showing the principle of differential genetic interaction analysis. (b) Differential analysis for yeast *SLT2*. Genetic interaction data from Bandyopadhyay *et al* (2010) [31] collected in either untreated or DNA-damage-treated conditions are compared to create (c) the differential interaction map underlying DNA damage. Adapted by permission from John Wiley and Sons: T. Ideker and N. J. Krogan, "Differential network biology", *Molecular Systems Biology* **8**:565 (2012) [29]. © (2012), EMBO and Macmillan Publishers Limited.

genes or proteins seldom act independently, and, as discussed early on in this book (§1.2.2), they are embedded in complex networks that orchestrate the various functions of every cell. Thus, beyond simple differential expression analyses, it is also imperative to study the *differences in the underlying networks* from one condition to the other. This principle underlies the central topic of this section, "differential network analysis" (DiNA).

Many approaches have been proposed to perform DiNA in the past; a landmark review was published by Ideker and Krogan in 2012 [29]. For a more recent review comparing the relative performance of 10 DiNA algorithms on coexpression networks, see reference [30].

Although there have been a few studies that have taken differential approaches to study various biological systems (for a timeline, see reference [29]), the classic study on differential network analysis was carried out by Bandyopadhyay *et al* in 2010 [31]. Their approach, which they called *differential epistasis mapping* (dE-MAP), unveiled widespread changes in genetic interaction among yeast kinases, phosphatases, and transcription factors in response to DNA damage in yeast. Their methodology is also illustrated in Figure 4.2.

First, static genetic interaction maps are computed in each of two conditions (Figure 4.2a) resulting in both positive (yellow[2]) and negative (blue) interactions. The network for condition 1 is "subtracted from" that for condition 2 to create a differential interaction map, in which the significant differential interactions are those that increase (green) or decrease (red) in score after the shift in conditions. Further, in the differential map, weak but dynamic interactions (dotted edges, viz. $A - B$) are magnified, and persistent 'housekeeping' interactions are removed (viz. $A - F, A - D$). Note that (A, E) and (A, C) are decreasing differential interactions achieved by different circumstances: (A, E) is a positive interaction that disappears after the shift in conditions, while (A, C) is a negative interaction that appears after the shift.

The differential interaction maps have been shown to uncover many gene functions that go undetected in static conditions. Bandyopadhyay et al identified DNA repair pathways, highlighting new damage–dependent roles for the Slt2 kinase, Pph3 phosphatase, and histone variant Htz1. They also showed that while protein complexes are generally stable in response to perturbation, the functional relations between these complexes are substantially reorganised. Figure 4.2c shows that the interactions with transcriptional machinery are present in both conditions and are thus downgraded in the differential map, while interactions with kinases and DNA damage response genes are highlighted.

Other approaches to DiNA have been since published [32, 33]. DiNA has the potential to unravel many interesting mechanisms that underlie the adaptation of an organism to changes in the environment or any other stress conditions, such as the introduction of a drug. Although DiNA has been widely applied to gene expression networks, the concepts are readily extended to other networks, such as host–pathogen networks or microbial association networks.

4.4   DISEASE SPREADING ON NETWORKS

Networks are also very useful to model the spread of disease in a population. We have already briefly discussed the SIR model for the spread of infectious diseases (§1.5.2). However, typical SIR models assume a well-mixed population, or at best some ad hoc models for the contact process. These are obviously approximations, and more realistic models assume an underlying network on which the disease propagates [34–36]. The study of the spread of disease on such networks has several parallels with percolation processes on graphs.

4.4.1   Percolation-based models

Moore and Newman [34] have studied simple models for the transmission of diseases on small-world networks. They focus on two key parameters—*susceptibility*, the probability of an exposed individual contracting disease, and *transmissibility*,

---

[2]Please consult the companion website for the book, for the colour image.

the probability that the contact between a susceptible individual and an infected individual will result in an infection. Moore and Newman showed that the spread of disease on this network is very similar to the standard percolation problems on graphs. Newman [35] showed that a large class of SIR models could be solved exactly on various kinds of networks, again using percolation theory and what are known as generating function methods.

4.4.2   Agent-based simulations

In 2004, Eubank and co-workers [36] developed a powerful agent-based simulation tool, EpiSims[3], to study disease outbreaks using social networks. The tool combines realistic estimates of population mobility, based on census and land-use data, with disease models that capture both the progress of disease within an individual as well as the transmission of disease between individuals. At the heart of their simulation is a large-scale dynamic contact network, which replaces the ODEs of the classic SIR models. EpiSims is based on the Transportation Analysis and Simulation System (TRANSIMS) developed at Los Alamos National Laboratory, which produces 'estimates' of social networks.

They modelled the social contact network as a bipartite graph. The bipartite graph G_{PL} has two types of vertices—people (P) and locations (L). Vertices in this graph are labelled with appropriate demographic or geographic information, and edges are labelled with arrival and departure times. As a case study, they presented a model of the city of Portland, USA, for which G_{PL} had ≈ 1.6 million vertices. Disease transmissions occur primarily between people who are co-located, while spread happens through the movement of people. Therefore, they considered two projections of G_{PL}, based on (i) people visiting the same location, and (ii) locations visited by the same people.

These projections are basically composed of just one type of nodes, with edges connecting nodes at a distance of two from each other on G_{PL}. This resulted in two disconnected graphs, G_P (containing only people vertices) and G_L (containing only location vertices). In G_P, the edges are labelled with the sets of time intervals during which the people were co-located. As a simplification, the authors made a static projection of G_P, which is a 'union' of all the time-resolved graphs, essentially a superset of all possible edges. A similar static projection was also made for G_L. Figure 4.3 illustrates a simple example to elucidate the nature of these contact maps and their various projections.

The people contact graph was seen to exhibit small-world properties, while the location network was scale-free. The modelling framework is very versatile and enables asking various kinds of questions: What is the maximum number of people at risk for a disease originating from a single infected person? This would translate to the size of the largest connected component in the people graph. Is mass vaccination a good strategy? Is shutting down key locations where people frequent (or vaccinating these people) a useful strategy to curtail spread? What are

---

[3]On a related note, recall the Vax (https://vax.herokuapp.com/) environment for simulating epidemic spreads on a network (§1.4.6).

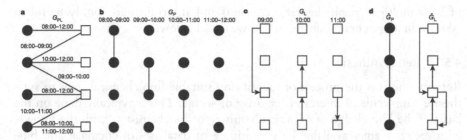

Figure 4.3 **An example of a small social contact network** (a) A bipartite graph G_{PL} with two types of vertices representing four people (P) and four locations (L). If person p visited location l, there is an edge in this graph between p and l. Vertices are labelled with appropriate demographic or geographic information, edges with arrival and departure times. (b, c) The two disconnected graphs G_P and G_L *induced* by connecting vertices that were separated by exactly two edges in G_{PL}. (d) The static projections $\hat{G}_P$ and $\hat{G}_L$ resulting from ignoring time labels in G_P and G_L. People are represented by filled circles, and locations, by open squares. Reproduced by permission from Springer Nature: S. Eubank, H. Guclu, V. S. A. Kumar, M. V. Marathe, A. Srinivasan, Z. Toroczkai, and N. Wang, "Modelling disease outbreaks in realistic urban social networks", *Nature* **429**:180–184 (2004) © Nature Publishing Group (2004).

the most important people to be vaccinated? What is the consequence of delays in response to an outbreak? The study presented several insights, which can be used to orchestrate a response to any epidemic outbreak.

4.5 MOLECULAR GRAPHS AND THEIR APPLICATIONS

The concept of molecular graphs was introduced by Arthur Cayley as early as 1874 [37], when he created *plerograms* and *kenograms* to represent molecular structures. Nodes in these graphs corresponded to molecules, while edges corresponded to covalent bonds. In plerograms, hydrogen atoms are explicitly included, unlike in kenograms, where they are suppressed because they can be readily inferred.

Chemical/molecular graphs have a variety of applications; see reference [38] for an excellent review. Classical graph problems such as graph isomorphism, sub-graph isomorphism, identification of maximum common sub-graphs have direct important applications in the matching of chemical molecules [39], identification of common sub-structures between molecules, and the identification of maximum common sub-structures . Heuristics have been developed on top of these basic algorithms for applications such as the identification of reaction atom mappings [40, 41] and other applications such as molecular classification [42]. One application, for identifying possible retrosynthetic pathways for the synthesis of molecules in metabolic networks [40], is a particularly illustrative example of how multiple graph-theoretic concepts and algorithms are highly useful in systems biology/chemoinformatics, as we detail below.

4.5.1 Retrosynthesis

Retrosynthesis is the process of identifying suitable (bio)chemical steps to synthesise a molecule of interest, *e.g.* a drug molecule. This is typically done on the basis of the knowledge of organic chemistry or biochemistry, with the knowledge being readily available in a multitude of databases of chemical and biochemical reactions. A nice review by Hadadi and Hatzimanikatis [43] overviews the various challenges in retrosynthesis, and several computational approaches to design retrosynthetic pathways. Although many computational approaches have been developed for retrosynthesis, many of them rely on hand-curated reaction rules/atom–atom mapping or cannot work on molecules not previously *seen* in the database. Raman and co-workers [40] developed an efficient method, Reaction-Miner, for predicting novel metabolic pathways for retrosynthesis by applying sub-graph mining.

Figure 4.4a illustrates the pipeline of the approach. ReactionMiner has two phases—the offline phase and the online phase. The offline phase involves graph mining of the KEGG reaction database (see Appendix C) to identify reaction *signatures*. Every molecule is represented as a graph (see Figure 4.4b for an example); every reaction involves the addition or removal of bonds between one or more atoms, and is captured by the addition or removal of edges in the molecular graphs.

The reaction signatures are essentially sub-graphs, *i.e.* sub-structures of the metabolites involved, around the *reaction centre*, where the addition or removal of bonds take place. Based on these reaction signatures, *reaction rules* are subsequently identified. These rules essentially capture the transformation of the various signatures in a reaction (Figure 4.4c) and comprise the reaction centre, reaction signature, the sub-graphs added or removed, as well as additional reactants (*e.g.* co-factors) necessary for the transformation to occur. Subsequently, the reaction rules so identified are embedded in a 'reaction rule network'. In this network, every node is a reaction rule, and edges connect reaction rules that can be successively applied. This representation of reaction rules on a network simplifies the problem of retrosynthetic pathway prediction to a simple path-finding on the rule network. In the online phase, the reaction rule network is searched to predict suitable pathways, on the arrival of a query A → B.

For the search on the network, a biologically motivated heuristic was used, which led to biologically relevant/natural pathways for biosynthesis being consistently ranked high amongst a number of possible alternatives. ReactionMiner also finds application in identifying pathways for xenobiotic degradation or uncovering pathways for the synthesis of known molecules, with unknown synthesis routes.

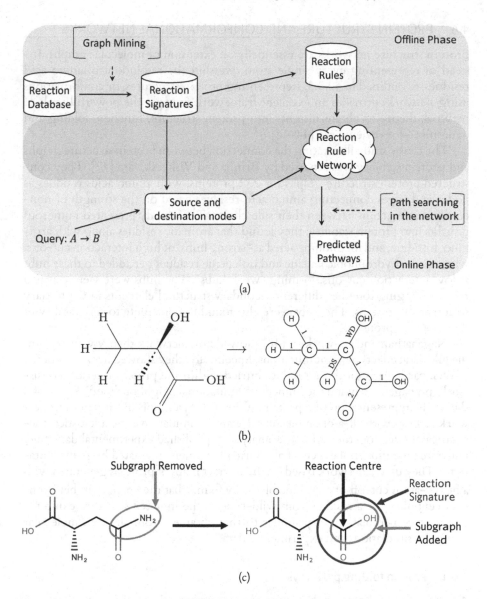

(a)

(b)

(c)

Figure 4.4 **Various aspects underlying the ReactionMiner algorithm.** (a) The ReactionMiner pipeline, outlining both the offline and online phases of the algorithm. (b) Conversion of an example molecule, D-lactate, to a graph. Note that double bonds are indicated by a changed edge label, as are wedges and dashes that represent bond stereochemistry. (c) Components of the reaction rule, for the conversion of L-Asparagine to L-Aspartic acid. Panels (a) and (b) are reprinted by permission from Oxford University Press: A. Sankar, S. Ranu, and K. Raman, "Predicting novel metabolic pathways through subgraph mining", *Bioinformatics* 33:3955–3963 (2017). © (2017) Oxford University Press.

4.6 PROTEIN STRUCTURE AND CONFORMATIONAL NETWORKS

Protein structure networks are essentially an extension of molecular graphs. Instead of representing every single atom, we shift the resolution to amino acid residues, as outlined in §2.5.3. Representing protein structures, or conformations using networks provides an excellent framework to apply the powerful tools of network theory to obtain insights into protein structure, function, folding and dynamics—for reviews, see [44–46].

The classic study illustrating the connection between protein structure graphs and protein stability was reported by Brinda and Vishveshwara [47]. They constructed protein structure graphs for 232 proteins, with amino acids residues as nodes and edges connecting amino acid residues based on the strength of non-covalent interaction between their side chains. Their study presented numerous insights into protein stability: they found that aromatic residues along with arginine, histidine, and methionine acted as "strong hubs" at high interaction cut-offs, whereas the hydrophobic leucine and isoleucine residues get added to these hubs at low interaction cut-offs, forming "weak hubs". The hubs were seen to play a role in bringing together different secondary structural elements in the tertiary structure of proteins. The hubs were also found to contribute to the stability of thermophilic proteins.

Naganathan and co-workers [48] showed how network theory well explains the allosteric effects of point mutations. Specifically, they showed how mutational effects, particularly of those residues buried within the protein interior, consistently propagated beyond the immediate 'mutational neighbourhood'. Figure 4.5 shows the representation of a protein, ubiquitin, as a graph. By integrating network analysis with all-atom molecular dynamics simulations, and a statistical mechanical model, together with a re-analysis of published experimental data, they illustrated the universal effects of allosteric phenomena mediated by point mutations. They used a network based on the contact maps of proteins generated with a heavy-atom cut-off of 6 Å. Notably, they found that the changes in betweenness centrality for a residue from wild-type to the mutated structure exhibited a pattern of percolation very similar to that obtained from all-atom molecular dynamics simulations and experimental data.

4.6.1 Protein folding pathways

An early study illustrating the power of protein networks to distinguish between conformations and their ability to fold was performed by Dokholyan et al [49]. Using two small proteins, CI2 and C-Src SH3, they showed how topological properties of the contact networks corresponding to different protein conformations could predict their kinetic ability to fold. They computed the average minimal distance of a node, L(i) as

$$L(i) = \frac{1}{n-1} \sum_{j=1}^{n} d_{ij} \qquad (4.1)$$

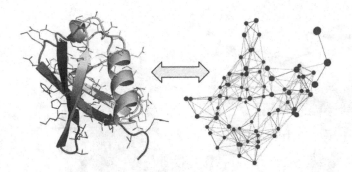

Figure 4.5 **Representation of a protein (ubiquitin) structure as a graph.** The left panel shows a cartoon representation of the structure, while the right panel shows the molecular graph representation. Every node in this network was an amino acid residue, and two residues were connected if they had an interaction energy corresponding to at least four atom–atom interactions. Reprinted with permission, from N. Rajasekaran, S. Suresh, S. Gopi, K. Raman, and A. N. Naganathan, "A general mechanism for the propagation of mutational effects in proteins", *Biochemistry* **56**:294–305 (2017). © (2017) American Chemical Society.

Note that this equation is practically the same as the measure of proximity in Equation 3.13, except that the denominator is $n - 1$ rather than n (which is inconsequential for very large graphs). Dokholyan *et al* showed that the L values of protein conformations can serve as a structurally reliable determinant of the pre- ($p_{fold} \approx 0$) and post-transition ($p_{fold} \approx 1$) fold states. Here, p_{fold} is the probability that a given conformation will fold to the native structure.

Another seminal study was carried out by Rao and Caflisch [50], where they performed a network analysis of the conformation space of a synthetic 20-residue anti-parallel β–sheet peptide, beta3s, sampled by molecular dynamics simulations. The conformation space 'network' was established with snapshots along the trajectory being grouped according to secondary structure into the nodes, and edges being the transitions between these conformations. They found the conformation space network to be scale-free. They also found that it contained hubs like the native state, which was the most populated free energy basin. The native basin of the structured peptide was also shown to have a hierarchical organisation of conformations. Their results also pointed towards the existence of two main average folding pathways, and an improved understanding of the heterogeneity of the "transition state ensemble"[4]. Lastly, they also showed that the average neighbour degree[5], k_{nn}, correlated with p_{fold}. Other studies have also exploited network theory to derive interesting insights into protein folding and dynamics [51, 52].

A very interesting study employed graph theory to shed light on the variety of microscopic routes that are accessible to a folding protein molecule [53].

---

[4]An ensemble of various conformations of the protein with varying abilities to fold/unfold/misfold.
[5]The average of the degrees of all nodes neighbouring a given node.

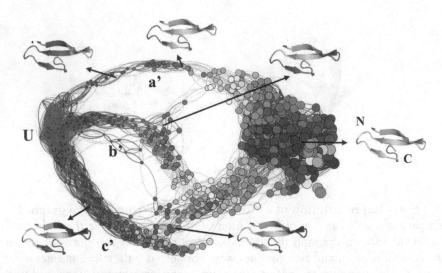

Figure 4.6 **Network of folding trajectories for WW domain.** The folding landscape is split into three distinct pathways, a′, b′, and c′, with varied fluxes. Reproduced by permission, from S. Gopi, A. Singh, S. Suresh, S. Paul, S. Ranu, and A. N. Naganathan, "Toward a quantitative description of microscopic pathway heterogeneity in protein folding", *Physical Chemistry Chemical Physics* **19**:20891–20903 (2017). © Royal Society of Chemistry (2017).

Naganathan and co-workers studied a graph of (a simplified subset of) possible conformational 'micro-states' for a globular protein. States that are reachable from one another, *i.e.* having nearby free energy values, are connected by edges. Performing a Markov clustering[6] on the above 'micro-state' graph, they derived a macro-state graph where nodes now represent communities, or clusters of micro-states. They also quantified the probable flux through the various transitions, thereby predicting several plausible folding pathways.

Figure 4.6 presents a simple visualisation of the complex folding network of the protein E3 ubiquitin-protein ligase NEDD4 WW domain ('WW') at 298 K. U represents the fully unfolded state of the protein, while N represents the native folded state. The size of every node is proportional to the number of structured ("folded") residues and the colour[7] highlights the micro-state free energy in the spectral scale (blue to red represents the linearly increasing free energy value in kJ/mol). The graph is organised such that the most unfolded and folded micro-states are close to the extreme left and right, respectively. Structural features of some representative paths are displayed as cartoons; the grey regions represent unfolded residues while coloured regions represent folded residues.

---

[6]An unsupervised clustering algorithm for community detection on graphs; see references [54, 55].

[7]Please consult the companion website for the book, for the colour image.

4.7 LINK PREDICTION

Another application of network theory stems from the prediction of false negative (*i.e.* missing) links, or false positive (*i.e.* spurious) links in complex networks. *Link prediction* can also involve the prediction of links that are most likely to form in the near future, given the current topology of the network. This problem has been studied on a variety of networks [56, 57] including protein interaction networks [58]. Clauset, Moore, and Newman [59] showed that it is possible to infer the hierarchical structure of a network and then exploit it to predict missing links. Link prediction has also been applied to predict cancer drug sensitivity [60, 61]. Drug response prediction can also be cast as a link prediction problem, as discussed in reference [61], where the authors illustrated their superior performance on the NCI-DREAM Drug Sensitivity Prediction Challenge dataset. This particular DREAM challenge [62] provided data on breast cancer cell lines and compiled 44 drug response prediction algorithms from the research community.

In social networks, the *triadic closure* principle is typically used to predict new 'friendships'—if A and B have a large number of common friends, then, there is a good chance that A and B are also friends. However, Barabási and co-workers showed that this does not typically apply to protein networks [58], where they show that paths of length three ('L3') predict neighbours with higher reliability. That is, two proteins P_1 and P_2 are expected to interact if they are linked by multiple paths of length 3 in the network. They also showed that longer paths (up to $L = 8$) are not as useful (predictive) as $L = 3$. Another useful albeit complex method is the "stochastic block model" approach of Guimerà and Sales-Pardo [57].

Importantly, link prediction methods can throw additional light on the underlying biology [57]. For instance, if a link prediction method predicts a very low probability for a particular interaction that is known to exist (or vice-versa) through experiments or otherwise, this would imply that the interaction is *surprising*, and has a biological significance. In other words, such an interaction is rare amongst nodes that are otherwise similar, and the interaction was perhaps evolutionarily or functionally important.

EXERCISES

4.1 Draw the molecular graphs for *cis*-but-2-ene and *trans*-but-2-ene. Are they isomorphic? What about D-glyceraldehyde and L-glyceraldehyde?

[LAB EXERCISES]

4.2 Write a simple piece of MATLAB code to perform DiNA. Your code must take in as input two networks, and output the 'differential' network.

4.3 In our 2014 study [2], we studied the centrality–lethality hypothesis using interactomes from STRING, by using a cut-off of 700. In the years since, many new interactions have been added to STRING. Can you plot and compare the degree distributions for essential and non-essential proteins, from the current interactome available in STRING? Use essentiality data from DEG (http://www.essentialgene.org/). What happens if you change the cut-off to 900? How do the distributions change?

4.4 Obtain the anonymised networks for the Disease Module Identification DREAM Challenge from https://www.synapse.org/#!Synapse:syn6156775. Use the MONET toolbox [23] to identify modules in this network using the 'R1' random-walk based algorithm. Also use a simple hierarchical clustering algorithm (see §3.3.2) to identify modules in this network, and compare the distributions of module sizes between the two approaches.

4.5 From the Therapeutic Target Database (http://db.idrblab.net/ttd/; [63]), obtain the file that provides "Target to drug mapping with mode of action". Construct a bipartite graph based on the information provided in this file. Now, can you identify *similar* drugs, *i.e.* how would you define the similarity between drugs? What possible strategies can you use? Can you cluster the drugs?

REFERENCES

[1] H. Jeong, S. P. Mason, A.-L. Barabási, and Z. N. Oltvai, "Lethality and centrality in protein networks", *Nature* **411**:41–42 (2001).

[2] K. Raman, N. Damaraju, and G. K. Joshi, "The organisational structure of protein networks: revisiting the centrality–lethality hypothesis", *Systems and Synthetic Biology* **8**:73–81 (2014).

[3] A. Franceschini, D. Szklarczyk, S. Frankild, et al., "STRING v9.1: protein-protein interaction networks, with increased coverage and integration", *Nucleic Acids Research* **41**:D808–D815 (2013).

[4] R. Zhang, H.-Y. Ou, and C.-T. Zhang, "DEG: a database of essential genes", *Nucleic Acids Research* **32**:D271–272 (2004).

[5] X. He and J. Zhang, "Why do hubs tend to be essential in protein networks?", *PLoS Genetics* **2**:e88 (2006).

[6] E. Zotenko, J. Mestre, D. P. O'Leary, and T. M. Przytycka, "Why do hubs in the yeast protein interaction network tend to be essential: reexamining the connection between the network topology and essentiality", *PLoS Computational Biology* **4**:e1000140 (2008).

[7] K. Song, T. Tong, and F. Wu, "Predicting essential genes in prokaryotic genomes using a linear method: ZUPLS", *Integrative Biology* **6**:460–469 (2014).

[8] X. Liu, B.-J. Wang, L. Xu, H.-L. Tang, and G.-Q. Xu, "Selection of key sequence-based features for prediction of essential genes in 31 diverse bacterial species", *PLoS ONE* **12**:e0174638 (2017).

[9]   Y.-C. Hwang, C.-C. Lin, J.-Y. Chang, H. Mori, H.-F. Juan, and H.-C. Huang, "Predicting essential genes based on network and sequence analysis", *Molecular BioSystems* **5**:1672–1678 (2009).

[10] K. Azhagesan, B. Ravindran, and K. Raman, "Network-based features enable prediction of essential genes across diverse organisms", *PLoS ONE* **13**:e0208722 (2018).

[11] A.-L. Barabási, N. Gulbahce, and J. Loscalzo, "Network medicine: a network-based approach to human disease", *Nature Reviews Genetics* **12**:56–68 (2010).

[12] K.-I. Goh, M. E. Cusick, D. Valle, B. Childs, M. Vidal, and A.-L. Barabási, "The human disease network", *Proceedings of the National Academy of Sciences USA* **104**:8685–8690 (2007).

[13] A. Hamosh, A. F. Scott, J. S. Amberger, C. A. Bocchini, and V. A. McKusick, "Online Mendelian Inheritance in Man (OMIM), a knowledgebase of human genes and genetic disorders", *Nucleic Acids Research* **33**:D514–D517 (2005).

[14] S. Choobdar, M. E. Ahsen, J. Crawford, et al., "Assessment of network module identification across complex diseases", *Nature Methods* **16**:843–852 (2019).

[15] E. L. Huttlin, R. J. Bruckner, J. A. Paulo, et al., "Architecture of the human interactome defines protein communities and disease networks", *Nature* **545**:505–509 (2017).

[16] A. Rzhetsky, D. Wajngurt, N. Park, and T. Zheng, "Probing genetic overlap among complex human phenotypes", *Proceedings of the National Academy of Sciences USA* **104**:11694–11699 (2007).

[17] X. Zhou, J. Menche, A.-L. Barabási, and A. Sharma, "Human symptoms–disease network", *Nature Communications* **5**:1–10 (2014).

[18] M. A. Yıldırım, K.-I. Goh, M. E. Cusick, A.-L. Barabási, and M. Vidal, "Drug-target network", *Nature Biotechnology* **25**:1119–1126 (2007).

[19] J.-U. Peters, "Polypharmacology—Foe or Friend?", *Journal of Medicinal Chemistry* **56**:8955–8971 (2013).

[20] L. H. Hartwell, J. J. Hopfield, S. Leibler, and A. W. Murray, "From molecular to modular cell biology", *Nature* **402**:C47–C52 (1999).

[21] S. D. Ghiassian, J. Menche, and A.-L. Barabási, "A DIseAse MOdule Detection (DIAMOnD) algorithm derived from a systematic analysis of connectivity patterns of disease proteins in the human interactome", *PLoS Computational Biology* **11**:e1004120 (2015).

[22] D. Lamparter, D. Marbach, R. Rueedi, Z. Kutalik, and S. Bergmann, "Fast and rigorous computation of gene and pathway scores from SNP-based summary statistics", *PLoS Computational Biology* **12**:e1004714 (2016).

[23]   M. Tomasoni, S. Gómez, J. Crawford, et al., "MONET: a toolbox integrating top-performing methods for network modularization", *Bioinformatics* **36**:3920–3921 (2020).

[24]   B. Tripathi, S. Parthasarathy, H. Sinha, K. Raman, and B. Ravindran, "Adapting community detection algorithms for disease module identification in heterogeneous biological networks", *Frontiers in Genetics* **10**:164 (2019).

[25]   M. Vidal, M. E. Cusick, and A.-L. Barabási, "Interactome networks and human disease", *Cell* **144**:986–998 (2011).

[26]   Q. Zhong, N. Simonis, Q.-R. R. Li, et al., "Edgetic perturbation models of human inherited disorders", *Molecular Systems Biology* **5**:321 (2009).

[27]   M. Dreze, B. Charloteaux, S. Milstein, et al., "'Edgetic' perturbation of a *C. elegans* BCL2 ortholog", *Nature Methods* **6**:843–849 (2009).

[28]   N. Sahni, S. Yi, M. Taipale, et al., "Widespread macromolecular interaction perturbations in human genetic disorders", *Cell* **161**:647–660 (2015).

[29]   T. Ideker and N. J. Krogan, "Differential network biology", *Molecular Systems Biology* **8**:565 (2012).

[30]   Y. Lichtblau, K. Zimmermann, B. Haldemann, D. Lenze, M. Hummel, and U. Leser, "Comparative assessment of differential network analysis methods", *Briefings in Bioinformatics* **18**:837–850 (2017).

[31]   S. Bandyopadhyay, M. Mehta, D. Kuo, et al., "Rewiring of genetic networks in response to DNA damage", *Science* **330**:1385–1389 (2010).

[32]   R. Gill, S. Datta, and S. Datta, "A statistical framework for differential network analysis from microarray data", *BMC Bioinformatics* **11**:95 (2010).

[33]   M. J. Ha, V. Baladandayuthapani, and K.-A. Do, "DINGO: differential network analysis in genomics", *Bioinformatics* **31**:3413–3420 (2015).

[34]   C. Moore and M. E. J. Newman, "Epidemics and percolation in small-world networks", *Physical Review E* **61**:5678–5682 (2000).

[35]   M. E. J. Newman, "Spread of epidemic disease on networks", *Physical Review E* **66**:016128+ (2002).

[36]   S. Eubank, H. Guclu, V. S. A. Kumar, et al., "Modelling disease outbreaks in realistic urban social networks", *Nature* **429**:180–184 (2004).

[37]   A. Cayley, "On the mathematical theory of isomers", *The London, Edinburgh, and Dublin Philosophical Magazine and Journal of Science* **47**:444–447 (1874).

[38]   T. Akutsu and H. Nagamochi, "Comparison and enumeration of chemical graphs", *Computational and Structural Biotechnology Journal* **5**:e201302004 (2013).

[39]   J. W. Raymond and P. Willett, "Maximum common subgraph isomorphism algorithms for the matching of chemical structures", *Journal of Computer-Aided Molecular Design* **16**:521–533 (2002).

[40] A. Sankar, S. Ranu, and K. Raman, "Predicting novel metabolic pathways through subgraph mining", *Bioinformatics* **33**:3955–3963 (2017).

[41] T. Akutsu, "Efficient extraction of mapping rules of atoms from enzymatic reaction data", *Journal of Computational Biology* **11**:449–462 (2004).

[42] S. Ranu and A. K. Singh, "Mining statistically significant molecular substructures for efficient molecular classification", *Journal of Chemical Information and Modeling* **49**:2537–2550 (2009).

[43] N. Hadadi and V. Hatzimanikatis, "Design of computational retrobiosyn-thesis tools for the design of *de novo* synthetic pathways", *Current Opinion in Chemical Biology* **28**:99–104 (2015).

[44] C. Böde, I. A. Kovács, M. S. Szalay, R. Palotai, T. Korcsmáros, and P. Cser-mely, "Network analysis of protein dynamics", *FEBS Letters* **581**:2776–2782 (2007).

[45] L. H. Greene, "Protein structure networks", *Briefings in Functional Genomics* **11**:469–478 (2012).

[46] M. Bhattacharyya, S. Ghosh, and S. Vishveshwara, "Protein structure and function: looking through the network of side-chain interactions", *Current Protein & Peptide Science* **17**:4–25 (2016).

[47] K. V. Brinda and S. Vishveshwara, "A network representation of protein structures: implications for protein stability", *Biophysical Journal* **89**:4159–4170 (2005).

[48] N. Rajasekaran, S. Suresh, S. Gopi, K. Raman, and A. N. Naganathan, "A general mechanism for the propagation of mutational effects in proteins", *Biochemistry* **56**:294–305 (2017).

[49] N. V. Dokholyan, L. Li, F. Ding, and E. I. Shakhnovich, "Topological determinants of protein folding", *Proceedings of the National Academy of Sciences USA* **99**:8637–8641 (2002).

[50] F. Rao and A. Caflisch, "The protein folding network", *Journal of Molecular Biology* **342**:299–306 (2004).

[51] M. Vendruscolo, N. V. Dokholyan, E. Paci, and M. Karplus, "Small-world view of the amino acids that play a key role in protein folding", *Physical Review E* **65**:061910 (2002).

[52] A. Ghosh, K. V. Brinda, and S. Vishveshwara, "Dynamics of lysozyme structure network: probing the process of unfolding", *Biophysical Journal* **92**:2523–2535 (2007).

[53] S. Gopi, A. Singh, S. Suresh, S. Paul, S. Ranu, and A. N. Naganathan, "Toward a quantitative description of microscopic pathway heterogeneity in protein folding", *Physical Chemistry Chemical Physics* **19**:20891–20903 (2017).

[54] S. Van Dongen, "Graph clustering via a discrete uncoupling process", *SIAM Journal on Matrix Analysis and Applications* **30**:121–141 (2008).

[55] A. J. Enright, S. Van Dongen, and C. A. Ouzounis, "An efficient algorithm for large-scale detection of protein families", *Nucleic Acids Research* **30**:1575–1584 (2002).

[56] D. Liben-Nowell and J. Kleinberg, "The link-prediction problem for social networks", *Journal of the American Society for Information Science and Technology* **58**:1019–1031 (2007).

[57] R. Guimerà and M. Sales-Pardo, "Missing and spurious interactions and the reconstruction of complex networks", *Proceedings of the National Academy of Sciences USA* **106**:22073–22078 (2009).

[58] I. A. Kovács, K. Luck, K. Spirohn, et al., "Network-based prediction of protein interactions", *Nature Communications* **10**:1–8 (2019).

[59] A. Clauset, C. Moore, and M. E. J. Newman, "Hierarchical structure and the prediction of missing links in networks", *Nature* **453**:98–101 (2008).

[60] T. Turki and Z. Wei, "A link prediction approach to cancer drug sensitivity prediction", *BMC Systems Biology* **11**:94 (2017).

[61] Z. Stanfield, M. Coşkun, and M. Koyutürk, "Drug response prediction as a link prediction problem", *Scientific Reports* **7**:1–13 (2017).

[62] J. C. Costello, L. M. Heiser, E. Georgii, et al., "A community effort to assess and improve drug sensitivity prediction algorithms", *Nature Biotechnology* **32**:1202–1212 (2014).

[63] H. Yang, C. Qin, Y. H. Li, et al., "Therapeutic target database update 2016: enriched resource for bench to clinical drug target and targeted pathway information", *Nucleic Acids Research* **44**:D1069–D1074 (2016).

FURTHER READING

O. Mason and M. Verwoerd, "Graph theory and networks in Biology", *IET Systems Biology* **1**:89–119 (2007).

G. A. Pavlopoulos, M. Secrier, C. N. Moschopoulos, et al., "Using graph theory to analyze biological networks", *BioData Mining* **4**:10 (2011).

K. Mitra, A.-R. Carvunis, S. K. Ramesh, and T. Ideker, "Integrative approaches for finding modular structure in biological networks", *Nature Reviews Genetics* **14**:719–732 (2013).

P. Csermely, K. S. Sandhu, E. Hazai, et al., "Disordered proteins and network disorder in network descriptions of protein structure, dynamics and function: hypotheses and a comprehensive review", *Current Protein & Peptide Science* **13**:19–33 (2012).

II

Dynamic Modelling

Introduction to dynamic modelling

CONTENTS

D YNAMIC MODELLING represents a departure from the static networks of the last three chapters. Of course, it is possible to study dynamic processes on networks, such as how a network evolves with time, or how the interacting partners of a protein change, or how the neighbours of an amino acid residue in a protein change as the protein folds or changes its conformation. Beyond a network-level perspective, however, it is often interesting to observe the dynamics of a system, in terms of the changes in levels of various species (metabolites, genes, or proteins, for instance).

Dynamic models are of utmost importance in biology, as they can capture key interactions and processes, such as the spread of infectious diseases in a population, the feedback inhibition of an enzyme, or the signalling cascade activated in response to glucose ingestion, or bacterial chemotaxis. More specifically, in this part of the book, we will focus on dynamic models built on the mathematical framework of differential equations. Such models are widely applied in biology, despite the various challenges arising from the size and complexity of the system, the availability of experimental data, and the difficulties in estimating the model parameters accurately and uniquely.

A very broad class of dynamic models, namely *kinetic models* can be expressed as

$$\frac{dx_i}{dt} = f_i(x_1, x_2, \ldots, x_n; \Theta) \quad i = 1, 2, \ldots, n \tag{5.1}$$

where x_i denotes the amount (*i.e.* concentration) of some biological molecule, say, an mRNA, a protein or a metabolite, and f_i is the rate *law* that captures how levels of x_i change in response to the levels of all the other x's. Θ represents the set of parameters, such as rate constants, which govern these rate equations.

Dynamic models outside biology

Dynamic models are ubiquitous, across disciplines. A very common dynamic model is one that captures the flow of water out of an orifice. In fact, it is one of the most common mathematical models taught in undergraduate courses on process control or mathematical modelling. Other models include those of planetary motion, which help predict eclipses, or those of current flow in electrical circuits. In chemical engineering, dynamic models are used to model chemical reactors to understand flow rates and conversion rates in chemical processes. In the financial world, dynamic models predict the behaviour of stocks. Many such models are also illustrated in Figure 1.2.

5.1 CONSTRUCTING DYNAMIC MODELS

How do we go about creating dynamic models? We first need to establish the scope, objectives, and feasibility. While discussing the scope of the model, as discussed in §1.4.1, we first need to decide the factors that the model will account for. Next, we need to ensure that the variable we desire to predict is well-captured in the model. Lastly, it must be feasible to build the model, with the data available—in terms of the kind of data, their quality, and quantity. It is essential to have *enough* data so that it can be used not only for model building, but also for validation (see also §1.4.5).

The Lotka–Volterra predator–prey model (§1.5.1) and the SIR model (§1.5.2) for the spread of infectious diseases are good examples of how one goes about constructing mechanistic dynamic models, based on first principles. In this chapter, however, we will focus on a different class of models, namely kinetic models.

5.1.1 Modelling a generic biochemical system

The archetypal dynamic model in systems biology is that of a biochemical system, which may comprise one or more processes, including metabolic reactions, gene transcription, protein translation, protein binding, allosteric inhibition, and so on. The first task is to correctly determine the topology of the system—the key species and their interactions. If the system is small enough, it is instructive to have a diagram[1]. Subsequently, for each interaction in the model, the right kind of kinetics must be chosen. Typically, for a ligand binding to an enzyme,

---

[1]Diagrams should be compliant with SBGN, see Appendix D.

or two proteins binding to one another, we may use mass–action kinetics; for an enzyme-catalysed reaction, we could use Michaelis–Menten kinetics; if there are inhibitions/activation reactions, suitable extensions of the Michaelis–Menten kinetic equations can be used; if there is co-operative binding between the enzyme and substrate, we could use Hill kinetics. We will discuss some of these kinetic models in the following sections.

5.2 MASS-ACTION KINETIC MODELS

The *law* of mass-action is the all–important *model* underlying most kinetic models. Building on the work of Guldberg and Waage in 1860s (which they reviewed in [1]), the law has been described in various forms. While Guldberg and Waage speak of the rate of chemical reactions being proportion to the "active masses" of the reactants, in its modern form, the law is taken to state that the rate of a chemical reaction is proportional to the probability of collision between the reactants in a given system. In a system where the concentrations are very low, of the order of a few molecules, such that the collisions are rare, stochastic effects will predominate. On the other hand, in a typical reactor, or in a cell, this probability will be proportional to the concentrations of the participating molecules (reactants) to the power of their respective *molecularities*. Molecularity refers to the number of colliding molecular entities that are involved in a single *elementary reaction* step. An elementary reaction is a chemical reaction where one or more chemical species react directly to form product(s) in a single reaction step, with a *single transition state*.

Consider a simple reaction,

$$A + B \rightleftharpoons 2C$$

The reaction rate can be written as

$$v = v_f - v_b = k_f[A][B] - k_b[C]^2$$

where v_f is the rate of the forward reaction, v_b is the rate of the backward reaction, $[\cdot]$ represents the concentration of a given species, and k_f and k_b represent the rate constants for the forward and backward reactions, respectively. The molecularities of A, B, and C are 1, 1, and 2, respectively. The equilibrium constant, K_{eq}, for this reaction is given as

$$K_{eq} = \frac{[C]_{eq}^2}{[A]_{eq}[B]_{eq}}$$

where $[\cdot]_{eq}$ denotes the equilibrium concentration of a given species. In a standard ordinary differential equation (ODE) form, we can write

$$-v_b = -\frac{dA}{dt} = -\frac{dB}{dt} = \frac{1}{2}\frac{dC}{dt} = v_f$$

Note that there are many tacit assumptions in mass–action, such as free diffusion. Due to its simplicity, mass–action has been widely applied in many scenarios, and can also be generalised to situations where the vanilla mass–action kinetics do not capture the dynamics correctly, as we will discuss in §5.4.1.

5.3   MODELLING ENZYME KINETICS

5.3.1   The Michaelis–Menten model

As with any modelling exercise, let us think about the scope, objectives, and feasibility. The choice of scope is crucial, as there are a large number of factors that affect the rate of an enzyme-catalysed reaction, such as temperature, pH, concentrations of reactants, and products, concentration of the enzyme itself, the levels of any potential inhibitors/activators, and so on. *In vivo*, in most scenarios, temperature and pH may not change—if they do, we can always extend our model. All classic Michaelis–Menten models focus on the relationship between the rate of the enzyme-catalysed reaction and the concentrations of the various species—enzyme, substrates, products, inhibitors, and activators. Note that the activators and inhibitors can be of different types as well, and differ in both the nature of their interactions with the enzyme, and consequently, their effect on the reaction and its rate.

The main objective of these models is to capture the rate of the reaction, typically at a given substrate concentration, at a given point in time. In terms of feasibility, it is essential to have the right kind of data for estimating the parameters of the Michaelis–Menten model. The data should be available at multiple time-points. Many of these points will be elaborated in later sections in this chapter.

Recall[2] that enzyme catalysis greatly accelerates the rates of a biochemical reaction—both the forward and backward reactions. A useful term to recall is the *turnover number*, k_{cat}, which is defined as the number of molecules of a substrate that an enzyme can convert to product per catalytic site per unit of time.

The reactions that are modelled in a standard Michaelis–Menten model are as follows:

$$E + S \underset{k_{-1}}{\overset{k_1}{\rightleftharpoons}} ES \underset{k_{-2}}{\overset{k_2}{\rightleftharpoons}} E + P \tag{5.2}$$

Now, assuming mass–action kinetics, we can write out ODEs for describing the change in concentration of each of the species:

$$\frac{d[S]}{dt} = -k_1[E][S] + k_{-1}[ES] \tag{5.3a}$$

$$\frac{d[ES]}{dt} = k_1[E][S] - k_{-1}[ES] - k_2[ES] + k_{-2}[E][P] \tag{5.3b}$$

---

[2]It is assumed here, that the reader is fairly familiar with the basics of enzyme catalysis, which is discussed at length in many a biochemistry textbook.

$$\frac{d[E]}{dt} = -k_1[E][S] + k_{-1}[ES] + k_2[ES] - k_{-2}[E][P] \tag{5.3c}$$

$$\frac{d[P]}{dt} = k_2[ES] - k_{-2}[E][P] \tag{5.3d}$$

Many textbooks assume that the second reaction is irreversible; this would mean that $k_{-2} = 0$, which can be over-simplifying at times, although it holds good in some cellular scenarios. A similar, but somewhat better approximation is to calculate the initial rate of the reaction v_0, where $[P] = 0$, and the terms involving k_{-2} similarly vanish, yielding the simpler equations:

$$\frac{d[S]}{dt} = -k_1[E][S] + k_{-1}[ES] \tag{5.4a}$$

$$\frac{d[ES]}{dt} = k_1[E][S] - (k_{-1} + k_2)[ES] \tag{5.4b}$$

$$\frac{d[E]}{dt} = -k_1[E][S] + (k_{-1} + k_2)[ES] \tag{5.4c}$$

$$\frac{d[P]}{dt} = k_2[ES] \tag{5.4d}$$

The rate of the reaction is basically the rate of product formation (or the negative of the rate of substrate consumption):

$$v_0 = \left.\frac{d[P]}{dt}\right|_{t=0} = -\left.\frac{d[S]}{dt}\right|_{t=0} \tag{5.5}$$

We also have another algebraic equation, that captures the conservation of all enzyme-containing species:

$$[E] + [ES] = [E]_0 \tag{5.6}$$

This also means that

$$\frac{d[E]}{dt} + \frac{d[ES]}{dt} = 0, \tag{5.7}$$

which can also be observed from Equations 5.4c and 5.4b. This system of ODEs cannot be easily solved analytically, but can be easily solved numerically[3], as we will see later in this chapter.

Either of two other approximations is popularly used to simplify the above equations—the original Michaelis–Menten[4] approximation [3], or the Briggs–Haldane approximation [4]. The original Michaelis–Menten assumption is that of *quasi-equilibrium*—they assumed that the substrate is present in much higher quantity, saturating the enzyme, *i.e.* $[S] \gg [E]_0$, and that the product formation reaction is the bottleneck, *i.e.* slow step ($k_2 \ll k_1, k_{-1}$). Together, these mean that

---

[3]Also see Lab Exercises.
[4]For a translation, see reference [2].

[S] does not appreciably change with time, i.e. $\frac{d[S]}{dt} = 0$. Applying this, along with Equation 5.7,

$$k_1[E][S] = k_{-1}[ES]$$
$$\Rightarrow k_1([E]_0 - [ES])[S] = k_{-1}[ES]$$
$$\Rightarrow [ES] = \frac{[E]_0[S]}{K_d + [S]}, \qquad (5.8)$$

where $K_d = \frac{k_{-1}}{k_1}$, the dissociation constant for ES. Ultimately,

$$v = k_2[ES] \qquad (5.9)$$
$$= \frac{k_2[E]_0[S]}{K_d + [S]} \qquad (5.10)$$

where k_2 is the same as k_{cat}, the turnover number of the enzyme.

Briggs and Haldane, on the other hand, made the *quasi-steady-state approximation*, assuming that [ES] does not change appreciably with time, particularly in the time-scale of product formation, i.e. $\frac{d[ES]}{dt} = 0$. Therefore,

$$k_1[E][S] = (k_{-1} + k_2)[ES]$$
$$\Rightarrow k_1([E]_0 - [ES])[S] = (k_{-1} + k_2)[ES]$$
$$\Rightarrow [ES] = \frac{k_2[E]_0[S]}{K_M + [S]}, \qquad (5.11)$$

where $K_M = \frac{k_{-1} + k_2}{k_1}$ is the Michaelis constant, and $k_{cat}[E]_0$ represents the maximum rate of the reaction, v_{max}. Combining Equations 5.9 and 5.11,

$$v = \frac{k_{cat}[E]_0[S]}{K_M + [S]} \qquad (5.12)$$
$$= \frac{v_{max}[S]}{K_M + [S]} \qquad (5.13)$$

Figure 5.1 illustrates a plot of rate of the reaction (v) versus substrate concentration ([S]). Note that at a substrate concentration equal to K_M, the reaction rate will be $0.5v_{max}$, as can be seen from Equation 5.13. The units of k_{cat} are s^{-1}, while the units for K_M are the same as that of (substrate) concentration, e.g. mM. Higher the K_M, lower the affinity of the enzyme for substrate, i.e. higher the [S] at which $0.5v_{max}$ is attained.

It is critical to remember the approximations used in deriving the Michaelis–Menten equation; in a regime where the underlying assumptions do not hold, the fit obtained is bound to be poor.

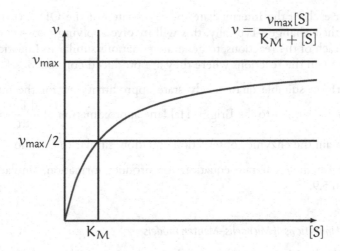

$$v = \frac{v_{max}[S]}{K_M + [S]}$$

Figure 5.1 **Plot of the Michaelis–Menten saturation kinetics.** Note that the value v_{max} is approached asymptotically, at very high [S], where the reaction rate becomes nearly independent of [S] (zero-order kinetics).

5.3.1.1 Extending the Michaelis–Menten model

The Michaelis–Menten model has been widely applied in other scenarios, where one or more inhibitors are present. For example, some common scenarios are characterised in biochemistry literature:

- Competitive inhibition, where S and I compete for E

- Uncompetitive inhibition, where I binds to ES to form ESI

- Non-competitive inhibition, where I binds well to both E and ES

- Mixed inhibition, which is similar to non-competitive inhibition, but the inhibitor has more affinity to one of E/ES

- Suicide inhibition, essentially an irreversible inhibition, where the EI complex is formed in an irreversible reaction

- Substrate inhibition, where S binds ES to form ESS, which cannot form a product directly

All of these cases can be systematically tackled in a fashion similar to our approach for deriving the standard Michaelis–Menten equation, as described in reference [5]:

1. Identify the topology of the system, from the nature of bindings and interactions: where does the inhibitor bind? Which species does it affect? What are all the intermediates? The topology is likely to be more complex than the standard Michaelis–Menten reactions (Equation 5.2).

2. For each of the intermediate species, write out the ODEs corresponding to their change—typically, this will involve applying mass-action kinetics to each of the reactions, to generate equations similar to Equations 5.4a–d, based on the reactions where they are produced/consumed.

3. Perform suitable quasi-steady-state approximations, for the intermediate species, similar to the Briggs–Haldane approximation of $\dfrac{d[ES]}{dt} = 0$.

4. Obtain the enzyme conservation equation, similar to Equation 5.6.

5. Derive the final rate equation for product formation, similar to Equation 5.9.

5.3.1.2 Limitations of Michaelis–Menten models

The Michaelis–Menten model considers only the initial rate, *i.e.* no product is present. Note that this is not the same as assuming $k_{-2} = 0$, although there are cases where the product formation reaction can be *practically* irreversible in cells, *e.g.* where the substrate concentration is much higher than product concentration, or when the product is continually removed. Of course, there are also the approximations arising from quasi-steady-state or quasi-equilibrium assumptions.

Also remember that the law of mass-action itself relies on free diffusion, whereas conditions inside the cell may be quite different, more resembling a *colloidal suspension* than just water. Further, spatial homogeneity is assumed, and high numbers of all molecules, so that we do not have to worry about stochasticity. Most importantly, we must remember that rate laws are approximate—and we must never lose track of the assumptions that have gone into building the model.

5.3.2 Co-operativity: Hill kinetics

Some systems do not display the hyperbolic Michaelis–Menten curve, but exhibit *sigmoidal* kinetics, *i.e.* an S-like curve, as shown in Figure 5.2. The classic example of co-operative binding kinetics is that of oxygen binding to haemoglobin. Deoxy-haemoglobin has a low affinity for oxygen. When oxygen molecules start binding the haeme group in haemoglobin, the affinity for oxygen increases. Haeme with three oxygen molecules bound has a very high affinity for oxygen, ≈ 300 times higher. The Hill equation [6] is of the form:

$$v = \frac{v_{max} S^n}{k' + S^n} \tag{5.14}$$

where n represents the Hill coefficient and k' represents the "Hill constant"[5]. v_{max} represents the maximum velocity of the reaction, as in the Michaelis–Menten

---

[5] k' may also be written as $K_{0.5}$, the "half-maximal concentration constant". Note, however, that the Hill equation was not originally derived in this form.

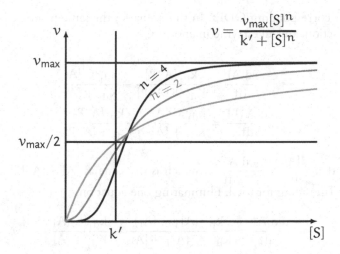

$$v = \frac{v_{max}[S]^n}{k' + [S]^n}$$

Figure 5.2 **Plot of sigmoidal Hill kinetics.** The dotted grey line indicates the Michaelis–Menten curve. The continuous grey and black curves represent Hill kinetics with $n = 2$ and $n = 4$, respectively.

equation. k' again provides an estimate of the affinity of the enzyme for the substrate, and is related to half-maximal substrate concentration $[S]_{0.5}$ as:

$$k' = [S]_{0.5}^n \qquad (5.15)$$

The Hill equation is the model of choice to capture the dynamics of systems where co-operativity is known to exist. The Hill coefficient for oxygen binding haemoglobin is close to 3, indicating a high positive co-operativity; a Hill coefficient of 1 signifies independent binding, *i.e.* no co-operativity. $n < 1$ signifies negative co-operativity.

5.3.3   An illustrative example: a three-node oscillator

Consider a simple network comprising three proteins, each of which can be phosphorylated. Protein A catalyses the dephosphorylation of B; B catalyses the dephosphorylation of C, while C catalyses the phosphorylation of A to its phosphorylated form, denoted $A^{\sim P}$. Let us build a dynamic model that captures the dynamics of each of the proteins. The reactions will be [7]:

$$A \underset{}{\overset{C}{\rightleftharpoons}} A^{\sim P} \qquad (5.16a)$$

$$B \underset{A}{\rightleftharpoons} B^{\sim P} \qquad (5.16b)$$

$$C \underset{B}{\rightleftharpoons} C^{\sim P} \qquad (5.16c)$$

One possibility is to assume that all three reactions follow Michaelis–Menten kinetics, and attempt to fit any available data. Here, let us merely study how to frame the corresponding ODEs. In this scenario, the same protein is a substrate in one reaction, and an enzyme in another.

$$\frac{d[A]}{dt} = \frac{k_{p_1}[A^{\sim P}]}{K_{d_1} + [A^{\sim P}]} - \frac{k_{d_1}[C][A]}{K_{d_1} + [A]}$$

$$\frac{d[A^{\sim P}]}{dt} = \frac{k_{d_1}[C][A]}{K_{d_1} + [A]} - \frac{k_{p_1}[A^{\sim P}]}{K_{d_1} + [A^{\sim P}]}$$

Note that $\dfrac{d[A]}{dt} = -\dfrac{d[A^{\sim P}]}{dt}$, which is the same as $[A] + [A^{\sim P}] = [A]_T$, the subscript T standing for total. Eliminating one equation,

$$\frac{d[A]}{dt} = \frac{k_{p_1}([A]_T - [A])}{K_{d_1} + ([A]_T - [A])} - \frac{k_{d_1}[C][A]}{K_{d_1} + [A]} \tag{5.17}$$

It is straightforward to write out similar rate equations for the other species:

$$\frac{d[B]}{dt} = \frac{k_{p_2}[A]([B]_T - [B])}{K_{d_2} + ([B]_T - [B])} - \frac{k_{d_2}[B]}{K_{d_2} + [B]} \tag{5.18}$$

$$\frac{d[C]}{dt} = \frac{k_{p_3}[B]([C]_T - [C])}{K_{d_3} + ([C]_T - [C])} - \frac{k_{d_3}[C]}{K_{d_3} + [C]} \tag{5.19}$$

A closer look at the equations reveals that each term is of the form

$$\frac{k[E][S]}{K_M + [S]},$$

much the same as the Michaelis–Menten equation, Equation 5.12. Of course, the [E] term is missing when the reaction is autocatalytic, such as B catalysing its own phosphorylation. These equations cannot be solved analytically but they can be easily integrated numerically, as we will see in §5.5.

5.4 GENERALISED RATE EQUATIONS

While mass-action kinetics, as well as Michaelis–Menten/Hill kinetics are very useful to model biochemical systems, there are many occasions where they fall short of capturing the actual dynamics. Also, there are many complexities introduced by the presence of multiple inhibitors and substrates. Further, in the case of large systems, it may be difficult to diligently identify mechanisms/kinetic laws for each and every reaction. Therefore, a more generalised approach, which can be systematically applied, directly from the topology of the system, can be advantageous.

5.4.1 Biochemical systems theory

Biochemical Systems Theory (BST) was originally proposed by Michael Savageau [8] and also developed and popularised by Eberhard Voit [9], as a canonical model for capturing biochemical system dynamics. Dynamic models in BST comprise systems of ODEs, where the change in each variable, *e.g.* the concentration of a metabolite, is modelled as a summation of multivariate products of power-law functions. One such canonical form is the *generalised mass action* (GMA) form:

$$\frac{dX_i}{dt} = \sum_{\cdot j} s_{ij} \cdot \gamma_j \prod_k X_k^{f_{ik}} \qquad (5.20)$$

where X_i represents the concentration of the ith species; s_{ij}, the stoichiometric coefficients; γ_j, the rate constants, and f_{ik}, the kinetic orders for each of the species X_k, capturing its effect on X_i. Unlike mass-action kinetics, the kinetic orders, *i.e.* f need not be integers; they can even be negative, for instance, in the case of inhibition.

Another canonical form of BST equations is the *S-system* form, with only two terms:

$$\frac{dX_i}{dt} = \alpha_i \prod_j X_j^{g_{ij}} - \beta_i \prod_j X_j^{h_{ij}} \qquad (5.21)$$

Here, α_i and β_i are non-negative rate constants, and g_{ij} and h_{ij} are kinetic orders (real numbers, not necessarily integers). These *power laws* can faithfully model the non-linearity of biological systems.

5.5 SOLVING ODEs

The dynamical systems of equations, such as the Michaelis–Menten equations cannot be solved analytically. So, how do we solve the ODEs numerically? As usual, we turn to Euler! In 1768, Euler came up with the first, and perhaps most elegant method for numerical integration. The ODEs we encounter in systems biology are commonly *initial value problems, i.e.* we have systems of differential equations, where the initial conditions, *e.g.* the starting concentration of various species (at t = 0), are known. These are different from *boundary value problems*, where the conditions are specified at the boundaries of the independent variable (*e.g.* both t = 0 and t = t_{final}).

Let us consider a simple rate equation, given by

$$\frac{dx}{dt} = f(x), \qquad x(t = 0) = x_0 \qquad (5.22)$$

where x possibly represents a substrate concentration, and f(x) is the rate law (model) that we have chosen to implement.

Recall the standard Taylor expansion for any function:

$$u(x) = u(x_i) + \frac{du}{dx}\bigg|_{x=x_i} (x - x_i) + \frac{1}{2!} \frac{d^2u}{dx^2}\bigg|_{x=x_i} (x - x_i)^2 + \cdots \quad (5.23)$$

$$\approx u(x_i) + \frac{du}{dx}\bigg|_{x=x_i} (x - x_i) \quad (5.24)$$

In our case, $u(x) \equiv x(t)$, $\frac{du}{dx} \equiv \frac{dx}{dt} = f(x)$, $x \equiv t + \Delta t$, and $x_i \equiv t$, giving:

$$x(t + \Delta t) \approx x(t) + f(x(t))\Delta t$$

Given that we know x_0,

$$x(\Delta t) = x_0 + f(x_0) \cdot \Delta t$$

Since we now know $x(\Delta t)$,

$$x(2\Delta t) = x(\Delta t) + f(x(\Delta t)) \cdot \Delta t$$
$$x(3\Delta t) = x(2\Delta t) + f(x(2\Delta t)) \cdot \Delta t$$
$$\cdots$$

Thus, $x(t)$ can be calculated at any desired t. Although the method seems straightforward, there is one pitfall—how do we choose the *correct* Δt? The choice of Δt has to be made carefully, as it affects the *numerical stability* of the method. The lower the value of Δt, the more accurate the answer, but it also demands far more computations. Therefore, there is a trade-off to be made here. A more detailed discussion of some of these interesting aspects of numerical methods for solving differential equations can be found in standard textbooks on numerical methods, such as reference [10].

The method described above is known as *forward* Euler method. Obviously, we made a somewhat extreme approximation in Equation 5.24, truncating the Taylor series to just the first two terms, which affects the accuracy of the computations. Euler also proposed another method, namely the *backward* Euler method, or implicit Euler method. This uses the backward difference formula, which can be obtained from Equation 5.24 using $u(x) \equiv x(t)$, $\frac{du}{dx} \equiv \frac{dx}{dt} = f(x)$ and $x_i \equiv t$ as before, but $x \equiv t - \Delta t$, giving

$$x(t - \Delta t) \approx x(t) + f(x(t))(-\Delta t) \quad (5.25)$$

$$\Rightarrow x(t) \approx x(t - \Delta t) + f(x(t))\Delta t \quad (5.26)$$

Note that this equation is implicit in $x(t)$, as the unknown, $x(t)$ is on both sides of the equation, and the (non-linear) equation must be solved to obtain the value of $x(t)$. Therefore, this method is much more computationally expensive, compared to the forward Euler method. However, this comes with a major advantage, of much-improved numerical stability!

A host of other methods, such as the Runge–Kutta method and Adams–Bashforth–Moulton methods are also available for solving ODEs numerically. The choice of method will be dictated by both the desired accuracy, as well as the inherent numerical characteristics of the system of ODEs, *i.e.* how numerically stable is the system of ODEs. §5.6.1 discusses more about how *stiff* systems of equations, which are difficult to solve, can be dealt with. A number of powerful ODE solvers, which can handle non-linear systems, as well as stiff systems of ODEs are available across a variety of programming languages, from C to Python to MATLAB.

5.6   TROUBLESHOOTING

Although dynamic modelling is probably the most mature of the methods discussed in this textbook, there still remain several challenges in building and simulating such models for the study of various biochemical systems. As in any modelling exercise, challenges stem from the kind of assumptions that need to be considered, the basic scope of the model and its objectives, as well as the availability of suitable data. Beyond these, there are specific challenges that are somewhat unique to dynamic modelling, which arise from the nature of the ODEs and the parameters, as also the uncertainty in the interactions, *i.e.* topology of the model.

5.6.1   Handling stiff equations

Some systems of ODEs are very difficult to solve using the simpler ODE solvers, such as the forward Euler method, without using very small step-sizes. Such systems of ODEs are called *stiff*; in stiff equations, stability considerations predominate accuracy considerations for choosing the Δt (step-sizes) [11]. Although there is no precise definition of stiffness, the above criterion is a generally good rule of thumb. Stiff systems are frequently encountered in biology, and the choice of appropriate solvers (numerical methods) to integrate them becomes rather important. Using a non–stiff solver on a stiff system of equations can produce unexpected and inaccurate results. Stiff ODE solvers perform more computations, but can work with larger time-steps than non-stiff ODE solvers. In MATLAB, ode15s and ode23s are examples of stiff solvers, while ode45 is the standard non-stiff solver. If a non–stiff solver such as ode45 takes too much time to compute the solution for a system of ODEs, it is normally advisable to switch to a stiff solver.

5.6.2   Handling uncertainty

Another major challenge while modelling biological systems, particularly in dynamic modelling, is the handling of various uncertainties, especially the uncertainty in the topology of the system. Many a time, models have to be built based on partial information, limited data, or conflicting reports in literature. As a consequence, we may find that multiple models (or topologies) can potentially describe the system, and the *best model* may have to be determined based on how well it can explain the data, as well as various other aspects of the real system that the model can capture. Importantly, the modelling exercise should help establish new hypotheses for testing, and also help in deciding the most informative experiments to be performed. Some other interesting approaches, such as ensemble modelling [12], have also been attempted to understand and resolve uncertainties in dynamic models.

5.7 SOFTWARE TOOLS

The companion code repository provides codes for the Euler forward method, as well as for solving ODEs in MATLAB using ode23s.

COPASI [13] (http://copasi.org/) is a versatile tool for dynamic modelling of SBML models using ODEs or stochastic simulation algorithms.

Tellurium [14] (http://tellurium.analogmachine.org/) is a Python environment for dynamic modelling; it comprises multiple libraries and plug-ins and can interface with powerful tools such as libRoadRunner (http://libroadrunner. org/) for simulations. Tellurium notebooks also provide a 'notebook environment' familiar to users of applications such as Jupyter Notebook.

There are many more popular tools for performing dynamic modelling, some of which are described in Appendix D. Appendix E illustrates how MATLAB solvers can be used to integrate systems of ODEs.

In addition to the tools, it is also important to mention the BioModels database [15] (http://www.ebi.ac.uk/biomodels/), which hosts hundreds of well-curated dynamic models. The repository provides the curated models in different formats, such as SBML and CellML, as well as model files that can be directly loaded and simulated in tools such as COPASI, Octave, Scilab, and XPP-Aut.

EXERCISES

5.1 What is the (mathematical) shape of the Michaelis–Menten curve? Can you represent it in a canonical form?

5.2 Consider a simple metabolic *system* where a metabolite A is imported into the cell, converted to B in an enzyme-catalysed reaction, and secreted out. Write out all the corresponding differential equations. State any assumptions that you make.

5.3 Consider the differential equation describing the rate of a biochemical reaction: $\frac{dx}{dt} = -x^3$, with $x_0 = 1$ mM. Illustrate the solution for this differential equation using Euler's method, for a step-size of 0.25. Can you solve the equation analytically, and compute the exact value of $x(2)$? What is the error in computing $x(t)$ at $t = 1$ for a step-size of 0.5?

[LAB EXERCISES]

5.4 For the above problem, plot a graph of error in computing $x(2)$ vs. step-size, for step-sizes 0.1, 0.2, 0.25, and 0.5. Also plot the errors when $x(2)$ is computed using ode23.

5.5 Consider the equation $\frac{dx}{dt} = -30x$, with $x(0) = 1$. Integrate it from $t = 0$ to $t = 1.5$ using the Euler forward method with $\Delta t = 0.001, 0.01,$ and 0.05, and compare with the results obtained from ode23. What do you observe?

5.6 Consider a simple enzymatic system, as discussed in §5.3.1. Can you solve these equations numerically (as discussed in §5.5), and plot the concentrations of [S], [E], [ES], and [P], against time. Assume $[S]_0 = [E]_0 = 1mM$, and that no product or enzyme–substrate complex is present initially. Plot these curves for different values of k_1, k_{-1}, and k_2, and comment.

5.7 Understand the model of Nielsen *et al* [16], describing the kinetics of glycolysis. This model can also be obtained from BioModels (Model ID: BIOMD0000000042).

    a. Write out an ODE file ode_model.m to encode the ODEs.

    b. Write out a file analyse_model.m to initialise the variables and perform the time-course simulation, using the ODEs from ode_model.m.

    c. Study the effect of changing parameters: Change V4, K4GAP, K4NAD, K8f, K8b individually to 50% and 150% of their original value, and study their effects on NAD and NADH concentrations.

REFERENCES

[1] C. M. Guldberg and P. Waage, "Ueber die chemische affinität", *Journal für Praktische Chemie* **19**:69–114 (1879).

[2] L. Michaelis, M. L. Menten, K. A. Johnson, and R. S. Goody, "The original Michaelis constant: translation of the 1913 Michaelis–Menten paper", *Biochemistry* **50**:8264–8269 (2011).

[3] L. Michaelis and M. L. Menten, "Kinetik der invertinwirkung", *Biochemische Zeitschrif* **49**:333–369 (1913).

[4] L. Briggs and J. B. S. Haldane, "A note on the kinetics of enzyme action", *The Biochemical Journal* **19**:338–339 (1925).

[5] E. Klipp, W. Liebermeister, C. Wierling, A. Kowald, H. Lehrach, and R. Herwig, *Systems biology*, 2nd ed. (Wiley-VCH, Germany, 2016).

[6] A. V. Hill, "The possible effects of the aggregation of the molecules of hæmoglobin on its dissociation curves", *The Journal of Physiology* **40**:iv–vii (1910).

[7]   A. Mogilner, R. Wollman, and W. F. Marshall, "Quantitative modeling in cell biology: what is it good for?", *Developmental Cell* **11**:279–287 (2006).

[8]   M. A. Savageau, *Biochemical systems analysis: a study of function and design in molecular biology* (Addison-Wesley, Reading, MA, USA, 1976).

[9]   E. O. Voit, *Computational analysis of biochemical systems: a practical guide for biochemists and molecular biologists* (Cambridge University Press, New York, USA, 2000).

[10] W. H. Press, S. A. Teukolsky, W. T. Vetterling, and B. P. Flannery, *Numerical recipes 3rd edition: the art of scientific computing*, 3rd ed. (Cambridge University Press, 2007).

[11] U. M. Ascher and L. R. Petzold, *Computer methods for ordinary differential equations and differential-algebraic equations*, 1st ed. (Society for Industrial and Applied Mathematics, Philadelphia, PA, USA, 1998).

[12] L. Kuepfer, M. Peter, U. Sauer, and J. Stelling, "Ensemble modeling for analysis of cell signaling dynamics", *Nature Biotechnology* **25**:1001–1006 (2007).

[13] S. Hoops, S. Sahle, R. Gauges, et al., "COPASI—a COmplex PAthway SImulator", *Bioinformatics* **22**:3067–3074 (2006).

[14] K. Choi, J. K. Medley, M. König, et al., "Tellurium: an extensible Python-based modeling environment for systems and synthetic biology", *Biosystems* **171**:74–79 (2018).

[15] C. Li, M. Courtot, N. Le Novère, and C. Laibe, "BioModels.net web services, a free and integrated toolkit for computational modelling software", *Briefings in Bioinformatics* **11**:270–277 (2010).

[16] K. Nielsen, P. G. Sorensen, F. Hynne, and H. G. Busse, "Sustained oscillations in glycolysis: an experimental and theoretical study of chaotic and complex periodic behavior and of quenching of simple oscillations", *Biophysical Chemistry* **72**:49–62 (1998).

FURTHER READING

K. A. Johnson, "A century of enzyme kinetic analysis, 1913 to 2013", *FEBS Letters* **587**:2753–2766 (2013).

E. A. Sobie, "An introduction to dynamical systems", *Science Signaling* **4**:tr6+ (2011).

C. B. Moler, "Ordinary differential equations", in *Numerical Computing with* MATLAB (Philadelphia, PA, USA, 2004) Chap. 7, pp. 185–234.

H.-M. Kaltenbach, S. Dimopoulos, and J. Stelling, "Systems analysis of cellular networks under uncertainty", *FEBS Letters* **583**:3923–3930 (2009).

Parameter estimation

CONTENTS

T HE CORNERSTONE of dynamic modelling in biology, especially *data-driven mechanistic modelling*, is the estimation of parameters for a given model topology, from experimental data. This is challenging, both in terms of the computation involved, as well as posing the right problem for parameter estimation. In this chapter, we will overview this problem of estimating parameters from data, and also survey some of the popular algorithms for parameter estimation.

6.1   DATA-DRIVEN MECHANISTIC MODELLING: AN OVERVIEW

Given a system of ODEs, such as those in §5.3.3, we have to know the parameters $K_{d_1}, K_{d_2}, \ldots$, to integrate the ODEs and make any sort of prediction. Typically, we have data available on a system, say, the concentration of various species. Even before we get started with the estimation exercise, we must ensure that we have data of sufficient quality and quantity. Also, are the data reliable, and potentially reproducible, if the experiment is repeated? Assuming the answers to these questions are all in the affirmative, we can embark on parameter estimation.

Parameter estimation, though, is preceded by several important steps in the modelling process. Figure 6.1 outlines a flowchart describing some of the most standard steps in a data-driven mechanistic modelling exercise[1]. Note that we predominantly pursue *mechanistic* modelling; therefore, model identification is an early and important step—we already have an understanding of the system in question, and consequently, some hypotheses about the mechanisms of the different reactions/interactions. These mechanisms/interactions are captured in the

---

[1] Also recall the 'systems biology cycle' (Figure 1.1), where we typically embark on the model building process with a bunch of hypotheses and/or experimental data.

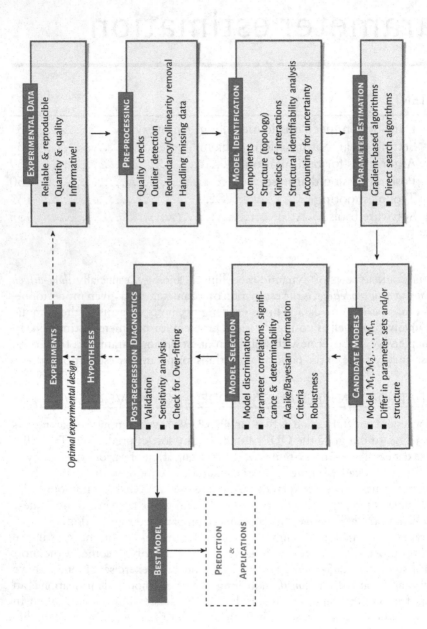

Figure 6.1 A flowchart for data-driven mechanistic modelling.

model via many mathematical equations. Of course, it is possible to try and build an empirical or semi-empirical model, purely based on the data, without mechanistic assumptions, but this is somewhat less common in systems biology, particularly in the modelling of biological networks. Empirical models, though, are more common in *quantitative systems pharmacology* (see Epilogue, p. 313), and PK/PD (pharmacokinetic/pharmacodynamic) modelling [1].

In the following sections, we will survey the important steps of data-driven mechanistic modelling.

6.1.1 Pre-processing the data

Once we have done the basic quality checks on the data, we need to pre-process them for modelling. It is vital to ensure that the data used for modelling/estimation is accurate, reliable, and informative. Every data point suffers from at least two sources of error: *experimental error*, arising from how the experiment or measurement was conducted (mitigated by performing technical replicates), or *biological variation/noise* (mitigated by performing multiple biological replicates). A good dataset is characterised by a *sufficient* number of technical and biological replicates.

As a first step, some simple diagnostics can be performed on the data, to understand their distribution, and determine if there are any obviously erroneous data points. While some extraordinary data points may be a consequence of true system variability and actually represent unexpected system behaviour, others may just be a consequence of large errors in experiment/measurement/instrumentation or even data entry/processing. It may be useful to find and eliminate the erroneous points early on; on the other hand, large variations that truly reflect system dynamics must be retained, so that the model may capture them.

Notably, the experimental data must be *informative*[2]. That is, the dataset should have *useful* data points from which inferences can be made. For example, in a Michaelis–Menten system, multiple measurements of the reaction velocity at very high values of K_M (where the enzyme is essentially saturated) are far less useful than measurements at lower values of K_M. Similarly, if we are studying how a signalling cascade responds following the addition of an effector molecule, we would again require many observations at the initial time-points, where a rapid initial response may be exhibited, rather than when the system has adapted to the stimulus and has reached a steady-state. Of course, we may not always have enough prior knowledge about a system to perform an optimal experimental design, but the systems biology cycle, and the flowchart of Figure 6.1 underline how iterative modelling can enable the design and conduct of more optimal and informative experiments. Any redundancies or collinearity in the data must also be eliminated, as they do not add any value to the estimation process.

---

[2]There are also systematic quantitative measures of information in literature, such as Fisher information, Shannon information, or Kullback–Leibler information. There are excellent books on these topics [2, 3].

We must also figure out how to handle missing data. It is common to have missing data for one or more species at one or more time-points. This may be a consequence of an experimental infeasibility, an instrumental malfunction/error and so on. At times, it may be useful to *impute* the missing data. However, it is perhaps best to go through the modelling exercise without imputation, then predict the missing data from the model at a later stage, and then finally judge whether the predictions are reasonable.

Lastly, it is very essential, yet rather difficult, to have a *sufficient* amount of data for parameter estimation. The data should be enough to build the model of desired complexity and uniquely determine its parameters. There are also issues about the *structural identifiability* of the model and other issues that we will address in the forthcoming sections. We may also want to set aside some data for cross-validation following parameter estimation, as it can help guard against over-fitting.

6.1.2 Model identification

Model identification involves identifying the components of the model, the structure or topology (wiring), as well as the kinetics that describe the interactions between the various components. Much of the previous chapter dwelt on these topics. An important aspect of model identification is the study of *structural identifiability*, which is essential before we go ahead with parameter estimation. A model is structurally *identifiable* if it is possible to determine the values of its parameters uniquely, by fitting the model to experimental observations [4]. In practice, structural identifiability requires that each parameter has a strong influence on at least one of the variables, and the effects of the parameters on the variables are not correlated with one another. When the parameter effects are correlated, we will have multiple feasible parameter sets, with the variation in one parameter counteracting the variation in another.

There are many challenges in assessing the structural identifiability of biological models; some useful approaches have been developed, and are discussed elsewhere [5, 6]. Villaverde *et al* [5] have reported a general approach to structural identifiability, along with a MATLAB toolbox. Gadkar *et al* [6] have proposed an iterative approach to model identification, for biological networks. Identifiability also has important consequences for *sloppiness*, as we will discuss in §6.5.2.

6.2 SETTING UP AN OPTIMISATION PROBLEM

Parameter estimation is at the heart of the entire modelling process, and how well this task is carried out, along with necessary preparatory tasks pre-estimation and diagnostic tasks post-estimation, determines the success of the modelling exercise. The problem of parameter estimation, is one of *regression*—we need to find the *best* parameters for a given model, *i.e.* the parameter set that will *best* capture, or fit the data. The notion of best hints that this will indeed be an *optimisation problem*. At the simplest level, parameter estimation can be a linear regression problem. Most problems are too complex for linear regression; therefore, another common method to perform regression, or parameter estimation, is to set up a *least-squares* problem. We will discuss these in the rest of this section.

6.2.1 Linear regression

Let us consider one of the simplest of systems, a single enzymatic reaction following Michaelis–Menten kinetics, where we need to estimate two parameters—v_{max} (or k_{cat}) and K_M. Indeed, this is a classic undergraduate biochemistry laboratory experiment. From time-course data corresponding to substrate concentrations and reaction velocities, it is common to estimate the parameter values via *linear regression*, made possible through transformations of the original Michaelis–Menten equations. Three transformations are widely popular:

$$\text{Lineweaver–Burk plot} \quad 1/v \text{ vs. } 1/[S] \quad \frac{1}{v} = \frac{K_M}{v_{max}} \frac{1}{[S]} + \frac{1}{v_{max}}$$

$$\text{Eadie–Hofstee plot} \quad v \text{ vs. } v/[S] \quad v = -K_M \frac{v}{[S]} + v_{max}$$

$$\text{Hanes–Woolf plot} \quad [S]/v \text{ vs. } [S] \quad \frac{[S]}{v} = \frac{[S]}{v_{max}} + \frac{K_M}{v_{max}}$$

These plots are commonly discussed in many textbooks; however, all three of them are plagued by a major problem: the transformation of the data also transforms the errors, *i.e.* the noise characteristics! This is undesirable, as the transformed errors confound the linear regression, and lead to incorrect estimates, especially in the double reciprocal plot. Although a knowledge of the standard deviation of the errors in measurement can be used to weight the errors suitably, this is seldom done. Furthermore, with the widespread availability of non–linear regression tools, there is really no compelling reason to use any of the transformations of the Michaelis–Menten equation for parameter estimation today.

6.2.2 Least squares

The least-squares problem, in its simplest form, captures the sum squared error (SSE) in the prediction of a variable x as:

$$SSE = \sum_{i=1}^{n} (x_{m,i} - x_{p,i})^2 \tag{6.1}$$

where x denotes the variable being modelled, *e.g.* the concentration of a species, and the subscripts m and p represent the measured and predicted values, at each of the n available data points. Note that $x_{p,i}$ is actually a function of the parameters, $\Theta = \{k_1, k_2, \ldots\}$. Therefore, the above equation is better written as

$$SSE(\Theta) = \sum_{i=1}^{n} (x_{m,i} - x_{p,i}(\Theta))^2 \tag{6.2}$$

We need to find the Θ that gives the least value of SSE, *i.e.* we need to solve an optimisation problem:

$$\hat{\Theta} = \arg\min_{\Theta} \text{SSE}(\Theta) \tag{6.3}$$

Before we solve the optimisation problem, of minimising the SSE, let us take a closer look at the SSE itself, the 'cost' function for our optimisation.

6.2.2.1 Modifications to the least-squares approach

Although the least-squares approach is most widely used, it has several obvious drawbacks. Firstly, it is very sensitive to scaling of the data. For instance, if we have $x_{m,1} = 10\text{mM}$ and $x_{m,2} = 0.5\text{mM}$, and the corresponding predictions are $x_{p,1} = 10.5\text{mM}$ and $x_{p,2} = 1.0\text{mM}$, then, the residual, $x_{m,i} - x_{p,i}$, at each point is the same, 0.5mM. But this is not how we would like to assess the model predictions—the prediction error at the first point is only 5%, whereas the prediction error for the second point is a whopping 100%! Therefore, the second point must ideally have a much higher contribution to the SSE; this can be achieved by a simple modification to Equation 6.2, to yield:

$$\text{SSE}(\Theta) = \sum_{i=1}^{n} \left(\frac{x_{m,i} - x_{p,i}(\Theta)}{x_{m,i}} \right)^2 \tag{6.4}$$

Now, the second point's contribution to the SSE is 400 times that of the first!

Further, the data may not be homoscedastic, *i.e.* that variance or the distribution of errors at different data points may not be the same. In such cases, it may also be advisable to weight the terms in the SSE based on the standard deviation at each of the individual data points:

$$\text{SSE}(\Theta) = \sum_{i=1}^{n} \left(\frac{x_{m,i} - x_{p,i}(\Theta)}{\delta_i \cdot x_{m,i}} \right)^2 \tag{6.5}$$

where δ_i is the weighting factor, based on the variance/error in measurement of the ith data point.

To intuit this, let us consider the curves shown in Figure 6.2, attempting to capture the kinetics underlying the data points shown. Both the grey curves shown have comparable SSE. Of course, we may be tempted to consider the dotted curve passing through all the data points to be the *truest* curve, corresponding to the most optimal parameter set. However, it might actually be an *unfair* ask of the model, pushing it to exactly predict the *mean* of the measured variable, at every data point, when, in reality, many points higher and lower than the mean were observed even in the experiment itself! To offset this, the δ_i term is helpful, downscaling the error by a factor proportional to the observed variance at the data point.

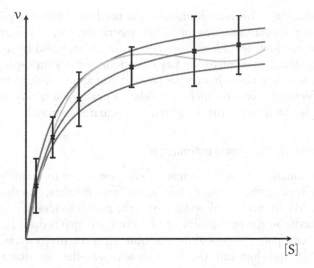

Figure 6.2 **Fitting the right curves for an example Michaelis–Menten system.** The (light and dark) grey curves correspond to predictions arising from different parameter sets/models. The dotted curve represents another parameter set, with *zero* SSE, as it passes through (the means of) all the experimental data points (denoted by crosses, with error bars).

Another important point to note is that we also have errors in the so-called independent variable—in this case, [S], or in many other cases, time. How accurately was variable this measured? We may also have error bars for this variable, adding another layer of complexity to our estimation process.

6.2.2.2 Formulating better cost functions

Note that the SSE is merely a proxy for the *best fit*; in reality, we want to identify the parameter set $\hat{\Theta}$ that can capture the key characteristics of the real system as closely as possible. Consider the light grey curve in Figure 6.2. Although it may have comparable SSE to the other dark grey curves, given it entirely lies within the error bars, it clearly cannot correctly describe a Michaelis–Menten system, as the reaction rate does not increase monotonically. Also, if we were to find a curve that does not pass through the origin, it would not be acceptable either, even if it were to display substantially lower SSE. Of course, neither of these possibilities are so worrisome, as we will first fix the *functional form* of the curve, to be Michaelis–Menten in this case, and only then proceed to estimate v_{max} and K_M.

Therefore, although the modified SSE of Equation 6.5, essentially a weighted least-squares formulation, can be a useful cost function, even it cannot help discriminate curves such as the light grey curve discussed above, which has low SSE but poor *behaviour*. This issue highlights the need for subjectivity in fitting, beyond objective measures such as the SSE. Importantly, every fit must be diligently examined to see how sensible the predictions are. It may also be useful to craft a cost function that lays emphasis on key behaviours, or key time-points where the predictions are important. It may be possible to capture/weight important features of a biological system, such as a delay, a peak, or a steady-state value, by even a simple extension of the weighted least-squares approach.

6.2.3   Maximum likelihood estimation

In the least-squares approach outlined above, we strove to identify a parameter set that minimises some measure of deviation from the data, like the SSE. In contrast, in the Maximum Likelihood scheme, the goal is to identify a parameter set that will maximise the probability of observing the given data. In other words, we wish to compute the probability of observing the given dataset for a given parameter set, and then find the *best parameter set*—the one that *maximises* this probability.

If we assume that the errors across the data points are not correlated, and that the errors follow a Gaussian distribution, then, the *likelihood* of a given parameter set Θ is expressed as [7]:

$$L \propto \prod_{i=1}^{n} P(x_{m,i} - x_{p,i}(\Theta)) \tag{6.6}$$

$$= \prod_{i=1}^{n} \frac{1}{\sqrt{2\pi\sigma_i^2}} \exp\left[-\frac{(x_{m,i} - x_{p,i}(\Theta))^2}{2\sigma_i^2}\right] \tag{6.7}$$

We can maximise the logarithm of this likelihood, equivalently, which means we must maximise

$$\sum_{i=1}^{n} \log \frac{1}{\sqrt{2\pi\sigma_i^2}} - \frac{1}{2} \sum_{i=1}^{n} \frac{(x_{m,i} - x_{p,i}(\Theta))^2}{\sigma_i^2}, \tag{6.8}$$

which is the same as *minimising*

$$\sum_{i=1}^{n} \frac{(x_{m,i} - x_{p,i}(\Theta))^2}{\sigma_i^2}, \tag{6.9}$$

a summation that very closely resembles Equation 6.5. Thus, we see that assuming a Gaussian distribution of errors leads to basically the least-squares formulation itself. Notably, if we (need to) assume a different distribution, it can be plugged back into Equation 6.6, to obtain a different formulation. Note that the least-squares formulation (and so also the maximum likelihood formulation, with Gaussian error distribution) is sensitive to the presence of outliers, and other alternatives exist for performing a more robust regression, such as the least median of squares [7, 8].

6.3 ALGORITHMS FOR OPTIMISATION

How do we solve complex non-linear optimisation problems? Remember, the least-squares cost function is quadratic in x but *highly non-linear* in Θ; $x(\Theta)$ is obtained by integrating the rate equations. The local minima for such functions are already very difficult to find—the global minimum, even harder! Therefore, to solve this harder-than-a-needle-in-a-haystack problem, we need robust algorithms. Many strategies exist, to sift through the massive parameter space, to find regions (and points!) of interest, *i.e.* parameter sets that fit the data well. The best of the algorithms rely on some sampling of the large parameter space, alongside various complex heuristics. In the following section, we will begin by looking at some of the desired features/behaviours for parameter estimation algorithms.

6.3.1 Desiderata

What are the desired characteristics of parameter estimation algorithms, especially in the context of biological systems? These desiderata are dictated by the complexities of our cost function and the biological system as well as the available data. The cost function is typically multi-modal, with many peaks and valleys, *i.e.* multiple local minima/maxima. The cost function is non-linear of course, and may not be differentiable, or, more commonly, too expensive to differentiate. Therefore, the use of gradient information is typically undesirable. The cost function computation is also computationally very expensive. Consequently, parameter estimation algorithms must be capable of handling non-differentiable non-linear cost functions. Also, they should ideally not rely on gradient information.

Since the cost function is expensive to compute, and cost function computations will happen very many times, the algorithm should be parallelisable, to speed up cost function computations. Other than these, we would obviously prefer that the method is easy to set up and use, in that it does not itself have too many parameters controlling its performance/behaviour, and lastly, show good convergence properties. In summary, it is highly desirable for optimisation algorithms to reference [9]:

1. possess the ability to handle non-linear multi-modal high-dimensional cost functions
2. be able to operate without gradient information
3. facilitate parallel cost function computations, for speed-up
4. have relatively few *hyperparameters*, *i.e.* the parameters that steer the optimisation algorithm
5. exhibit good (and rapid) convergence to the global minimum

6.3.2 Gradient-based methods

Gradient-based optimisation methods are highly popular in several disciplines of science and engineering and there are many good textbooks that cover these algorithms in much detail [7, 10–13].

Some of the popular methods include the steepest descent method, the conjugate gradient method, the Levenberg–Marquardt algorithm, Gauss–Newton algorithm, Newton's method, and quasi-Newton methods such as the Broyden–Fletcher–Goldfarb–Shanno algorithm. All these methods use the gradient of the Hessian[3] of the cost function in different ways, including making efficient approximations for the gradient/Hessian.

Although these methods are useful for a variety of problems, they are often of limited use in biology, notably in the context of data-driven mechanistic modelling. While they may be useful in solving simple parameter estimation problems, they are rarely applicable to more complex problems, as the cost function may be non-differentiable or too expensive to differentiate. Therefore, we must look for methods that do not compute/use gradient information. The direct search methods we discuss below, as the name suggests, fulfil this requirement.

Nonetheless, some methods based on Levenberg–Marquardt [14] and LSQNONLIN SE (a local gradient-based search algorithm with Latin hypercube restarts; [15]) have been found to outperform direct search algorithms in terms of both accuracy and function evaluations. Another method based on non-linear programming has been developed, which was shown to be highly efficient and scalable to very large systems having thousands of ODEs and tens of thousands of data points [16].

6.3.3 Direct search methods

Direct search methods follow a simple template: there has to be (a) a strategy to vary Θ_n, the parameter vector, and (b) given a new parameter vector Θ_{n+1}, there must be a strategy to accept or reject it. Each of the methods we discuss below has some unique strategies for either case. Suitable stopping conditions, for terminating the optimisation must also be decided. A nice review of direct search methods has been published elsewhere [17]. It is worth noting that some of these methods were originally devised to solve (discrete) combinatorial optimisation problems, and were later ported to solve (continuous) real-valued problems such as parameter estimation, which we discuss here.

6.3.3.1 Classic methods

There are many classic methods of direct search, and they broadly fall into three categories [17]—pattern search, simplex methods, and adaptive methods. These classic methods were developed between 1960–1971. The pattern search methods are characterised by a serious of exploratory moves to understand how the cost function varies in different parts of the parameter space. The step-size for the exploration is decided based on the cost function values seen during the exploratory moves. Simplex methods are based on a simple device, the simplex—a set of $n + 1$ points in $\mathbb{R}^n$. MATLAB `fminsearch` also uses the classic Nelder–Mead simplex

---

[3]The Hessian matrix of a function is defined as $\mathbf{H}_{i,j} = \frac{\partial^2 f}{\partial x_i \partial x_j}$.

method. The adaptive methods seek to improve the search by figuring out better directions based on the points already explored during the search. Examples of adaptive methods include Powell's method and Rosenbrock's method. Rosenbrock also came up with a classic 'banana' function[4], which is commonly used as a benchmark to evaluate direct search algorithms.

6.3.3.2 Hill climbing

One of the simplest methods for direct search is called hill climbing, a *greedy* algorithm for optimisation. The algorithm is greedy in the sense it makes locally optimal decisions, continuously moving in the direction of increasing 'elevation', to scale the peak of the mountain. The mountain, in this case, is the cost function landscape, and the peak of course is *one* optimal solution to the problem. The term hill *climbing* comes from the notion of maximisation; its minimisation counterpart is hill descending, a term less frequently used. The algorithm starts at an arbitrary point (Θ_0), and then picks an arbitrary *neighbour* (in discrete optimisation cases), or an arbitrary point close by, in some direction (in continuous cases), as the next point (Θ_1). The algorithm accepts this point and moves, if and only if the cost function has a lower value at that point. Otherwise, a new point must be found, going back to the original point.

Hill climbing converges fairly fast, reaching a peak, where no neighbouring point has a better value for the cost function. However, it has the obvious drawback of getting stuck at a local maximum, a challenge even more pronounced in the case of curves with many maxima (or minima), such as the one in Figure 6.3. At a local minimum, the cost function increases in all directions—therefore, the algorithm has converged, or misconverged, rather, to a local minimum. For multi-modal cost functions, there are far too many local minima to have a useful estimation using hill climbing. A simple variant is hill climbing with *random restart*. This essentially involves restarting the hill climbing algorithm with different initial values (Θ_0's). This allows for finding many local minima and better solutions, than a single run of hill climbing.

The other methods that we will discuss in this Chapter have better safeguards and novel heuristics to avoid such misconvergence. Interestingly, many direct search methods have been inspired by biological processes in nature, *e.g.* evolutionary algorithms and particle swarm algorithms, or physical processes, such as the annealing of metals. The excellent book by Brownlee [18] discusses many such algorithms in detail.

6.3.3.3 Simulated annealing

Simulated annealing was one of the first algorithms to take inspiration from a physical process and codify it as a search algorithm [19]. The algorithm is a

---

[4]$f(x,y) = (a-x)^2 + b(y-x^2)^2$, with global minimum at $(x,y) = (a,a^2)$, where $f(x,y) = 0$. Normally, $a = 1$ and $b = 100$ are used. The global minimum of this function is within a long parabolic valley, and is difficult to find.

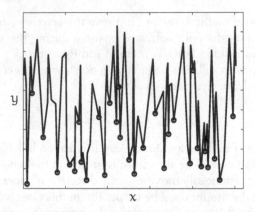

Figure 6.3 **A curve with multiple local minima.** Shown is an arbitrary 1D-curve, $y = f(x)$, which has many local minima (and maxima). The circles represent various local minima on the curve; the first circle (from left) is the global minimum.

beautiful adaptation of the Metropolis–Hastings Monte Carlo algorithm [20, 21]. Simulated annealing is inspired by the metallurgical process of annealing, in the process of which a metal is heated and gradually cooled under controlled conditions to obtain larger crystals and fewer defects in the final material. At higher temperatures, the atoms have higher energy and move about freely, and the slow cooling allows the atoms to attain new low-energy configurations.

It is important to understand all the metaphors here, and relate them to the computational algorithm. The possible *configurations* of the system represents the search space (*i.e.* values of Θ), each having a different *energy*, *i.e.* cost function value, $E(\Theta)$. The *temperature* T (or rather, the *annealing schedule*) is a parameter that affects how the exploratory the algorithm is, at different stages of the iterations during the optimisation. Different annealing schedules are possible, representing the 'cooling programme' for the system, capturing how the temperature is lowered as the iterations progress *e.g.* linearly, or exponentially.

The key defining aspect of simulated annealing is the acceptance–rejection strategy it uses to avoid getting stuck in local minima. Moves or steps that reduce energy (cost) are always accepted. On the other hand, unlike a greedy algorithm, moves that increase cost are not outright rejected, but occasionally accepted, albeit with a low probability! This enables the algorithm to climb out of local minima, searching for a better, hopefully global, minimum. Simulated annealing employs the Metropolis–Hastings algorithm or the *Metropolis criterion* to make a decision on the acceptance of the new point (Θ_{n+1}).

Mathematically, the probability of accepting the new state, Θ_{n+1} (generated randomly from the previous state, Θ_n) is computed as

$$P(E(\Theta_n), E(\Theta_{n+1}), T) \propto \exp\left(-\frac{E(\Theta_{n+1}) - E(\Theta_n)}{k_B T}\right) \qquad (6.10)$$

where k_B represents the Boltzmann's constant, a constant of nature that relates temperature to energy.

This probability is compared with $u(0, 1)$, a random number uniformly distributed in $(0, 1)$. If $P(E(\Theta_n), E(\Theta_{n+1}), T) \geqslant u(0, 1)$, Θ_{n+1} is accepted. Of course, when $E(\Theta_{n+1}) < E(\Theta_n)$, the above probability is greater than unity; therefore, P is arbitrarily set as 1, and this means that we will always accept moves that reduce cost. Importantly, moves that increase energy are not always rejected—the decision is made on the basis of the current temperature, as well as energy, *i.e.* how much poorer is the new cost function value. At higher temperatures, *i.e.* earlier iterations, the algorithm is more exploratory, permitting more moves that increase cost function, whereas, at lower temperatures, the algorithm only locally explores a particular region of the search space.

The key hyperparameters for simulated annealing are the initial temperature, cooling (annealing) schedule, the length of the run (or the number of iterations) as well as stopping conditions for the algorithm. These affect the performance of the algorithm, and need to be chosen carefully for any given optimisation problem, often by trial-and-error.

Various extensions to simulated annealing have also been proposed. Similar to hill climbing, it is also possible to restart simulated annealing from a significantly superior solution obtained along the way, and re-explore from that solution, searching for better minima.

6.3.3.4 *Other algorithms*

Many other direct search algorithms exist, such as particle swarm optimisation, ant colony optimisation, artificial bee colony algorithm, tabu search, immune algorithms, and neural network–based algorithms. A number of these have been adapted to solve parameter estimation problems in systems biology as well. A large selection of such optimisation algorithms are available from the PyGMO Python package (see Appendix D). The interested reader is referred to the books by Brownlee [18] and Michalewicz and Fogel [22], and the detailed review by Kolda [23]. A nice overview of global optimisation methods for parameter estimation has been published by Banga and co-workers [24].

6.3.4 Evolutionary algorithms

Evolutionary Algorithms (EAs) are a broad class of algorithms, attributed to John Holland [25][5], inspired by and modelled after biological evolution. In essence, they are algorithms that probabilistically apply search operators to a set of points in the search space of an optimisation problem. The preceding sentence already alludes to several interesting and distinguishing aspects of EAs, which set them apart from many other direct search algorithms:

- They are *stochastic* direct search algorithms

---

[5]These ideas were originally propounded in the first edition of Holland's book, in 1975.

- They use *multiple strategies* (search operators) to search for new solutions, *i.e.* generate new Θ's
- They simultaneously operate on a *set of points* (*i.e.* a population of solutions) in the search space

EAs encompass a wide variety of algorithms, and many different flavours exist: genetic algorithms (GAs), evolution strategies (ES), differential evolution (DE), evolutionary programming, gene expression programming, and so on. In the rest of this section, we will overview some key concepts and some key examples of EAs.

6.3.4.1 Key concepts and metaphors

The first important thing about evolution is, it happens only in *populations*. Every population, of course, is made up of *individuals*. How does evolution occur? Rather, what are the processes that drive evolution? Genetic variation is fundamental to evolution and is achieved by means of processes ('genetic operators') such as *mutations* and *crossovers*. A key mechanism of evolution is natural *selection*, which acts on the phenotype of the organism, leading organisms with better *fitness* or beneficial phenotypes to survive with higher probability (*survival of the fittest*).

The paragraph above is a succinct summary of evolutionary processes—all the biological terms discussed above have new connotations in the context of EAs. We now proceed to understand these various metaphors, and how they come handy in the context of optimisation:

Chromosome/Individual A single solution (Θ) to the optimisation problem.

Population A collection of solutions to the optimisation problem.

Generation Each iterative step of an EA is a generation, where a new population replaces the previous population.

Mutation A small change effected on a single solution, to produce a new solution ('child').

Crossover A combination of two (or more) solutions to create two new solutions ('children'), somewhat akin to sexual reproduction.

Fitness (function) The cost function, in fact its inverse, as EAs strive to find solutions with the highest fitness, which must correspond to least cost.

Selection The process of selecting the best or the most promising solutions from one generation to the next.

It is worth emphasising that the EAs follow the general recipe for direct search algorithms but afford better performance by virtue of mixing global and local exploration, improved search diversity offered by multiple operators operating on populations of solutions, and a rich variety of selection strategies. All the above concepts are leveraged by different EAs in different ways, as we will see below.

Another interesting point of note is that while non-Darwinian theories of evolution are often dismissed by biologists, algorithms based on Lamarckian evolution exist [26] and are useful as well, in optimisation!

We will now look at three different flavours of EAs, viz. genetic algorithms, evolution strategies, and differential evolution. The key differences between these three approaches lie in how they represent the individuals (solutions) and apply various operators.

6.3.4.2 Genetic algorithms

Genetic algorithms are perhaps the most widely used family of EAs. GAs were originally established to solve discrete optimisation problems.

Representation The first challenge in any EA is the choice of an appropriate representation of a solution to the problem. Typically, in GAs, the representation is a bit vector, or an array of bits, although other representations are also possible. Figure 6.4 shows a simple example, where a bit representation is used, with the first five bits representing x, and the next five representing y, two variables that are being optimised. Therefore, the first bit vector listed (top left, Figure 6.4), 1100100100, corresponds to $x = 25, y = 4$. The problem considered involves the maximisation of $f(x, y) = x^2 - y^2$, for $0 \leqslant x < 32$ and $0 \leqslant y < 32$, which can be taken as the fitness function[6].

Mutations and Crossovers Mutations in bit vector representations are effected by 'bit flips', where a zero is made into a one, or vice-versa. Crossovers involve an exchange of bits between chromosomes, and can happen at one or more points. Figure 6.4 illustrates two types of crossovers: a single-point crossover, where one segment of bits from one vector is exchanged with the second, and a multi-point crossover, where multiple segments of bits are exchanged. All the new chromosomes created by mutation and crossover are 'children', and from this population of children and parents, a fixed number of chromosomes (decided by the 'population size', a hyperparameter) are selected for the next generation, based on fitness. The rates/probabilities of mutations and crossovers are decided by specific hyperparameters.

Beyond classic mutations and crossovers, many EAs employ enhancements, that afford performance improvements. Special macro-mutation operators effect large changes in the chromosomes without recombination. Hybrid operators, which have nothing to do with evolutionary processes, have also been used, *e.g.* hill climbing, which involves searching for small changes to a chromosome that improve fitness, and implementing these changes.

---

[6]Strictly speaking, the fitness function should be non-negative, and this can be achieved by appropriate transformations.

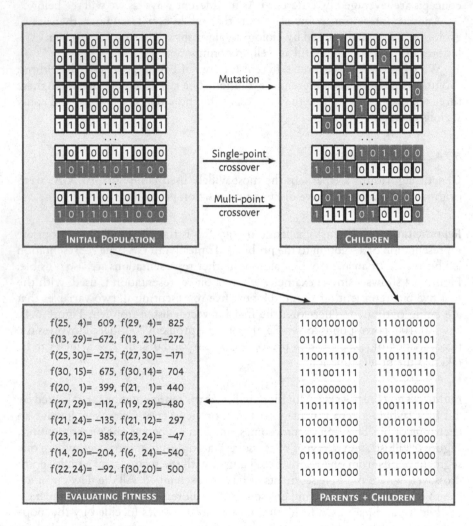

Figure 6.4 **An overview of a GA for a simple problem.** Illustrated are the initial population ('parents'), represented as bit vectors (see text). These individuals undergo mutation and crossovers to form children. Mutations involve single bit flips, marked grey in the children bit vectors. Crossovers involve the exchange of bits between two vectors, shown in different colours. The interim population has both parents and children, as shown. Following this, fitness is computed for all the individuals, based on the x and y values they encode, and then, by applying a suitable selection, the next generation can of the population is arrived at, completing one iteration of the GA.

Although we are predominantly interested in solving parameter estimation problems using EAs/GAs, it must be noted that EAs can be used to solve many different types of optimisation problems, including discrete optimisation problems, such as scheduling exam timetables, or the classic travelling salesperson problem, to which many other complex problems can be readily mapped. While the general approach to solving these problems is similar, in terms of mutations, crossovers and selection, the representation could be markedly different, necessitating different approaches to mutation. For example, in a travelling salesperson problem, where every solution could be a vector of cities to be visited in order, a mutation, where one city is arbitrarily replaced with another, may not result in a valid solution, as the same city may occur twice and one city may be left out. On the other hand, a mutation operation could be defined as performing a swap of two cities, which would yield valid solutions. Similar approaches can be devised for different kinds of problems, which demand different kinds of representations.

Selection How do we perform this selection? Selection is one of the most important steps in the GA, and the selection strategy directly impacts how well the GA explores various parts of the search space and consequently, the convergence of the algorithm as well. A common selection strategy in EAs is denoted $(\mu + \lambda)$, where λ children are created from μ parents, and the entire population of $(\mu + \lambda)$ individuals is evaluated, to select the μ *best* individuals. Here, best is a notion of how well the selected individuals will enable the GA to identify solutions with the highest fitness. Alternatively, there are (μ, λ) strategies, where μ best children are selected from the λ children $(\lambda \geqslant \mu)$. Selection is contingent on fitness, and many strategies exist:

Fitness proportionate selection Also known as 'Roulette wheel selection', this strategy involves picking individuals from the population in proportion to their fitnesses. Individuals with higher fitness have a greater probability of being selected into the next generation. Note that this means that even individuals with very low fitness have a finite probability of making it to the next generation, somewhat reminiscent of the Metropolis criterion, providing some safeguard against getting stuck in local minima. The name Roulette wheel selection comes from the analogy of putting all the solutions on a roulette wheel, with sector angles proportional to their fitness, and then spinning the wheel to make selections.

Tournament selection In (deterministic) tournament selection, a specific number of individuals (t, 'tournament size') are uniformly randomly picked from the population to play a 'tournament', and the fittest individual from the tournament (t individuals) is selected. This process is repeated to pick the entire next generation. t = 1 essentially corresponds to random selection. The value of t is turns out to be another hyperparameter that can control the convergence of the algorithm.

Elitist selection In elitist selection, a fixed number of top (fittest) individuals are retained without performing mutations/crossovers, *i.e.* selected directly, for the next generation.

6.3.4.3 Evolution strategies

Evolution Strategies (ES) typically use real number representations, and although they were also originally designed for discrete optimisation problems, they have long been used for continuous optimisation problems as well. A notable addition to the representation of every solution is a *strategy parameter*, which can be thought of as a gene that affects the evolutionary processes (*e.g.* mutation) for a given individual. These parameters themselves evolve during the course of the EA, consequently altering the evolutionary process itself.

A classic ES is the $(1 + 1) - ES$, which is essentially the same as greedy hill climbing, as a single child is derived from a single parent, and one of these two is selected based on their fitness [27]. The child is arrived at by performing a mutation, by adding a random number $r \sim N(0, \sigma^2)$ to the initial value, where σ^2 is the mutation variance (strategy parameter). This is a departure from how GAs represent solutions and perform mutations. Every individual is represented as a vector, *e.g.* $\Theta_i = \{\theta_{i1}, \theta_{i2}, \dots, \theta_{in}\}$, and a set of standard deviations $\{\sigma_{i1}, \sigma_{i2}, \dots, \sigma_{in}\}$.

By performing a crossover between Θ_p and Θ_q, a single child can be generated in different ways. Consider individuals (problem dimension = 4) described as

$$\Theta_1 = \{\theta_{11}, \theta_{12}, \theta_{13}, \theta_{14}\}; \{\sigma_{11}, \sigma_{12}, \sigma_{13}, \sigma_{14}\}$$
$$\Theta_2 = \{\theta_{21}, \theta_{22}, \theta_{23}, \theta_{24}\}; \{\sigma_{21}, \sigma_{22}, \sigma_{23}, \sigma_{24}\}$$

with grey and black colours being used to help discern the various elements coming from different parents as they mix during a crossover. Different types of crossovers can yield different children [27]. In a *discrete sexual crossover*, each element of the solution and the standard deviation is picked at random from a parent, to yield children like:

$$\{\theta_{11}, \theta_{22}, \theta_{23}, \theta_{14}\}; \{\sigma_{11}, \sigma_{22}, \sigma_{23}, \sigma_{14}\}$$

In *intermediate sexual crossover*, each element of the solution and the standard deviation is computed as the midpoint of the corresponding values for either parent:

$$\left\{\frac{\theta_{11} + \theta_{21}}{2}, \frac{\theta_{12} + \theta_{22}}{2}, \frac{\theta_{13} + \theta_{23}}{2}, \frac{\theta_{14} + \theta_{24}}{2}\right\};$$
$$\left\{\frac{\sigma_{11} + \sigma_{21}}{2}, \frac{\sigma_{12} + \sigma_{22}}{2}, \frac{\sigma_{13} + \sigma_{23}}{2}, \frac{\sigma_{14} + \sigma_{24}}{2}\right\}$$

Crossovers can also be performed by picking elements of a child from across the entire population. A popular variant of ES algorithm is the Covariance Matrix Adaptation-ES (CMA-ES) algorithm [28].

6.3.4.4 Differential evolution

Unlike other EAs, differential evolution (DE) is not biologically motivated [9], but nevertheless, performs very well. DE represents solutions as populations of real vectors, which can be independently manipulated. An initial vector population is chosen to cover the entire parameter space. The key operators of mutation and crossover are implemented somewhat differently from ES/GAs.

For mutation, parameter vector variations are created as follows, by adding some weighted fraction of the difference between two other solutions (parameter vectors) to a third:

$$\Theta_{k,G+1} = \Theta_{k,G} + \alpha(\Theta_{i,G} - \Theta_{j,G})$$

where the subscripts G and $G+1$ denote the generation.

Crossover is implemented as *parameter mixing*, in a fashion somewhat similar to the discrete sexual crossover in ES, discussed above. The child vector is created by randomly picking elements from two different parameter vectors, *e.g.* $\Theta_{k,G+1}$, the mutant vector and $\Theta_{i,G}$.

The selection procedure used is generally greedy; if the new vector has a lower cost function than the old, *i.e.* if $f(\Theta_{k,G+1}) < f(\Theta_{k,G})$, then the vector is replaced in the population; if not, the old value of $\Theta_{k,G}$ is retained.

6.3.4.5 Summary

The main applications of EAs are in complex optimisation problems, or even multi-objective optimisation problems, which are difficult to solve using more conventional methods. Difficult scheduling problems or routing problems, such as the scheduling of matches in a tournament, or the time-table for a university exam may have many difficult constraints, and these can be tackled by EAs by using suitable fitness functions.

In computational biology, we can use EAs to compute phylogenetic trees, multiple sequence alignments, clustering microarray data, and of course, parameter optimisation. Outside biology, EAs/GAs also find applications in the design of electric circuits, or 'evolvable' hardware. A famous example of an evolved object is the 2006 NASA ST5 spacecraft antenna, which was designed by means of an EA [29], so as to achieve many complex and unusual design requirements. Many other optimisation problems can also be solved using EAs, such as general k-means clustering.

When to use EAs? EAs can be particularly useful in hard-to-solve optimisation problems, where the cost functions are non-linear, non-differentiable, and they can handle multiple objectives as well. They are especially useful when the structure of the search space is poorly understood, and cannot really be exploited by other kinds of (deterministic) algorithms. Of course, many algorithms that are gradient-based cannot be used in a number of scenarios, as discussed at the beginning of this chapter, and we may have no other choice than to use EAs. At times,

EAs and direct search algorithms may help unravel important insights into the structure of the search space itself, although there is no strong theory to suggest that EAs would be better than other optimisation techniques. Nevertheless, various types of EAs have been widely used to tackle difficult optimisation problems from circuit design in electronics to parameter estimation problems in biology.

Challenges There are also a number of challenges one encounters while employing EAs. In general, the choice of representation is tricky, although, in the context of parameter estimation, it is reasonably straightforward. A major issue with EAs and many other direct search algorithms is that they can still converge to local minima, and there is often no strong theory about the convergence. There is also the issue of 'black-box' behaviour—it is difficult to characterise *how* an EA searches through a given space. Lastly, they are computationally very expensive: for a population size of 100 and 10,000 generations, numbers not uncommon in GAs/ES, a huge number of fitness function evaluations are necessary.

6.4 POST-REGRESSION DIAGNOSTICS

6.4.1 Model selection

At the end of the parameter estimation (regression) exercise, it is likely we have not one, but many parameter sets that capture (or reproduce) the experimental data *reasonably well*. We may also have more than one topology, which is able to capture the experimental data, with parameters that are different both in nature (*i.e.* the processes or interactions that they capture), number, and value. Thus, given a bunch of candidate models, $M_1, M_2, \ldots, M_n$, we are tasked with selecting the *best* model. This is quite a challenging task, given the question itself is somewhat ill-posed. What do we truly mean by *best*? Are we merely looking for the model that gives the least SSE with respect to the available data? Are we looking for the simplest model that can explain the data? Surely, we want to avoid the pitfalls of over-parametrisation and over-fitting.

During the process of model selection, parameter diagnostics are also performed [8]. For example, are the parameters statistically *significant*? That is, given the uncertainty in parameter estimates owing to noise, is there a significant probability that the parameter must be different from zero? How *determinable* are the parameters? That is, how well can we infer the value of the parameter from the data, independent of other parameter values? This is easily studied by computing the cross-correlation between the parameters. For two parameters θ_i and θ_j, the cross-correlation is defined as:

$$\kappa_{ij} = \frac{\hat{V}_{ij}}{\sqrt{\hat{V}_{ii}}\sqrt{\hat{V}_{jj}}} \tag{6.11}$$

where $\hat{V}$ is the covariance matrix corresponding to the parameter estimates. This cross-correlation is normalised between -1 and $+1$. Values of ± 1.0 indicate a complete dependency, while κ = 0 signifies no dependency between the parameters. Large values of cross-correlation between parameter estimates, say $|\kappa| > 0.95$, means that the two parameters are weakly determinable owing to their strong influence on one another [8].

For statistical models, it is common to use measures such as the Akaike Information Criterion (AIC) or Bayesian Information Criterion (BIC). These approaches are based on likelihood, and can be extended to the dynamical models used in systems biology, as detailed elsewhere [30]. A rule of thumb, or heuristic, that is commonly used is Occam's razor, as described below.

Occam's razor

Occam's razor, also spelt Ockham's razor, is a commonly used heuristic for selecting theories to explain phenomena, or in our case, models to explain data. William of Ockham (1285–1347/49 CE) was a philosopher who employed this principle— essentially a principle of parsimony or economy—which he stated as *pluralitas non est ponenda sine necessitate*, i.e. *"plurality should not be posited without necessity."* The idea of the principle is that, when confronted with two competing theories, the simpler theory must be picked ahead of the more complex one. In the context of systems biology, it would mean picking a model with fewer parameters, but only if it explains the model as well as another model with more parameters.

Model discrimination or selection remains a challenging task, and is central to the success of any modelling exercise. In the context of biology, it is also possible to summon other domain knowledge to aid in this discrimination. For instance, the models should not be too sensitive to minor changes in parameters, as biological systems are seldom so sensitive. Rather, they are robust, displaying (nearly) unchanged behaviour, across a wide array of environmental conditions, or parameters. Thus, while models must be *responsive* to change in parameters (otherwise, the parameter could well be eliminated), they must not be overly sensitive, and mimic the general robust nature of biological systems.

6.4.2   Sensitivity and robustness of biological models

Sensitivity analysis is a common step in any modelling process—to determine how various perturbations and uncertainties can have an effect on the 'model output'. In the context of biological models, sensitivity analysis bears greater significance, as it also enables us to understand and characterise the robustness of a model. Studies of sensitivity and robustness are critical in systems biology, as biological systems are uniquely robust, and therefore, the models that capture them must also mirror this behaviour. Robustness has even been used as a measure to evaluate the plausibility of alternate models [31]. See §13.3 (box on p. 303) for an interesting discussion on how model robustness may be quantified, and used in model selection.

6.4.2.1 Local

In a general dynamical system, captured by ODEs such as in Equation 5.1, sensitivity analysis can be broadly performed by studying the variation of specific outputs with respect to parametric perturbations [32]. Very simply, one can capture parametric state sensitivity in an $m \times n$ matrix, capturing the sensitivity of each of the m states in the model with respect to isolated perturbations of each of the n parameters [33]:

$$S_{ij}(t) = \frac{dx_i(t)}{d\theta_j} \tag{6.12}$$

capturing how the state x_i at a specific time-point t changes with variation in θ_j. Such an analysis is *local*, because it is generally restricted to studying the behaviour of a system near a *single* operation point; rather than *globally* across different parameter sets/operating points.

6.4.2.2 Global

Global methods strive to characterise the entire parameter space, notably in terms of regions of desired/acceptable behaviour. It is conceivable that, the larger the volume of this region of acceptable behaviour, the more robust the system, *i.e.* there are more parameter sets, or configurations, for which the system displays the desired behaviour. How do we estimate this volume? This is typically done through sampling various parameter sets from the parameter space and examining viability [31, 34, 35].

It is important to note that the robustness of a system depends not only on the mere volume of this parameter space, but also on its topology and geometry [34]. For instance, even if the total volume is high, but if there exist certain narrow regions in this space, where minor changes (in parameter values) can push the system to outrun the boundary of the region (*i.e.* lose desired behaviour), robustness can be poor. Thus, it is necessary to examine both the size and the shape/-geometry of the viable parameter spaces (again, see §13.3, p. 303).

6.4.2.3 'Glocal' sensitivity analyses

Methods have also been developed to combine both global and local approaches (hence the name 'glocal') [36]. The key idea here is to obtain a large sample of 'viable' parameter sets from across the space, where the system shows desirable behaviour, through an iterative Gaussian sampling. Subsequently, a local sensitivity analysis is performed around each of these individual parameter sets. Together, these represent a global–local measure of robustness. However, a challenge remains, in that it is extremely difficult to obtain a statistically representative sample of viable points. Another algorithm developed by Zamora-Sillero *et al* combines global and local explorations of the parameter space [37]. Specifically, it can tackle parameter spaces where the regions of viability are many, potentially disconnected, and also individually non-convex. Of course, these methods are computationally much more expensive, which is unsurprising, as we seek to characterise the entire parameter space, even if approximately.

6.4.2.4 'Topological' sensitivity

Beyond the above measures of sensitivity, it is also possible to examine the sensitivity of model or system behaviour to *topology*; that is, is a system, or a model able to retain its behaviour following the loss or gain of certain interactions. Such an analysis can also guide model selection. For instance, is a particular interaction, phosphorylation of feedback loop essential for capturing the function of a system? Are there other model topologies that can explain observations (experimental data) equally well? Some of these concepts also underlie the notion of ensemble modelling [38].

6.5 TROUBLESHOOTING

A number of challenges arise while building robust dynamic mathematical models and performing parameter estimation—practically at every step of the flowchart in Figure 6.1. In parameter estimation itself, we are confronted with the choice of a suitable objective function as discussed in §6.2.2.2.

6.5.1 Regularisation

A number of ills in modelling biological systems arise from the lack of identifiability, owing to both model structure and lack of *informative* data. This typically results in the parameter estimation problem being ill-posed[7] and ill-conditioned[8].

One useful strategy to counter these problems is regularisation, which is related to the parsimony principle (recall Occam's razor discussed above, p. 151). The key aim of regularisation is to replace an ill-posed problem by a family of neighbouring well-posed problems [39], by including an additional term that typically penalises model complexity or widely varying behaviour. A variety of regularisation schemes exist, and the interested reader is referred to the excellent review by Engl *et al* [39]. Regularisation can also help overcome over-fitting [40].

6.5.2 Sloppiness

Sloppiness [41, 42] is another pitfall that one must be aware of, during parameter estimation of systems biology models. Intimately connected with identifiability [43], sloppiness essentially qualifies the parameter sensitivities of a model. Gutenkunst *et al* showed that, for a large number of systems biology models,

---

[7]The solution may not exist, or it may not be unique or stable.

[8]Crudely, a small change in the parameters can cause a very large change in the behaviour, *i.e.* cost function value. The Hessian of the objective function is typically ill-conditioned, having a very large spread of eigenvalues, and consequently, a large "condition number". The condition number of a matrix $\kappa(\mathbf{A}) = \|\mathbf{A}\|\|\mathbf{A}^{-1}\|$.

there exist directions in parameter space, where large variations in the parameter values produce feeble variation in the cost function ("sloppy" parameter sensitivities) and fewer directions that are "stiff", where changes in parameter values cause changes in the cost function. Sloppiness can be determined from the eigenvalues (conditioning) of the Hessian of the cost function, $\frac{\partial^2 C}{\partial\theta_1\partial\theta_2}$, as detailed in [42]; a model can be considered sloppy if the ratio of the minimum eigenvalue to the maximum, $\frac{\lambda_{min}}{\lambda_{max}} \lesssim 10^{-3}$. It is also possible to use the Fisher Information Matrix to approximate the Hessian (see [43]).

6.5.3 Choosing a search algorithm

Another tricky point in a parameter estimation exercise is the choice of a search algorithm. Several studies have benchmarked a variety of algorithms and present interesting results [5, 14, 15, 44]. Depending upon the size, complexity, and overall nature of the parameter estimation problem, a number of algorithms can be used effectively to estimate parameters. Tuning the hyperparameters for these algorithms can be another major challenge—typical benchmarking studies do not carefully tune these hyperparameters to avoid bias [15]. Therefore, one may have to explore a variety of optimisation methods and hyperparameter combinations to identify a potent algorithm to solve a given problem.

6.5.4 Model reduction

An automatic consequence of mechanistic modelling is the fact that the models tend to be quite large and complex. It is quite likely that certain portions of a model, for example, certain reactions in a network, have little or no impact on the variables of interest. A number of approaches have been developed for model reduction, based on sensitivity analysis, optimisation, parameter lumping, etc., and are reviewed in [45].

6.5.5 The curse of dimensionality

In large parameter spaces, finding the global minimum is inherently a challenge—typically called the *curse of dimensionality*. It is a *curse*, in the sense it is an inherent problem of very large spaces. For even a simple enzymatic system with five reactions, we are looking at a space of $\mathbb{R}^{10}$, assuming two parameters (*e.g.* K_M and v_{max}) for each of the reactions. Finding a single point, the minimum, in n dimensions is much much harder than finding it in a single dimension; with every increasing dimension, the task becomes *exponentially harder*[9].

Suppose we want to estimate parameters in $\mathbb{R}^{10}$. To even merely examine simple combinations that belong in different 'corners' of the space, like "low k_1, low k_2, low k_3, ..., high k_{10}", "high k_1, low k_2, high k_3, ..., high k_{10}", etc., we would need $2^{10} = 1,024$ samples. In 20 dimensions, this explodes to over

---

[9]Imagine searching for a single point of interest on a number line, vs. a 2D square, and a 3D cube.

a million! Thus, sampling algorithms and parameter estimation algorithms are highly disadvantaged in large spaces, and have to surmount major computational challenges, to find biologically meaningful optima.

Being a challenge inherent in the size of the space, it is really not possible to effectively surmount this. However, the availability of good parameter estimates can make this process much more computationally efficient. These estimates could be obtained from experiments specifically designed to estimate few parameters, or from previous knowledge available on the same species currently being modelled, or even related species.

6.6 SOFTWARE TOOLS

Multiple tools are available for parameter estimation and all the other steps of data-driven mechanistic modelling; some of these are listed below, and others are detailed in Appendix D.

AMIGO2 toolbox [46] is a versatile MATLAB-based toolbox for performing various steps of data-driven modelling, including identifiability analysis, parameter estimation, local and global sensitivity analyses, and optimal experimental design.

PESTO toolbox [47] is a parameter estimation toolbox for MATLAB and supports various optimisation algorithms as well as identifiability and uncertainty analyses.

PyGMO (*Python Parallel Global Multiobjective Optimizer*) [48] has efficient implementations of various kinds of optimisation algorithms, especially bio-inspired algorithms. The companion code repository contains simple snippets of PyGMO codes.

EXERCISES

6.1 Consider a biological system that follows Michaelis–Menten kinetics with competitive inhibition. For the data given below,

$v\,(mM/s)$	$S\,(mM)$	$I\,(mM)$	δ
1.0	1.8	12.0	0.2
5.0	7.1	12.0	0.1
10.0	10.6	12.0	0.1
15.0	12.2	12.0	0.1

a. Compute which of the following parameter sets gives the least error.

$$\theta_1 = \{V_m = 20mM/s, K_m = 5mM, K_I = 4mM\}$$

$$\theta_2 = \{V_m = 20mM/s, K_m = 4mM, K_I = 3mM\}$$

$$\theta_3 = \{V_m = 20mM/s, K_m = 10mM, K_I = 12mM\}$$

b. Will you require any additional experimental data to evaluate the parameter sets? How will you then identify the 'true' parameters of the system?

6.2 If 010101 and 100110 are individuals in a genetic algorithm, which of the following are possible children, following a single-point crossover?

a. 010100, 100111
b. 010110, 100101
c. 010111, 100001
d. 000110, 110101
e. 011000, 101110
f. 010101, 100110

6.3 Consider a simple integer optimisation problem, to be solved using Genetic Algorithms: maximise $x^3 + y^2 - 3x^2y$, in the range $0 < x < 32, 0 < y < 32$. One can use a ten-bit representation of the chromosome, where the first five bits represent x, and the next five bits represent y.

a. What would be the fitness of the chromosome represented as 1111000010?
b. If you mutated the first and the last bits of the above chromosome, what is the new fitness?

[LAB EXERCISES]

6.4 Similar to the Rosenbrock function, there exist many other 'test functions' for optimisation. One of them is the Rastrigin function, proposed by Rastrigin in 1974 and later generalised to n-dimensions

$$f(x) = 10n + \sum_{i=1}^{n} \left[x_i^2 - 10\cos(2\pi x_i) \right]$$

with $x_i \in \{-5.12, 5.12\}$. This function has a large number of local minima, with the global minimum of zero, at $x = 0$. This function is available in MATLAB's global optimisation toolbox (rastriginsfcn). Try finding the minimum of this function using (a) MATLAB fminsearch beginning at random starting points and (b) evolutionary algorithms.

6.5 Consider a simple Michaelis–Menten system, with the following available measurements.

v (mM/s)	0.15	0.35	0.65	1.2	2.0	3.0	4.0	5.0	6.0
S (mM)		0.35	0.70	1.26	1.62	1.5	2.21	2.167	2.33	2.12

Can you obtain the optimal values of v_{max} and K_M based on (a) a Lineweaver–Burk double reciprocal plot followed by linear regression, (b) an Eadie–Hofstee plot followed by linear regression, and (c) non-linear regression (e.g. fminsearch in MATLAB, for a simple SSE objective function.) Can you see how the errors vary between the three methods?

6.6 The DREAM6 "Estimation of Model Parameters Challenge" (https://www.synapse.org/#!Synapse:syn2841366/) provided a fully characterised gene regulatory network comprising nine genes, modelled using ODEs. For each gene, both the mRNA level and protein level are modelled, giving rise to 18 states, and 45 parameters in all. The challenge also included a 'budget' for performing experiments, which could be used to generate new data.

Create the parameter estimation problem in PyGMO to solve it using any of the several built-in algorithms in PyGMO. Try out the parameter estimation using two different sets of initial values: (a) a random initial guess of parameters and (b) the winning team's parameter estimates, and compare your performance.

REFERENCES

[1] N. Benson, "Quantitative systems pharmacology and empirical models: friends or foes?", *CPT: Pharmacometrics & Systems Pharmacology* 8:135–137 (2019).

[2] C. Arndt, *Information measures: information and its description in science and engineering* (Springer-Verlag, Berlin Heidelberg, 2001).

[3] S. Kullback, *Information theory and statistics*, New edition (Dover Publications Inc., Mineola, N.Y, 1997).

[4] R. Bellman and K. Åström, "On structural identifiability", *Mathematical Biosciences* 7:329–339 (1970).

[5] A. F. Villaverde, A. Barreiro, and A. Papachristodoulou, "Structural identifiability of dynamic systems biology models", *PLoS Computational Biology* 12:e1005153+ (2016).

[6] K. G. Gadkar, R. Gunawan, and F. J. Doyle, "Iterative approach to model identification of biological networks", *BMC Bioinformatics* 6:155 (2005).

[7] W. H. Press, S. A. Teukolsky, W. T. Vetterling, and B. P. Flannery, *Numerical recipes 3rd edition: the art of scientific computing*, 3rd ed. (Cambridge University Press, 2007).

[8] K. Jaqaman and G. Danuser, "Linking data to models: data regression", *Nature Reviews Molecular Cell Biology* 7:813–819 (2006).

[9] R. Storn and K. Price, "Differential Evolution —a simple and efficient heuristic for global optimization over continuous spaces", *Journal of Global Optimization* 11:341–359 (1997).

[10] S. Boyd and L. Vandenberghe, *Convex optimization* (Cambridge University Press, 2004).

[11] R. Fletcher, *Practical methods of optimization*, 2nd ed. (Wiley, 2000).

[12] J. Nocedal and S. J. Wright, *Numerical optimization* (Springer-Verlag, New York, USA, 1999).

[13] M. Kochenderfer, *Algorithms for optimization* (The MIT Press, Cambridge, Massachusetts, 2019).

[14] A. Degasperi, D. Fey, and B. N. Kholodenko, "Performance of objective functions and optimisation procedures for parameter estimation in system biology models", *npj Systems Biology and Applications* 3:20 (2017).

[15] A. Raue, M. Schilling, J. Bachmann, et al., "Lessons learned from quantitative dynamical modeling in systems biology", *PLoS ONE* 8:e74335 (2013).

[16] S. Shin, O. S. Venturelli, and V. M. Zavala, "Scalable nonlinear programming framework for parameter estimation in dynamic biological system models", *PLoS Computational Biology* 15:e1006828 (2019).

[17] R. M. Lewis, V. Torczon, and M. W. Trosset, "Direct search methods: then and now", *Journal of Computational and Applied Mathematics* 124:191–207 (2000).

[18] J. Brownlee, *Clever algorithms: nature-inspired programming recipes* (Lulu, 2011).

[19] S. Kirkpatrick, C. D. Gelatt, and M. P. Vecchi, "Optimization by simulated annealing", *Science* 220:671–680 (1983).

[20] N. Metropolis, A. W. Rosenbluth, M. N. Rosenbluth, A. H. Teller, and E. Teller, "Equation of state calculations by fast computing machines", *The Journal of Chemical Physics* 21:1087–1092 (1953).

[21] W. K. Hastings, "Monte Carlo sampling methods using Markov chains and their applications", *Biometrika* 57:97–109 (1970).

[22] Z. Michalewicz and D. B. Fogel, *How to solve it: modern heuristics* (Springer, Berlin, Germany, 2004).

[23] T. G. Kolda, R. M. Lewis, and V. Torczon, "Optimization by direct search: new perspectives on some classical and modern methods", *SIAM Review* 45:385–482 (2003).

[24] C. G. Moles, P. Mendes, and J. R. Banga, "Parameter estimation in biochemical pathways: a comparison of global optimization methods", *Genome Research* 13:2467–2474 (2003).

[25] J. Holland, *Adaptation in natural and artificial systems: an introductory analysis with applications to biology, control, and artificial intelligence* (MIT Press, Cambridge, Mass, 1992).

[26] G. M. Morris, D. S. Goodsell, R. S. Halliday, et al., "Automated docking using a Lamarckian genetic algorithm and an empirical binding free energy function", *Journal of Computational Chemistry* 19:1639–1662 (1998).

[27] D. Simon, *Evolutionary optimization algorithms* (Wiley-Blackwell, New Jersey, USA, 2013).

[28] N. Hansen and A. Ostermeier, "Adapting arbitrary normal mutation distributions in evolution strategies: the covariance matrix adaptation", in Proceedings of IEEE International Conference on Evolutionary Computation (1996), pp. 312–317.

[29] G. Hornby, A. Globus, D. Linden, and J. Lohn, "Automated antenna design with evolutionary algorithms", in *Space 2006* (2006).

[30] P. Kirk, T. Thorne, and M. P. Stumpf, "Model selection in systems and synthetic biology", *Current Opinion in Biotechnology* **24**:767–774 (2013).

[31] M. Morohashi, A. E. Winn, M. T. Borisuk, H. Bolouri, J. Doyle, and H. Kitano, "Robustness as a measure of plausibility in models of biochemical networks", *Journal of Theoretical Biology* **216**:19–30 (2002).

[32] J. Stelling, E. D. Gilles, and F. J. Doyle, "Robustness properties of circadian clock architectures", *Proceedings of the National Academy of Sciences USA* **101**:13210–13215 (2004).

[33] N. Bagheri, J. Stelling, and I. Doyle Francis J., "Quantitative performance metrics for robustness in circadian rhythms", *Bioinformatics* **23**:358–364 (2006).

[34] A. Dayarian, M. Chaves, E. D. Sontag, and A. M. Sengupta, "Shape, size, and robustness: feasible regions in the parameter space of biochemical networks", *PLoS Computational Biology* **5**:1–12 (2009).

[35] A. Wagner, "Circuit topology and the evolution of robustness in two-gene circadian oscillators", *Proceedings of the National Academy of Sciences USA* **102**:11775–11780 (2005).

[36] M. Hafner, H. Koeppl, M. Hasler, and A. Wagner, "'Glocal' robustness analysis and model discrimination for circadian oscillators", *PLoS Computational Biology* **5**:1–10 (2009).

[37] E. Zamora-Sillero, M. Hafner, A. Ibig, J. Stelling, and A. Wagner, "Efficient characterization of high-dimensional parameter spaces for systems biology", *BMC Systems Biology* **5**:142 (2011).

[38] L. Kuepfer, M. Peter, U. Sauer, and J. Stelling, "Ensemble modeling for analysis of cell signaling dynamics", *Nature Biotechnology* **25**:1001–1006 (2007).

[39] H. W. Engl, C. Flamm, P. Kügler, J. Lu, S. Müller, and P. Schuster, "Inverse problems in systems biology", *Inverse Problems* **25**:123014 (2009).

[40] A. Gábor and J. R. Banga, "Robust and efficient parameter estimation in dynamic models of biological systems", *BMC Systems Biology* **9**:74 (2015).

[41] K. S. Brown and J. P. Sethna, "Statistical mechanical approaches to models with many poorly known parameters", *Physical Review E* **68**:021904 (2003).

[42] R. N. Gutenkunst, J. J. Waterfall, F. P. Casey, K. S. Brown, C. R. Myers, and J. P. Sethna, "Universally sloppy parameter sensitivities in systems biology models", *PLoS Computational Biology* **3**:e189 (2007).

[43] O.-T. Chis, A. F. Villaverde, J. R. Banga, and E. Balsa-Canto, "On the relationship between sloppiness and identifiability", *Mathematical Biosciences* **282**:147–161 (2016).

[44] M. Ashyraliyev, Y. Fomekong-Nanfack, J. A. Kaandorp, and J. G. Blom, "Systems biology: parameter estimation for biochemical models", *The FEBS Journal* **276**:886–902 (2009).

[45] T. J. Snowden, P. H. van der Graaf, and M. J. Tindall, "Methods of model reduction for large-scale biological systems: a survey of current methods and trends", *Bulletin of Mathematical Biology* **79**:1449–1486 (2017).

[46] E. Balsa-Canto, D. Henriques, A. Gábor, and J. R. Banga, "AMIGO2, a toolbox for dynamic modeling, optimization and control in systems biology", *Bioinformatics* **32**:3357–3359 (2016).

[47] P. Stapor, D. Weindl, B. Ballnus, et al., "PESTO: Parameter EStimation TOolbox", *Bioinformatics* **34**:705–707 (2018).

[48] F. Biscani and D. Izzo, "A parallel global multiobjective framework for optimization: pagmo", *Journal of Open Source Software* **5**:2338 (2020).

FURTHER READING

A. F. Villaverde and J. R. Banga, "Reverse engineering and identification in systems biology: strategies, perspectives and challenges", *Journal of the Royal Society Interface* **11**:20130505 (2014).

F. Fröhlich, B. Kaltenbacher, F. J. Theis, and J. Hasenauer, "Scalable parameter estimation for genome-scale biochemical reaction networks", *PLoS Computational Biology* **13**:1–18 (2017).

Z. Zi, "Sensitivity analysis approaches applied to systems biology models", *IET Systems Biology* **5**:336 (2011).

M. Renardy, C. Hult, S. Evans, J. J. Linderman, and D. E. Kirschner, "Global sensitivity analysis of biological multi-scale models", *Current Opinion in Biomedical Engineering* **11**:109–116 (2019).

I.-C. Chou and E. O. Voit, "Recent developments in parameter estimation and structure identification of biochemical and genomic systems", *Mathematical Biosciences* **219**:57–83 (2009).

D. V. Raman, J. Anderson, and A. Papachristodoulou, "Delineating parameter unidentifiabilities in complex models", *Physical Review E* **95**:032314 (2017).

A. G. Busetto, A. Hauser, G. Krummenacher, et al., "Near-optimal experimental design for model selection in systems biology", *Bioinformatics* **29**:2625–2632 (2013).

CHAPTER 7

Discrete dynamic models: Boolean networks

CONTENTS

B EYOND CONTINUOUS dynamic models, the dynamic nature of biological systems can also be captured by discrete dynamic models. Discrete-time models have been popularly applied in biology to model population growth, where the population of a given species at time-step t is modelled as a function of the population in the previous time-step(s), $\{t-1,\ldots\}$. In this chapter, however, we will focus on models where the variables are also assumed to have a finite set of possible states. In particular, we will focus on *logical models*, as captured by Boolean networks, where these states are only two, namely ON and OFF. Boolean networks, which are thus discrete both in time and state, consequently provide a coarser and somewhat qualitative high-level view of the dynamics of a system. This is in contrast to the detailed insights into system dynamics provided by ODE models. The origins of Boolean networks can be traced to as early as 1969, when Stuart Kauffman [1] introduced "random genetic nets" to study cellular processes. Later work by Kauffman [2] and René Thomas [3] further established the utility of Boolean network formalisms to study regulatory circuits in biology.

7.1 INTRODUCTION

While ODE-based dynamic models provide detailed insights into the dynamics of a given system, they also demand large amounts of data for effective parameter estimation, as well as accurate measurements of various variables, *e.g.* concentrations of metabolites or expression levels of different genes. In many cases, notably in the case of gene regulatory networks, both the available measurements and

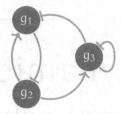

Figure 7.1 **A simple gene regulatory network.** Three nodes, for genes g_1, g_2, and g_3 are shown, with normal arrowheads on the edges indicating activation, and bar-headed arrows indicating inhibition/repression.

the required readouts, such as the activity of a gene, may be restricted to discrete values such as high, medium, and low, or, more simply, just ON and OFF. In gene regulatory networks, it is common to capture the activity of a given gene (or regulatory protein) as the net consequence of various effectors/inhibitors. For example, "*Gene* A *is active in the absence of Gene* B *or Gene* C *and low levels of metabolite* M" is a typical description of regulatory relationships in a biological system.

In the following section, we will understand the basics of constructing discrete models using Boolean networks and simulating them.

7.2   BOOLEAN NETWORKS: TRANSFER FUNCTIONS

As in any modelling exercise, a very vital step is the accurate reconstruction of the network from known hypotheses and available literature. The key assumptions in a Boolean network model are that the system evolves in discrete time-steps, and that the variables take on discrete values, namely OFF and ON, or $\{0, 1\}$. Boolean networks can be represented similar to the graphs we studied in Chapter 2, for example Figure 2.7. A similar but simpler figure is shown in Figure 7.1. The interactions (*i.e.* edges) shown in the figure can be translated to Boolean transfer functions.

Transfer functions relate the output or response of a variable to the inputs. Boolean transfer functions capture the relationship between the various variables in a model, as follows:

$$X_i(t) = F_i(X_1(t-1), X_2(t-1), \ldots, X_n(t-1)) \tag{7.1}$$

where X_i represents the various state variables (*e.g.* genes) and F_i represents the transfer function for the ith variable. The above equation represents a *synchronous* update, as all the state variables are updated simultaneously. F_i is composed using standard Boolean operators such as AND, OR, and NOT. For the simple network in Figure 7.1, the transfer functions can be described as:

$$g_1(t) = g_2(t-1) \text{ AND } (\text{NOT } g_3(t-1)) \tag{7.2a}$$

$$g_2(t) = \text{NOT } g_1(t-1) \tag{7.2b}$$

$$g_3(t) = (\text{NOT } g_3(t-1)) \text{ OR } g_2(t-1) \tag{7.2c}$$

In a more compact notation, it is common to write out the above transfer functions as

$$g_1^* = g_2 \text{ AND } (\text{NOT } g_3) \tag{7.3a}$$

$$g_2^* = \text{NOT } g_1 \tag{7.3b}$$

$$g_3^* = (\text{NOT } g_3) \text{ OR } g_2 \tag{7.3c}$$

Note how the transfer functions *composes* the different inputs—for g_1^*, AND is used, while for g_3^*, OR is used to compose the two inputs. This is not apparent from the figure; it must be inferred from the information used to build the network, as to which of the inputs/interactions predominate, and whether the effects are individual or cumulative. It is easy to simulate this system of equations, beginning with the initial values for the states of expression of all three genes, *i.e.* $g_1(0)$, $g_2(0)$, and $g_3(0)$.

The transfer functions given below represent a more complex system, that of the mammalian cell cycle [4]:

$$\text{CycD}^* = \text{CycD}$$

$\quad \text{Rb}^* = (\text{NOT CycA AND NOT CycB AND NOT CycD AND NOT CycE})$

$\qquad \text{OR } (\text{p27 AND NOT CycB AND NOT CycD})$

$\quad \text{E2F}^* = (\text{NOT Rb AND NOT CycA AND NOT CycB})$

$\qquad \text{OR } (\text{p27 AND NOT Rb AND NOT CycB})$

$\text{CycE}^* = (\text{E2F AND NOT Rb})$

$\text{CycA}^* = (\text{E2F AND NOT Rb AND NOT Cdc20 AND NOT (Cdh1 AND Ubc)})$

$\qquad \text{OR } (\text{CycA AND NOT Rb AND NOT Cdc20 AND NOT (Cdh1 AND Ubc)})$

$\quad \text{p27}^* = (\text{NOT CycD AND NOT CycE AND NOT CycA AND NOT CycB})$

$\qquad \text{OR } (\text{p27 AND NOT (CycE AND CycA) AND NOT CycB AND NOT CycD})$

$\text{Cdc20}^* = \text{CycB}$

$\text{Cdh1}^* = (\text{NOT CycA AND NOT CycB}) \text{ OR } (\text{Cdc20}) \text{ OR } (\text{p27 AND NOT CycB})$

$\quad \text{Ubc}^* = \text{NOT Cdh1 OR } (\text{Cdh1 AND Ubc AND (Cdc20 OR CycA OR CycB)})$

$$\text{CycB}^* = \text{NOT Cdc20 AND NOT Cdh1} \tag{7.4}$$

To discuss one example of how such complex transfer functions are composed, let us take the example of the protein UbcH10 (denoted Ubc above). This protein is active when Cdh1 is absent—this contributes the first part of the transfer function. Alternatively, in the presence of CdH1, this activity can be maintained when at least one of its other targets, namely CycA, Cdc20, or CycB, is present—giving rise to the second part of the transfer function.

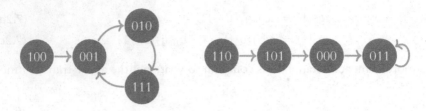

Figure 7.2 **State transition map for the simple gene regulatory network, with synchronous updates.** The three binary digits indicate the states of the three genes g_1, g_2, and g_3, respectively. Arrows indicate transitions.

7.2.1 Characterising Boolean network dynamics

Given a Boolean network, we need to understand its behaviour and characterise its dynamics. Beyond a simple simulation of the Boolean network, it is often helpful to characterise the various possible states of the network, which can be reached from different states. For a Boolean network with n variables, there are obviously 2^n possible states of the system, where every variable can be either ON or OFF. A simple representation of the state of the system can be done using a Boolean vector of dimensions $n \times 1$. From a given starting state, by applying synchronous updates, only a single state can be reached. From a given state, it is possible that the system remains in the same state, following the update, indicating that a steady-state has been reached. Alternatively, the system might continue to *cycle* through one or more states. The transitions between these different states are governed by the Boolean transfer functions.

For the system illustrated in Figure 7.1, the following are the transitions, from each of the eight starting states:

$000 \Rightarrow 011$	$001 \Rightarrow 010$	$010 \Rightarrow 111$	$011 \Rightarrow 011$
$100 \Rightarrow 001$	$101 \Rightarrow 000$	$110 \Rightarrow 101$	$111 \Rightarrow 001$

It is very helpful to visualise these transitions in a *map*, which helps clearly identify how the system cycles through the various states. Figure 7.2 illustrates the state transition map for the simple network in Figure 7.1. Note that there are two disconnected sets of state transitions. That is, it is not possible to reach certain states from certain other states. For example, there is no way to reach from the state 000 to 111, or vice-versa. Starting from the state 100, the system transits to 001, 010 and 111, and then again to 001, resulting in a *cycle* of *period* three. Such sets of repeating states are known as *attractors*. These attractors are further classified based on their period. The cycle, where 011 transits to itself, is a cycle of period one and is called a point attractor or a singleton attractor. On the other hand, cycles with larger periods, such as the cycle of period three mentioned above, are *cycle attractors*.

7.2.2 Synchronous vs. asynchronous updates

Recall equation 7.1, which describes the synchronous update for a given system:

$$X_i(t) = F_i(X_1(t-1), X_2(t-1), \ldots, X_n(t-1)) \tag{7.1}$$

Such an update is synchronous, as all states are simultaneously updated; an important consequence of this update is that all processes are assumed to be equally fast or slow, spanning exactly one time-step. For example, the repression of one gene by another, or the activation of two genes by a third, are all assumed to take place within one time-step. This may be an over-simplifying assumption at times, but it is straightforward to extend the model to account for some more stochasticity in process times. The *asynchronous* update involves using $X(t_i)$ values instead of $X(t-1)$ [3, 5, 6]:

$$X_i(t) = F_i(X_1(t_1), X_2(t_2), \ldots, X_n(t_n)), \tag{7.5}$$

where $t_i \in \{t, t-1, \ldots\}$, *i.e.* the value of X_i is taken at the current (t) or a previous time-step (*e.g.* $t-1$, $t-2$, etc.) This allows for significant stochasticity in the durations of various processes, and also allows for accommodation of aspects such as delays.

Conway's Game of Life

Created by British mathematician John Conway in 1970, the "Game of Life" is a classic example of a discrete simulation.

The rules of the game are as follows: for a cell that is currently 'populated', if it has one or fewer neighbours, it dies. It also dies if it has four or more neighbours, as if by overcrowding. Cells with two or three neighbours continue to survive. An empty cell with three neighbours becomes populated. The two examples shown below end in 'point attractor' and a 'cycle attractor', respectively:

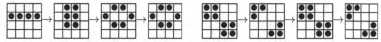

Such simulations can be carried out for large grids, as available on the website https://bitstorm.org/gameoflife/.

Sontag and co-workers [7] have proposed additional ways of performing asynchronous updates. For example, they propose that in each time interval, proteins can be updated first, followed by mRNAs, which allows the separation of post-translation and pre-translational processes. They demonstrate their approaches on the regulatory network of the segment polarity genes in *Drosophila melanogaster*, to study the development of gene expression patterns.

7.3   OTHER PARADIGMS

7.3.1   Probabilistic Boolean networks

As with every biological system, there is always a cloud of uncertainty over the exact nature of the regulatory functions that can capture the exact behaviour of a system. At times, multiple candidate Boolean functions may exist, for describing a given interaction. Shmulevich and co-workers [8] tackled this challenge by modifying the Boolean network to admit more than one transfer function, each of which has a particular probability, assigned based on prior data. Such networks are termed probabilistic Boolean networks (PBNs).

At every time-step of the simulation, one of the admissible Boolean transfer functions is picked at random, based on a probability distribution. Since we are making random choices at each step, the outcome of the simulation is decidedly stochastic. Therefore, from a single initial state, unlike in the case of regular Boolean networks, more than one final state/sets of transitions are possible.

7.3.2   Logical interaction hypergraphs

Proposed by Steffen Klamt and co-workers, logical interaction hypergraphs were shown to be a useful formalism to model and understand signalling and regulatory networks [9]. As discussed in §2.3.5, simple graphs can only capture binary interactions. To represent AND relationships between species in their influence on say, a particular protein in a signalling network, it is therefore necessary to use other formalisms such as hypergraphs. Logical interaction hypergraphs can be implemented and simulated using CellNetAnalyzer (see §7.6).

7.3.3   Generalised logical networks

Generalised logical networks build on the Boolean network formalism, but extend it further to allow the variables to have more than two values, as well as permit asynchronous transitions [10, 11]. *Multi-level* variables are particularly useful in biology since they provide for a way to naturally capture different levels of concentration of species, such as non-existent, basal, medium, or high. Suitable thresholds can be used to convert continuous variables to discrete logical variables.

7.3.4   Petri nets

Petri Nets are named after Carl Adam Petri, who proposed a graphical and mathematical formalism for the modelling of concurrent/distributed systems, such as chemical reactions[1]. Petri nets are characterised by *states* and *transitions*, which correspond to substances (*e.g.* metabolites, proteins, or genes) and reactions/regulatory interactions. Petri nets are indeed directed bipartite graphs where states

---

[1]Petri proposed this formalism in 1939, at the age of 13!

and transitions form nodes, with edges connecting only states to transitions and transitions to other states.

The first biological application of Petri nets was proposed by Reddy *et al* in 1993 [12], who used Petri nets to represent and study metabolic networks. More recent reviews capture the breadth of applications of Petri nets to biology [13–15]. Despite great potential, Petri nets have been somewhat under-explored in systems biology and metabolic network analysis, compared to other approaches.

7.4   APPLICATIONS

Many studies have illustrated the utility of qualitative Boolean network models to understand regulatory networks, ranging from the simple *lac* operon [16], to the cell cycle [4] or more complex signalling networks in plants [5] or *Drosophila* [17], and even the complex interplay between pathogens and the human immune system during respiratory infections and tuberculosis infection (see §11.2.2.1). Most importantly, such systems-level models of regulation/signalling enable an integrated study of a multitude of factors, rather than studying various factors in isolation, without considering various cross-talks and interactions across pathways [5].

7.5   TROUBLESHOOTING

The remarkable simplicity of logical models means that there are relatively fewer challenges to surmount while building and simulating logical models/Boolean networks. The key challenge is in the accurate reconstruction of the model, such that the underlying biology is faithfully captured by the Boolean transfer functions. In particular, the precedence of various interactions have to be carefully considered to compose the various input variables into the transfer functions using AND, OR, AND NOT, and so on.

Also, a considered choice has to be made, as to whether a Boolean model suffices, or multi-level variables are necessary to capture the dynamics. Another key choice is whether a simple synchronous update suffices, or an asynchronous update is necessary to capture the differences in durations of various processes. A very recent approach, Most Permissive Boolean Networks (MPBNs), attempts to mitigate many of the shortcomings of Boolean networks, and attempts to capture a broader range of dynamics [18].

7.6   SOFTWARE TOOLS

BooleanNet   [19] is a Python-based package for performing simulations of Boolean networks. It accepts simple text encodings of the Boolean model, and can perform simulations for a specified number of time-steps. It also supports knock-out studies, evaluating the effect of removing a given state from the model. BooleanNet is available from https://github.com/ialbert/booleannet.

BooleSim [20] is an excellent JavaScript-based online tool for Boolean network simulations, particularly for simpler networks, and for pedagogical purposes. It can also be downloaded for offline use, and runs within a browser itself. BooleSim is available from https://matthiasbock.github.io/BooleSim/ or https://rumo.biologie.hu-berlin.de/boolesim/.

CellNetAnalyzer (https://www2.mpi-magdeburg.mpg.de/projects/cna/cna.html) is a MATLAB toolbox that can perform logical steady-state analysis of Boolean networks (*e.g.* signalling networks). It also has various algorithms for simulating metabolic networks (as we will see in the following chapter).

EXERCISES

7.1 Consider a gene network with the following regulatory interactions: A is activated by B and C. B is repressed by D. C is activated by A or B. D is activated in an XOR fashion by A and C. Sketch the network and compute all the possible state transitions, identifying the attractors. Assume synchronous updates.

7.2 Consider a gene network containing four genes, a through d, which are transcribed to mRNA. The mRNAs serve as a template for the proteins A through D, synthesised by translation. The proteins B and C dimerise to form a complex, which activates the expression of gene d. The protein D represses the expression of genes a and b (which are thus co-regulated) *instantly*. Protein D activates the translation of protein B. Sketch the network and compute all the possible state transitions. How many attractors are there in this system?

7.3 Consider the classic three-node repressilator (see §12.2.1). The transcriptionally regulated oscillatory circuit involves three different proteins, LacI, λcI and TetR, wired such that LacI represses TetR, TetR represses λcI and λcI represses LacI. What are the Boolean transfer functions for this circuit? If you simulate this circuit, you will observe oscillations. If the repression is *delayed* by a few cycles, does the oscillatory behaviour remain?

[LAB EXERCISES]

7.4 Implement Conway's game of life in MATLAB. Take as input an $n \times n$ matrix, and return the series of matrices up to a pre-specified number of iterations, k. Assume cells beyond the grid boundary to be all dead.

7.5 Construct Boolean transfer functions for the neurotransmitter signalling pathway described in reference [21] and simulate them. How sensitive is the system, to changes in the regulatory topology? A simple way to study this would be to change some of the AND 'interactions' to OR and simulate the network again—these essentially can account for the uncertainties in our understanding of the regulatory interactions.

7.6 Study the Boolean model of the cell cycle as described in reference [4]. The same model is readily imported into BooleSim (available as an example). Simulate the model for different initial conditions to understand the dynamics of the system. Systematically knock out each node from the network and examine the effect of the knock-out on the system dynamics. Based on the knock-out analyses, what would you conclude as the most important components of the network? Note that knock-out analyses are easier to perform with BooleanNet.

REFERENCES

[1] S. A. Kauffman, "Metabolic stability and epigenesis in randomly constructed genetic nets", *Journal of Theoretical Biology* 22:437–467 (1969).

[2] L. Glass and S. A. Kauffman, "The logical analysis of continuous, nonlinear biochemical control networks", *Journal of Theoretical Biology* 39:103–129 (1973).

[3] R. Thomas, "Boolean formalization of genetic control circuits", *Journal of Theoretical Biology* 42:563–585 (1973).

[4] A. Faure, A. Naldi, C. Chaouiya, and D. Thieffry, "Dynamical analysis of a generic boolean model for the control of the mammalian cell cycle", *Bioinformatics* 22:e124–e131 (2006).

[5] S. Li, S. M. Assmann, and R. Albert, "Predicting essential components of signal transduction networks: a dynamic model of guard cell abscisic acid signaling", *PLoS Biology* 4:e312+ (2006).

[6] J. Thakar, M. Pilione, G. Kirimanjeswara, E. T. Harvill, and R. Albert, "Modeling systems-level regulation of host immune responses", *PLoS Computational Biology* 3:e109 (2007).

[7] M. Chaves, R. Albert, and E. D. Sontag, "Robustness and fragility of Boolean models for genetic regulatory networks", *Journal of Theoretical Biology* 235:431–449 (2005).

[8] I. Shmulevich, E. R. Dougherty, S. Kim, and W. Zhang, "Probabilistic Boolean networks: a rule-based uncertainty model for gene regulatory networks ", *Bioinformatics* 18:261–274 (2002).

[9] S. Klamt, J. Saez-Rodriguez, J. A. Lindquist, L. Simeoni, and E. D. Gilles, "A methodology for the structural and functional analysis of signaling and regulatory networks", *BMC Bioinformatics* 7:56 (2006).

[10] R. Thomas, "Regulatory networks seen as asynchronous automata: a logical description", *Journal of Theoretical Biology* 153:1–23 (1991).

[11] A. Ghysen and R. Thomas, "The formation of sense organs in *Drosophila*: a logical approach", *Bioessays* 25:802–807 (2003).

[12] V. N. Reddy, M. L. Mavrovouniotis, and M. N. Liebman, "Petri net representations in metabolic pathways", *Proceedings of the First International Conference on Intelligent Systems for Molecular Biology* **1**:328–336 (1993).

[13] C. Chaouiya, "Petri net modelling of biological networks", *Briefings in Bioinformatics* **8**:210–219 (2007).

[14] I. Koch, "Petri nets in systems biology", *Software & Systems Modeling* **14**:703–710 (2015).

[15] M. Heiner, D. Gilbert, and R. Donaldson, "Petri nets for systems and synthetic biology", in Formal Methods for Computational Systems Biology (2008), pp. 215–264.

[16] A. Veliz-Cuba and B. Stigler, "Boolean models can explain bistability in the *lac* operon", *Journal of Computational Biology* **18**:783–794 (2011).

[17] R. Albert and H. G. Othmer, "The topology of the regulatory interactions predicts the expression pattern of the segment polarity genes in *Drosophila melanogaster*", *Journal of Theoretical Biology* **223**:1–18 (2003).

[18] L. Paulevé, J. Kolčák, T. Chatain, and S. Haar, "Reconciling qualitative, abstract, and scalable modeling of biological networks", *Nature Communications* **11**:4256 (2020).

[19] I. Albert, J. Thakar, S. Li, R. Zhang, and R. Albert, "Boolean network simulations for life scientists", *Source Code for Biology and Medicine* **3**:16+ (2008).

[20] M. Bock, T. Scharp, C. Talnikar, and E. Klipp, "BooleSim: an interactive boolean network simulator", *Bioinformatics* **30**:131–132 (2014).

[21] S. Gupta, S. S. Bisht, R. Kukreti, S. Jain, and S. K. Brahmachari, "Boolean network analysis of a neurotransmitter signaling pathway", *Journal of Theoretical Biology* **244**:463–469 (2007).

FURTHER READING

S. A. Kauffman, *The origins of order: self-organization and selection in evolution* (Oxford University Press, U.S.A., 1993).

H. de Jong, "Modeling and simulation of genetic regulatory systems: a literature review", *Journal of Computational Biology* **9**:67–103 (2002).

R. Laubenbacher and A. S. Jarrah, "Algebraic models of biochemical networks", *Methods in Enzymology* **467**:163–196 (2009).

III

Constraint-based Modelling

CHAPTER 8

Introduction to constraint-based modelling

CONTENTS

I N THE previous part, we covered a significant amount of ground in dynamic modelling. While dynamic modelling is potent and highly useful, the modelling exercise itself is fraught with the *curse of dimensionality*, greatly limiting our ability to work with such models, especially for large systems. Despite brave efforts to construct genome-scale kinetic models [1], many challenges remain. Given the inherent difficulties in constructing, analysing, and simulating large-scale kinetic models, it is important to look for other modelling paradigms that can come to our rescue, if we are to understand biological networks from a systems perspective. Amongst these paradigms are a set of powerful, versatile, and popular techniques, collectively termed "constraint-based modelling" methods.

Constraint-based modelling is focussed on metabolic networks and is most useful to predict the growth rate of a cell, under specific environmental conditions, such as given glucose uptake rate, or following perturbations, such as the knock-out of certain genes from an organism.

More specifically, the key goal of constraint-based methods is to predict the 'flux distribution' or flux configuration of the cell, *i.e.* the fluxes[1], or velocities of each and every reaction in the metabolic network. Of course, we start with a genome-scale metabolic reconstruction (Appendix B), which has a comprehensive list of reactions happening within a cell, replete with details on stoichiometry, reaction direction, the corresponding catalysing enzymes, the gene–protein–reaction (GPR) associations[2] and so on. Constraint-based methods were originally developed by Watson [2, 3], who brought about the idea of using linear programming (LP) to minimise a cost in the system. Subsequently, Fell and Small (1986) [4] studied stoichiometric constraints in adipose tissue, demonstrating the utility of constraints to study metabolic models with 50+ reactions.

However, constraint-based modelling gained much wider popularity with the applications to *Escherichia coli* metabolic models in the lab of Bernhard Palsson [5, 6] at the University of California, San Diego; his lab remains a major contributor to the field of systems biology and constraint-based modelling. Constraint-based methods have been extremely popular in the last two decades, to predict microbial phenotypes and growth rates, relying on only a few experimental measurements. In the following sections, we will study the key principles underlying constraint-based modelling.

8.1 WHAT ARE CONSTRAINTS?

Constraints are essentially limitations or restrictions; in the present context, constraints refer to the behaviours that are (not) possible. It is easy to imagine that constraints pervade all biological systems, and life in general. The box below gives a very macroscopic view of constraints; nevertheless, constraints in microbial systems can also be conceived easily.

"On being the right size" ...

This box takes its title from the famous essay by the evolutionary biologist J. B. S. Haldane in 1926 [7]. In this essay, Haldane discusses the relationships between the size and structure of different animals. For instance, if a human were to be sixty feet tall, instead of six, and consequently ten times broader and wider, the total weight would be a 1000 times, roughly 80–90 tonnes, against a human's weight of 80–90 kg. The cross-section of the bones, on the other hand, are only hundred times larger, leaving them to support ten times the weight supported by an ordinary human bone, an impossibility! This principle is enshrined in the "square-

---

[1]Fluxes generally denote the net flow of a material per unit area of cross-section; in the context of metabolism, however, it commonly refers to the net flow of metabolite through a reaction—the *reaction flux*, or rate of a reaction.

[2]GPR associations carefully catalogue the many-to-many mapping between genes, proteins (including isozymes and protein complexes), and reactions. For example, *glk* in *E. coli* encodes for glucokinase, which catalyses the first reaction of glycolysis. In general, one or more genes may encode for one or more proteins, which catalyse one or more reactions. Also see §10.1.1 and Appendix B.

cube law", originally attributed to Galileo (1638), although Haldane never referred to it explicitly in his essay. These observations emphasise that *life*, in general, is subject to various constraints, which ultimately govern the shape, structure and size of organisms.

Notably, in the context of metabolism/metabolic networks, several constraints exist. For instance, no reaction in the body, or any living cell for that matter, can violate the laws of thermodynamics or stoichiometry. Such restrictions form the core of the constraints that we will study in this section.

8.1.1 Types of constraints

Physico-chemical constraints All cells are subject to the laws of physics and chemistry. Physico-chemical laws, therefore, place 'hard' inviolable constraints on the cell, in terms of thermodynamics, conservation of mass, energy and momentum, kinetics, and so on. Accumulation of metabolites within a cell can prove to be toxic—therefore, these lead to constraints on how the cell produces and eliminates various metabolites. Mass balance and elemental/charge conservation in chemical reactions also impose constraints on the cell. Notably, these manifest as stoichiometric constraints. From one mole of glucose, at the end of glycolysis, a cell produces exactly two moles of pyruvate[3]. The stoichiometric constraints are so powerful that constraint-based methods are able to make accurate quantitative predictions, based merely on these constraints and only a few additional strain-specific parameters [8, 9].

The directionality of reactions within the cell is governed by thermodynamics. Some reactions in the cell are reversible, while others are irreversible and proceed only in one direction. Enzymes can only convert a limited number of substrate molecules to product per unit time ("turnover number", k_{cat}). The maximum rate, v_{max}, of an enzyme-catalysed reaction is $k_{cat}[E_0]$ (see §5.3.1), which gives rise to *capacity* constraints. A recent study has also shown the existence of other constraints such as cellular Gibbs energy dissipation rate [10]. Other constraints include those arising from the kinetics of reactions within cells and osmotic pressure [8].

Environmental constraints The environmental conditions of a cell—both intracellular and extracellular—have a profound effect on its growth and metabolism. Both the intracellular and extracellular environments have a role to play. External environments, such as the presence of various nutrients—*e.g.* carbon, oxygen, nitrogen, sulphur, and phosphate sources—or pH, temperature, osmolarity, and the availability of electron acceptors, influence cellular behaviour. For instance, it is easy to imagine that the cellular metabolism in a rich medium will be

---

[3]More exactly, $C_6H_{12}O_6$ (glucose) $+ 2\,NAD^+ + 2\,ADP + 2\,P_i \rightarrow 2\,CH3(CO)COOH$ (pyruvic acid) $+ 2\,ATP + 2\,NADH + 2\,H^+$.

very different from that in a minimal medium, where it has to synthesise many more metabolites, such as amino acids, for survival. An important consequence for modelling is that it is essential to know the environmental conditions (*i.e.* medium composition) for an experiment, to accurately predict growth rates.

Within the cell, too, the presence or absence of various proteins and enzymes, as well as small molecule effectors, significantly affect cellular behaviour. Therefore, integrating omics data, such as transcriptomic or proteomic data, can account for important constraints, limiting the space of possible phenotypes to those that are more likely.

Regulatory constraints   These constraints are more intriguing—as they are often self-imposed. For instance, an organism may *choose* not to grow under certain conditions, when growth is indeed possible. This may be a decision driven by increasing survivability. Sporulation is a good example of this—as conditions for growth deteriorate (albeit growth still remaining possible), organisms may shut down their functions and form spores, which are very stable (to heat and other exigencies) and can germinate when conditions are much more favourable. A simpler example is that of catabolite repression: *E. coli* will not use lactose as a source of energy in the presence of glucose, while it could very well use both sources concurrently and potentially grow at a higher rate. Such regulatory constraints enable the cell to make the best use of its environment, eliminate sub-optimal phenotypes and seek out behaviours with increased fitness [11].

Spatial constraints   Constraints also arise out of the spatial organisation of the cell, notably the crowding of an extraordinary variety of molecules including macromolecules such as DNA, which could not exist within the cell unless tightly packed. Such constraints are harder to quantify and integrate into models, although a few studies have begun to address such issues [12].

8.1.2   Mathematical representation of constraints

All constraints discussed above are represented as *balances* or *bounds*. For instance, balances can be represented as

$$v_1 + 2v_2 - v_3 - v_4 - 2v_5 = 0$$

where the v_i represent fluxes through some five reactions. This balance likely arises from the mass balance of a metabolite that is produced in reactions 1 and 2, and consumed in reactions 3, 4, and 5 (with different stoichiometries). The other class of constraints are bounds; Table 8.1 represents common bounds used in constraint-based models of metabolism. Typically, every reaction has a specified lower and upper bound:

$$v_{lb} \preceq v \preceq v_{ub} \tag{8.1}$$

where v_{lb} and v_{ub} denote the lower and upper bounds for each of the fluxes in v.

TABLE 8.1 **Examples of constraints.** It is common to use 'default' bounds of -1000 and 1000 mmol/gDW/h (millimoles per gram dry weight per hour), whenever constraints are unknown. Other constraints are specified based on the knowledge of thermodynamic constraints, environmental conditions, and enzyme capacities.

Constraint	Nature of constraint	Example scenario
$-1000 \leqslant v_1 \leqslant 1000$	Thermodynamic	reversible reaction
$0 \leqslant v_2 \leqslant 1000$	Thermodynamic	irreversible reaction
$-20 \leqslant v_3 \leqslant -10$	Environmental (extracellular)	uptake of a nutrient
$0 \leqslant v_4 \leqslant 10$	Physico-chemical	enzyme capacity
$0 \leqslant v_5 \leqslant 0$	Environmental (intracellular)	enzyme expression
$10 \leqslant v_6 \leqslant 1000$	Environmental (intracellular)	ATP maintenance

8.1.3 Why are constraints useful?

From the discussions in the previous section, it is evident that every living cell must obey many constraints, to survive and grow. In the context of modelling, every predicted 'configuration' of the cell must obey these constraints. A configuration, in this context, refers to a flux state of the cell, or a "flux distribution", which describes the rate of every single reaction within a given cell. Satisfying the various stoichiometric, thermodynamic, environmental, regulatory, and spatial constraints is, therefore, a pre-requisite for cellular survival and growth.

In a sense, the constraints limit the range of observable phenotypes; or, the knowledge of constraints can be used to prune the phenotype space, eliminating impossible behaviours. Constraint-based models are able to predict cellular phenotypes on the basis of stoichiometric and thermodynamic constraints, along with the knowledge of key environmental (medium) constraints such as substrate and oxygen uptake rates.

8.2 THE STOICHIOMETRIC MATRIX

The defining mathematical object, for constraint-based models is the stoichiometric matrix, which captures the entire metabolic network of a cell in a mathematical form. The stoichiometric matrix $\mathbf{S} \in \mathbb{R}^{m \times r}$ connects the reaction fluxes ($\mathbf{v}_{r \times 1}$) in a given metabolic network to the rate of change of metabolite concentrations ($\mathbf{z}_{m \times 1}$) as follows

$$\frac{d\mathbf{z}}{dt} = \mathbf{S}\mathbf{v} \qquad (8.2)$$

Every row in a stoichiometric matrix corresponds to a metabolite in the network; every column in the stoichiometric matrix corresponds to a reaction—either an internal reaction, or an exchange with the outer environment. The entry s_{ij} in the $m \times r$ stoichiometric matrix denotes the stoichiometry of the ith metabolite participating in the jth reaction, with reactants having negative entries and products in a reaction having positive entries.

Consider a simple metabolic network with eight metabolites and ten reactions, as shown in Figure 8.1. The reactions are also listed in Table 8.2. All the reactions are stoichiometrically balanced, save for the exchange fluxes. These exchange fluxes can account for metabolites that are taken up from or secreted into the extracellular medium where the organism grows.

TABLE 8.2 **A simple metabolic network.** The arrows indicate the reversibility of the reactions. Fluxes of internal reactions are represented by v_i's, and the fluxes of the exchange reactions E_1 and E_2 are represented by b_1 and b_2, respectively. The flux of the 'biomass reaction', is represented as v_{bio}.

	Reaction	Flux
R_1:	$A \rightarrow ATP + B$	v_1
R_2:	$A \rightarrow C$	v_2
R_3:	$C \rightarrow ATP + B$	v_3
R_4:	$B \Leftrightarrow 2\,D$	v_4
R_5:	$B \rightarrow F$	v_5
R_6:	$D \Leftrightarrow E$	v_6
R_7:	$D + ATP \rightarrow G$	v_7
E_1:	$\rightarrow A$	b_1
E_2:	$E \Leftrightarrow$	b_2
Biomass:	$F + 2\,G + 10\,ATP \rightarrow$	v_{bio}

The stoichiometric matrix ($S_{8 \times 10}$) and the flux vector ($v_{10 \times 1}$) for the metabolic network are represented as follows:

$$
S = \begin{array}{c} \\ A \\ B \\ C \\ D \\ E \\ F \\ G \\ ATP \end{array}
\begin{array}{c} \begin{matrix} v_1 & v_2 & v_3 & v_4 & v_5 & v_6 & v_7 & b_1 & b_2 & v_{bio} \end{matrix} \\
\begin{pmatrix}
-1 & -1 & 0 & 0 & 0 & 0 & 0 & 1 & 0 & 0 \\
1 & 0 & 1 & -1 & -1 & 0 & 0 & 0 & 0 & 0 \\
0 & 1 & -1 & 0 & 0 & 0 & 0 & 0 & 0 & 0 \\
0 & 0 & 0 & 2 & 0 & -1 & -1 & 0 & 0 & 0 \\
0 & 0 & 0 & 0 & 0 & 1 & 0 & 0 & -1 & 0 \\
0 & 0 & 0 & 0 & 1 & 0 & 0 & 0 & 0 & -1 \\
0 & 0 & 0 & 0 & 0 & 0 & 1 & 0 & 0 & -2 \\
1 & 0 & 1 & 0 & 0 & 0 & -1 & 0 & 0 & -10
\end{pmatrix}
\end{array}
\quad v = \begin{bmatrix} v_1 \\ v_2 \\ v_3 \\ v_4 \\ v_5 \\ v_6 \\ v_7 \\ b_1 \\ b_2 \\ v_{bio} \end{bmatrix}
\quad (8.3)
$$

8.3 STEADY-STATE MASS BALANCE: FLUX BALANCE ANALYSIS (FBA)

At metabolic steady-state, the right-hand side of Equation 8.2 goes to zero, giving

$$Sv = 0 \qquad (8.4)$$

the key *flux balance* equation underlying constraint-based modelling. Observe that this is actually a system of linear equations in v_i. Any flux configuration of

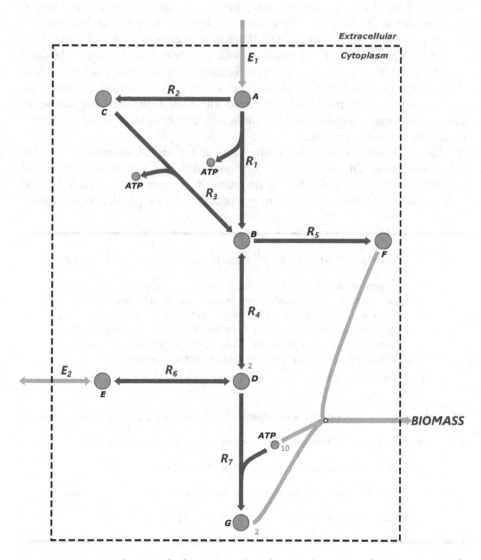

Figure 8.1 **A simple metabolic network.** The stoichiometry (when not unity) is indicated next to the arrow. The figure was generated using Escher [13] (see Appendix D).

the cell, at steady-state, must satisfy the above equation; *i.e.* any **v** that is a possible flux distribution, lies in the null space of the stoichiometric matrix.

What is the shape of the stoichiometric matrix, for a typical metabolic network? *Tall* or *fat*? That is, are there more metabolites than reactions, or vice-versa? In other words, are there more unknown variables (reaction fluxes) than flux balance equations (for each metabolite)? While both scenarios are possible, it is very common to find that $m < r$ in real metabolic networks. This is because the same metabolites mix and match in different reactions; further, many sink reactions that represent the exchange of intracellular metabolites are always present in the model. Thus, we commonly are confronted with a situation where the number of unknowns is much higher than the number of equations, *i.e.* the system of linear equations is *under-determined*.

Also note that $\mathbf{v} = \mathbf{0}$ is almost always a valid, if trivial, solution to the flux balance equation. Of course, if $v_i = 0$ is precluded by a non-zero lower bound for some reaction i, (*e.g.* the ATP maintenance flux mentioned in Table 8.1 and detailed in §8.8.6), then the solver may report that the LP problem is infeasible.

Stoichiometric matrices, adjacency matrices and sub-spaces

The binarised version of the stoichiometric matrix, generally denoted $\hat{\mathbf{S}}$, where every non-zero entry is replaced by one, has interesting properties. $\hat{\mathbf{S}}\hat{\mathbf{S}}^{\top}$ is an $m \times m$ matrix, which captures the coupling of metabolites across reactions [14]. It also happens to be the adjacency matrix for the substrate graph (see §2.5.5.1), although the diagonal entries are non-zero, and represent the number of reactions a metabolite participates in. Can you now think of what $\hat{\mathbf{S}}^{\top}\hat{\mathbf{S}}$ represents?

It is also important to know the four fundamental sub-spaces of the stoichiometric matrix, **S**:

(Right) null space contains all the steady-state flux distributions allowable in the network

Left null space contains all the conservation relationships, or time invariants, in the network

Row space contains all the *dynamic flux distributions* of the network

Column space contains all the possible time derivatives of the concentration vector **z**

A more detailed treatment of the above is available in [15].

For the metabolic network discussed above (Figure 8.1), we obtain the following constraints, or flux balances, by multiplying out **S** and **v** from Equation 8.3 and equating it to zero, as in Equation 8.4:

$$-v_1 - v_2 + b_1 = 0 \tag{8.5a}$$

$$v_1 + v_3 - v_4 - v_5 = 0 \tag{8.5b}$$

$$v_2 - v_3 = 0 \tag{8.5c}$$

$$2v_4 - v_6 - v_7 = 0 \qquad (8.5d)$$

$$v_6 - b_2 = 0 \qquad (8.5e)$$

$$v_5 - v_{bio} = 0 \qquad (8.5f)$$

$$v_7 - 2v_{bio} = 0 \qquad (8.5g)$$

$$v_1 + v_3 - v_7 - 10v_{bio} = 0 \qquad (8.5h)$$

Now, how do we solve these equations? As is evident, there are more variables than equations; therefore, a unique solution is not possible. In the following section, we will look at a number of approaches to predict these unknown fluxes, and consequently, the phenotype of the cell, by solving these equations.

8.4 THE OBJECTIVE FUNCTION

A key aspect of FBA involves picking one of the infinitely many possible solutions to the flux balance equation. Which of these solutions is biologically the most plausible? What do the other solutions mean, anyway? These are some of the interesting questions that we will address in this section.

From the classic papers of Watson [3] and Fell and Small [4], one solution to the problem of infinitely many solutions is to employ an objective criterion, often a linear one. For instance, Fell and Small examined fat synthesis in adipose tissue, under conditions of minimum glucose utilisation, or maximum NADPH production [4]. For larger genome-scale metabolic models, a common strategy has been to maximise the growth of the cell, or *biomass*, as we will discuss in the following section.

Objective functions can be chosen in many ways, and serve to interrogate a metabolic model in multiple ways. For instance, the objective function can:

1. *probe the solution space*—to simply characterise the overall metabolic capabilities of the cell; this includes methods such as flux variability analysis (FVA; §8.7).
2. *represent a physiologically likely/plausible objective*—to accurately predict cellular phenotype; *e.g.* the biomass objective function (§8.4.1).
3. *represent a bio-engineering design objective*—to examine possible states where a particular cellular configuration can be achieved, *e.g.* the over-production of a vitamin.

It is easy to imagine that no one objective function can capture cellular behaviour under all possible conditions. Therefore, a careful choice of the objective function is the cornerstone of constraint-based modelling. For example, while the cell may favour maximum biomass yield in certain conditions, it might work towards maximal energetic efficiency in other conditions. The choice of the objective function is amongst the most critical choices a modeller has to make, while performing constraint-based modelling.

A detailed discussion of many objective functions, along with the rationales for applying them is available elsewhere [16, 17]. Table 8.3 illustrates a few of them.

TABLE 8.3   **Examples of objective functions.**

Objective	Reference(s)
Maximisation of growth rate/biomass yield	[19–21]
Maximisation of product formation	[22]
Maximisation of maintenance energy expenditure	[23]
Minimisation of ATP production	[19]
Minimisation of ATP usage	[22]
Minimisation of substrate uptake	[22]
Minimisation of redox metabolism/NADPH production	[19]

While many approaches employ linear objective functions, there are a number of approaches that have explored other alternatives, as we discuss in the next chapter.

While maximisation of growth or maximisation of ATP production could conceivably be traits evolutionarily selected for, other objectives like the maximisation of product formation, say that of a commercially valuable vitamin or antibiotic, are bio-engineering design objectives, and are unlikely to be aligned with the physiological goals of a cell. In such cases, what do the fluxes actually mean? For instance, if we maximise the production of lycopene, an important secondary metabolite, we will obtain a sub-optimal growth rate [18]—an incorrect prediction of the wild-type cell's actual growth rate. However, the importance of this simulation lies in its ability to predict the theoretical maximum yield of lycopene. Further, this simulation provides a platform to explore cellular configurations that support a higher production of lycopene, and consequently predict possible perturbations that can enhance its production (also see §10.3).

8.4.1   The biomass objective function

The biomass objective function has been heavily used in constraint-based modelling since the late 1990s [20, 21]. The biomass objective function generally corresponds to a biomass reaction that has been curated into the model; this is essentially a fictitious reaction, which captures all the metabolites necessary for cellular growth, in their correct proportions. Table 8.4 exemplifies a complex biomass objective function, one that is used in the latest genome-scale metabolic reconstruction of *E. coli*, *i*ML1515 [24]. Curating this reaction in detail is central to being able to make reliable and accurate phenotypic predictions from a genome-scale metabolic model.

At the *basic* level, the biomass objective function captures the macromolecular content of the cell (*e.g.* weight fraction of RNA, lipids, proteins, etc.) as well as the building blocks (*e.g.* amino acids and nucleotide triphosphates). At the next *intermediate* level, one typically integrates information on energy requirements (ATP for growth and non-growth related maintenance of the cell). Finally, *advanced* objective functions capture vitamins, elements, and co-factors essential for growth, as well as other core metabolites necessary for cellular viability. As more and more information is included, it increases the prediction accuracy of growth

TABLE 8.4 **Biomass objective function definition in the *E. coli* *i*ML1515 model [24].** The reaction corresponding to hydrolysis of ATP—growth-associated maintenance of 75.55 mmol/gDW/h—is part of biomass, and has been indicated in bold font; the coefficients are not all same, since some of these metabolites are involved in other reactions as well. Note that the reaction has been abbreviated, yet clearly emphasises the complexity; it shows only some example metabolites, including some amino acids, key central carbon metabolism precursors, carbohydrates, lipids, co-factors, cell-wall components, DNA, RNA etc. All abbreviations follow the conventions used in the BiGG/ModelSEED databases (see Appendix C).

...+ **75.55223 atp_c** + ...+ 0.000674 cu2_c + 0.088988 cys__L_c + 0.024805 datp_c + 0.025612 dctp_c + 0.025612 dgtp_c + 0.024805 dttp_c + 0.000223 fad_c + 0.006388 fe2_c + 0.007428 fe3_c + 0.255712 gln__L_c + 0.255712 glu__L_c + 0.595297 gly_c + 0.154187 glycogen_c + 0.000223 gthrd_c + 0.209121 gtp_c + **70.028756 h2o_c** + 0.000223 hemeO_c + ...+ 0.001787 nad_c + 4.5×10^{-5} nadh_c + 0.000112 nadp_c + 0.000335 nadph_c + 0.012379 nh4_c +...+ 0.000223 ribflv_c + ...+ 0.005448 murein4px4p_p + ...→ **75.37723 adp_c + 75.37723 h_c + 75.37323 pi_c** + 0.749831 ppi_c

rate, as well as essentiality of genes/reactions. For a detailed discussion about the biomass objective function, and how it can be built incrementally, see the excellent review by Feist and Palsson [17].

Note that both the components of the biomass reaction and their stoichiometric coefficients, *i.e.* proportions are important to make accurate predictions. The components themselves are of critical importance while making predictions of gene essentiality from a given metabolic model, while their exact proportions are more critical in making accurate predictions of growth rates [25]. Note that the units of the objective are h^{-1} since all the coefficients have units of mmol/gDW. Thus, the biomass reaction essentially sums the *mole fraction* of each precursor necessary to produce 1 g dry weight of cells [25]. The reaction coefficients are scaled so that the reaction flux is equal to the exponential growth rate of the organism (μ). The level of detail in biomass reactions/objective functions has continued to grow, with our increasing ability to reconstruct in detail, the various parts of cellular machinery. Another recent paper reiterated the importance of the biomass equation, and also illustrates the sensitivity of flux predictions to biomass composition [26].

8.5 OPTIMISATION TO COMPUTE FLUX DISTRIBUTION

Having identified a suitable objective function, as discussed above, we now need to solve an optimisation problem. When the objective function, **c**, is linear, the optimisation is an LP problem:

$$\max \sum_{j=1}^{r} c_j v_j \tag{8.6a}$$

$$subject\ to\ \sum_{j=1}^{r} s_{ij}v_j = 0 \qquad \forall i \in \{1, 2, \ldots, m\} \qquad (8.6b)$$

$$\alpha_j = v_{lb,j} \leqslant v_j \leqslant v_{ub,j} = \beta_j \qquad \forall j \in \{1, 2, \ldots, r\} \qquad (8.6c)$$

Linear Programming (LP)

LP is commonly taught in high schools, with a problem on the following lines. There is a factory that manufactures tables and chairs. Chairs fetch a profit of $80, while tables fetch a profit of $100. Each chair takes 4 units of metal and 2 units of wood to make, while each table takes 1 unit of metal and 3 units of wood. The factory has 50 units of metal and 70 units of wood available. How many tables and chairs should the factory manufacture, to maximise its profit?

This can be posed as follows:

$$max\ 80c + 100t$$
$$subject\ to\ 4c + t \leqslant 50$$
$$2c + 3t \leqslant 70$$
$$c \geqslant 0, t \geqslant 0$$

Computing the objective at each of the 'corner' points, the maximum is obtained at $c = 8, t = 18$, a profit of $2,440. Of course, strictly speaking, this is an integer programming problem, as solutions like 12.5 chairs do not make sense. In MATLAB, this problem is easily solved as:

```
>> [x,profit]=linprog(-[80; 100], [4 1; 2 3], [50; 70])
Optimal solution found.

x =
    8
   18

profit =
    -2440
```

linprog assumes a minimisation problem by default; the negative sign for the objective function transforms the problem to a maximisation problem.

In matrix/vector notation, this can be written as[4]:

$$max\ \mathbf{c}^\top \mathbf{v} \qquad (8.7a)$$
$$subject\ to\ \mathbf{Sv} = \mathbf{0} \qquad (8.7b)$$
$$\mathbf{v}_{lb} \preceq \mathbf{v} \preceq \mathbf{v}_{ub} \qquad (8.7c)$$

---

[4]Note that Equations 8.7b and 8.7c are the same as Equations 8.4 and 8.1, respectively.

where $c_{r \times 1}$ is the objective function, $v_{r \times 1}$ is the unknown flux vector ('flux distribution') being computed, $S_{m \times r}$ is the stoichiometric matrix, v_{lb} (or α_j's) denote the lower bounds for each of the fluxes in v, and v_{ub} (or β_j's) denote the upper bounds for each of the fluxes in v.

Note that there are $2r$ inequality constraints arising from the r upper and r lower bounds, as well as m equality constraints arising from flux balance. There are many algorithms to solve LP problems, from the classic simplex method to more complex interior-point algorithms [27]. A number of solvers are available, both free and commercial, which can efficiently solve the kind of LP problems encountered in FBA (see §8.10).

It is useful to understand the canonical form of an LP:

$$\max \ c^T x \quad subject \ to \ Ax \preceq b \qquad (8.8)$$

where the objective is c, and the constraints are captured in the inequalities $Ax \preceq b$. In practice, most solvers allow a separate specification of the lower and upper bounds—so, only the equalities arising from flux balance need be described as $Ax = 0$, i.e. $Sv = 0$. In MATLAB, the LP solver linprog has several arguments, which are mapped to the terms of Equation 8.7 in Table 8.5. Note that linprog solves a *minimisation* problem by default; therefore, to *maximise* growth rate, the objective function should be negated, since the maximum of $f(x)$ is the minimum of $-f(x)$ (as can also be observed in the box on the previous page).

TABLE 8.5 **FBA as an LP problem, from the perspective of MATLAB linprog.** The output will be x, the optimal value of the flux v_{opt}.

Argument	LP term	FBA term
f	objective function	$\pm c$
A	matrix corresponding to inequality constraints	—
b	column vector corresponding to RHS of the inequality constraints	—
Aeq	matrix corresponding to equality constraints	S
beq	column vector corresponding to RHS of the equality constraints	0
lb	column vector corresponding to lower bounds on the solution	v_{lb}
ub	column vector corresponding to upper bounds on the solution	v_{ub}

8.6 AN ILLUSTRATION

The broad steps in carrying out an FBA are as follows:

0. Curate the metabolic network (see Appendix B)
1. Obtain the stoichiometric matrix $S_{m \times r}$ for the network (see §8.2)
2. Compute the flux balance constraints $Sv = 0$ (see §8.3, Equation 8.4)

3. Identify the constraints on each reaction, *e.g.* lower and upper bounds (see §8.1, Equation 8.1). Usually, these lower and upper bounds are respectively set to some values like −1000 and 1000 (for reversible reactions), and 0 and 1000, for irreversible reactions (units being mmol/gDW/h). As mentioned earlier, it is important to obtain accurate bounds at least for the limiting substrate (see §8.1.1).

4. Choose a suitable objective function, *e.g.* maximisation of biomass (see §8.4)

5. Solve the optimisation problem, to identify the optimal solution (see §8.5, Equation 8.7)

Now, let us revisit the toy network of Figure 8.1. The stoichiometric matrix is shown in Equation 8.3. Applying steady-state mass balance, we obtain the flux balance constraints as shown in Equation 8.5. For simplicity, let us assume 'default' bounds: $\mathbf{v_{lb}} = \begin{bmatrix} 0 & 0 & 0 & -1000 & 0 & -1000 & 0 & 0 & -1000 & 0 \end{bmatrix}^T$, and $\mathbf{v_{ub}} = \begin{bmatrix} 1000 & 1000 & 1000 & 1000 & 1000 & 1000 & 1000 & 1000 & 1000 & 1000 \end{bmatrix}^T$, all in mmol/gDW/h, with the order of fluxes same as in Equation 8.3. Thus, the medium for growth includes only A, which is fed by the exchange flux b_1.

For this network, the 'biomass' objective is simply the reaction represented as v_{bio}, F + 2G + 10ATP →, as shown in Table 8.2. Note that this is very different from maximising an objective like $b_F + 2b_G + 10b_{ATP}$, assuming exchange fluxes for each of these metabolites. The latter optimisation can possibly provide optimal solutions with one or two of the individual fluxes going to zero, while the former works in an 'all-or-none' fashion—the biomass reaction cannot carry flux if any of F, G or ATP are not produced, and in the required proportions, which is a more realistic requirement in the context of cellular growth. It is also possible to have other objectives, such as say, maximising b_2, or even an internal reaction such as v_7.

Maximising v_{bio}, alongside the flux balance constraints (Equation 8.5) and bounds as indicated above results in the following solution:

$$v = \begin{bmatrix} 600 & 0 & 0 & 550 & 50 & 1000 & 100 & 600 & 1000 & 50 \end{bmatrix}^T \qquad (8.9)$$

from which v_{bio} can be seen to be 50 (mmol/gDW/h). Another, flux configuration, with the same 'growth rate' is

$$v = \begin{bmatrix} 0 & 600 & 600 & 550 & 50 & 1000 & 100 & 600 & 1000 & 50 \end{bmatrix}^T \qquad (8.10)$$

It is easy to verify that both the above solutions satisfy the flux balance constraints of Equation 8.5.

Further, it is common for such optimisation problems to have multiple optimal solutions, all having the same optimal value for the objective function. However, in the context of FBA, this has a potential downside—while the prediction of optimal growth rate is unique, there are potentially multiple flux configurations that FBA can predict, for a given cell, under a given set of conditions (constraints). Therefore, it is important to remember that FBA predicts the theoretical capabilities of a cell, and this is captured more faithfully by performing an FVA, as we discuss in the following section.

8.7 FLUX VARIABILITY ANALYSIS (FVA)

From the two alternative solutions to the FBA problem discussed above (equations 8.9 and 8.10), it is clear that there are likely multiple optimal solutions to FBA LP problems. How do we make sense of these multiple solutions, and understand the *variability* in the flux space better? FVA captures the possible variability in every flux in a network, under a given objective [28].

The two different solutions for the toy network (Figure 8.1) indicate that for the same growth rate, the first three reactions can carry fluxes in the range of $[0, 600]$. Depending on the partition of flux through reactions R_1 and R_2, various flux solutions, such as $v = [100 \ 500 \ 500 \ 550 \ 50 \ 1000 \ 100 \ 600 \ 1000 \ 50]^\top$ are possible.

How to perform an FVA? FVA essentially involves performing a maximisation and minimisation of flux, through every single reaction in the network, but with an additional constraint on the objective value achieved, over and above the normal FBA formulation (Equation 8.7), as follows:

$$\max \text{ (or min) } v_j \qquad \forall j \in \{1, 2, \ldots, r\} \tag{8.11a}$$

$$\textit{subject to } \mathbf{Sv} = 0 \tag{8.11b}$$

$$\mathbf{v_{lb}} \preceq \mathbf{v} \preceq \mathbf{v_{ub}} \tag{8.11c}$$

$$\mathbf{c}^\top v \geqslant f^* \tag{8.11d}$$

where all terms have their usual meanings; additionally f^* represents a minimum value of the objective $\mathbf{c}^\top v$ (or some fraction of it, say 50%), obtained by solving the original FBA LP. Equation 8.11b translates to a total of $2r$ optimisation problems—a maximisation and a minimisation each, for $j = 1, 2, \ldots, r$. For the toy network, we obtain the minimum and maximum fluxes through each reaction as

$$
\begin{aligned}
\mathbf{v_{min}} &= [\ \ \ 0 \quad \ \ \ 0 \quad \ \ \ 0 \ \ \ 550 \ \ \ 50 \ \ \ 1000 \ \ \ 100 \ \ \ 600 \ \ \ 1000 \ \ \ 50]^\top \\
\mathbf{v_{max}} &= [600 \ \ \ 600 \ \ \ 600 \ \ \ 550 \ \ \ 50 \ \ \ 1000 \ \ \ 100 \ \ \ 600 \ \ \ 1000 \ \ \ 50]^\top
\end{aligned} \tag{8.12}
$$

by solving 18 LP problems, maximising and minimising separately, each of $v_1, v_2, \ldots, v_7, b_1, b_2$, constraining v_{bio} to 50 all along. Obviously, we do not maximise or minimise v_{bio} separately, as it has already been constrained. Thus, FVA provides us with the boundaries of the allowable flux space, while maintaining a given objective.

Figure 8.2 illustrates geometrically, the solution space for a constraint-based problem, how the solution space changes following the application of constraints, and finally, the FBA solutions, as well as the FVA solutions (*i.e.* boundaries for v_1, v_2) that correspond to maximum v_{obj}.

8.8 UNDERSTANDING FBA

In this section, we will deal with several of the practical issues faced while working with constraint-based models.

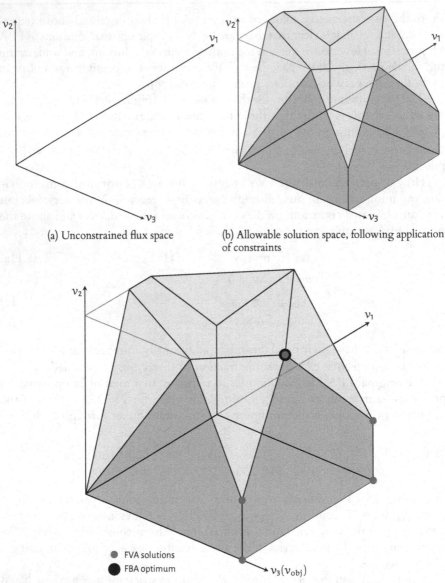

(a) Unconstrained flux space

(b) Allowable solution space, following application of constraints

(c) Maximisation of an objective function, along with FBA/FVA solution points

Figure 8.2 **A geometric picture of the flux space, allowable solution space and FBA/FVA solutions.** This figure is adapted by permission from Springer Nature: A. Bordbar, J. M. Monk, Z. A. King, and B. Ø. Palsson, "Constraint-based models predict metabolic and associated cellular functions", *Nature Reviews Genetics* **15**:107–120 (2014). © Macmillan Publishers Limited (2014).

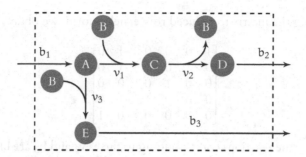

Figure 8.3 **An example network to illustrate blocked reactions.** Labels on the arrows indicate the fluxes. The dashed box represents the boundary of the system, across which metabolites A, D, and E are exchanged. This example is adapted by permission from Springer Nature: I. Thiele and B. Ø. Palsson, "A protocol for generating a high-quality genome-scale metabolic reconstruction", *Nature Protocols* **5**:93–121 (2010). © Nature Publishing Group (2010).

8.8.1 Blocked reactions and dead-end metabolites

Not all reactions in a metabolic network will carry a flux in all conditions. This is easily conceivable since it is likely that some reactions will be useful only in certain environments. However, it is common for metabolic networks to contain reactions, which do not carry a flux under *any* condition. Such reactions are known as *blocked reactions*. Typically, these reactions involve metabolites that are only produced (and never consumed) or only consumed (and never produced) in the metabolic network, *i.e. dead-end metabolites*. Such metabolites may also arise from 'gaps' in the metabolic network, as we will discuss later (§8.8.2). Blocked reactions can be easily identified by performing an FVA: $v_{min} = v_{max} = 0$ for such reactions.

It is also possible to deduce blocked reactions from the stoichiometric matrix of the network, directly. By simplifying the mass-balance constraints, for example through row operations on the stoichiometric matrix, we may arrive at 'simplified' constraints. Simplified constraints of the form $v_i = 0$ represent reactions that cannot carry flux, and are thus blocked. For example, consider the metabolic network in Figure 8.3, represented by the following stoichiometric matrix:

$$
\begin{array}{c}
\quad\quad v_1 \quad v_2 \quad v_3 \quad b_1 \quad b_2 \quad b_3 \\
\begin{array}{c} A \\ B \\ C \\ D \\ E \end{array}
\left(
\begin{array}{cccccc}
-1 & 0 & -1 & 1 & 0 & 0 \\
-1 & 1 & -1 & 0 & 0 & 0 \\
1 & -1 & 0 & 0 & 0 & 0 \\
0 & 1 & 0 & 0 & -1 & 0 \\
0 & 0 & 1 & 0 & 0 & -1
\end{array}
\right)
\end{array}
\quad (8.13)
$$

Converting the matrix to reduced row echelon form, we obtain

$$
\begin{bmatrix}
1 & 0 & 0 & 0 & -1 & 0 \\
0 & 1 & 0 & 0 & -1 & 0 \\
0 & 0 & 1 & 0 & 0 & 0 \\
0 & 0 & 0 & 1 & -1 & 0 \\
0 & 0 & 0 & 0 & 0 & 1
\end{bmatrix}
\tag{8.14}
$$

indicating that the reactions corresponding to fluxes v_3 and b_3 are blocked (multiplying out rows 3 and 5 with $\mathbf{v}$ yield $v_3 = 0$ and $b_3 = 0$), despite the fact that there is no dead-end metabolite involved. The matrix also additionally illustrates the constraints $v_1 = b_2$, $v_2 = b_2$, $b_1 = b_2$, from rows 1, 2, and 4.

Note that if v_2 and b_2 did not exist, B would be only consumed in the network, and C will be only produced in the network, causing both of them to become dead-end metabolites. On the other hand, if v_3 and b_3 were not to exist, the system can carry a flux—*i.e.* there are no blocked reactions. In such a case, however, how is the system able to carry flux, despite no explicit supply of B? This is an important artefact of metabolic modelling. Metabolites such as B can be internally *cycled*—B is produced by v_2, and consumed in v_1, to produce C, which subsequently produces B! More discussion on such cycles can be found in §8.8.4. Note that the reactions corresponding to v_1 and v_2 can be combined to A → D, eliminating both B and C.

Now, since B is being cycled in v_1 and v_2, the input of B in v_3 is 'unbalanced', leading to the 'block'. This is also evident from the balance equations that can be derived from 8.13, where rows 2 and 3 give rise to constraints $-v_1 + v_2 - v_3 = 0$ and $v_1 - v_2 = 0$—immediately showing that $v_3 = 0$. The functions pFBA and findBlockedReactions in the COBRA toolbox are useful to identify blocked reactions.

8.8.2 Gaps in metabolic networks

Somewhat related to the concept of dead-end metabolites is the concept of a *gap* in a metabolic network. Gaps in a metabolic network can arise owing to gaps in our knowledge of metabolism; many reactions, even in well-curated metabolic networks, remain unknown till date [30]. In some cases, an intermediate reaction in a pathway may be unknown. Gaps can also arise from incorrectly curated biomass components (*e.g.* wrongly included a component not synthesised in the network), or dead-end metabolites that should have been supplied in the medium via an exchange reaction. An analysis of gaps in a metabolic network is one of the first and crucial steps in debugging the network [25]. Many algorithms also exist for gap-filling, and have enabled improved predictions from genome-scale metabolic networks and have also led to the discovery of new metabolic gene functions [31, 32].

8.8.3 Multiple solutions

While the objective function value for a linear program is guaranteed to be unique, there are still many solutions that are possible. As we have seen earlier, FVA can be used to understand the various flux solutions that are possible for a given set of conditions. Beyond this, what do these multiple optimal solutions mean? First, the number of constraints imposed on the metabolic network may be insufficient. Alternatively, these multiple optima can actually represent various feasible flux states that a cell can actually take, such as alternative metabolic pathways. Note that FBA provides a handle to understand the theoretical capabilities of a cell. In a sense, the growth rates (objective function values) predicted by FBA are more valuable than the exact flux configurations predicted. Nevertheless, across FBA, FVA, and a suite of constraint-based methods that we will also discuss in the next chapter, these techniques overall provide a reasonably accurate and detailed quantitative picture of metabolism within a cell. The COBRA toolbox function `enumerateOptimalSolutions` can easily enumerate multiple optimal solutions for a given model. Reed and Palsson (2004) [33] present a detailed discussion on multiple alternate optima in metabolic networks.

8.8.4 Loops

Another important artefact in constraint-based metabolic modelling is the existence of loops, or cycles, in metabolic networks. Different types of loops are possible in metabolic networks; they arise from a combination of lack of sufficient knowledge on reaction reversibility, and a lack of thermodynamic constraints on the model. Some of the cycles may also be thermodynamically impossible and erroneously generate energy [34]. A few different types of loops/cycles are possible in metabolic networks [34, 35]:

1. actual metabolic cycles like the TCA cycle
2. reactions that involve transport of non-currency metabolites from the environment, and drive the cycling of currencies such as ATP inside the cell
3. futile cycles, which consume energy from a co-factor pool (may occur in cells under certain conditions)
4. futile cycles running in reverse, which generate energy, erroneously
5. stoichiometrically balanced cycles, consisting only of internal reactions, without exchanging metabolites with the environment (see Figure 8.4)

Methods exist to remove futile and energy generating cycles, as discussed in references [25, 34]. In practice, the use of the minimum norm solution (as discussed in the following section) can be used to eliminate the net flux around stoichiometrically balanced cycles. For example, the system in Figure 8.4 can carry fluxes (consistent with balances) such as $b_1 = 10, v_1 = 1000, v_2 = 1000, b_2 = 10, v_3 = 990, v_4 = 990$, or $b_1 = 10, v_1 = 10, v_2 = 10, b_2 = 10, v_3 = 0, v_4 = 0$.

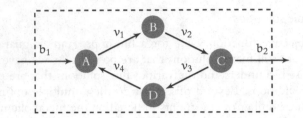

Figure 8.4 **An example network to illustrate stoichiometrically balanced cycles.** Labels on the arrows indicate the fluxes. The dashed box represents the boundary of the system, across which metabolites A, and C are exchanged.

8.8.5 Parsimonious FBA (pFBA)

One way around the multiple solutions for FBA is to pick a solution that minimises the number of active reaction fluxes. This is known as parsimonious enzyme usage FBA or parsimonious FBA (pFBA). pFBA assumes that under exponential growth, selection favours faster growing strains—strains that use relatively fewer reactions (or equivalently, enzymes) to channel flux through the metabolic network [36]. In a sense, pFBA applies Occam's razor (see §6.4.1, p. 151) to flux distributions that can capture the growth of an organism equally well.

Vector Norms

The 'length' of a vector $\mathbf{x}$ is typically captured using a vector *norm*. Typically written as $\|\mathbf{x}\|$, a vector's norm most commonly refers to its ℓ_2-norm, $\|\mathbf{x}\|_2 = \sqrt{\sum_i |x_i|^2}$. More generally, for $p = 1, 2, \ldots$, the p-norm $\|\mathbf{x}\|_p$ is defined as:

$$\|\mathbf{x}\|_p = \left(\sum_{i=1}^{n} |x_i|^p\right)^{1/p}$$

Important special cases are the
- ℓ_1-norm (or) 1-norm (or) city block distance (or) Manhattan distance (or) taxicab norm, defined as $\|\mathbf{x}\|_1 = \left(\sum_{i=1}^{n} |x_i|\right)$
- ℓ_1-norm (or) 2-norm (or) Euclidean norm discussed above,
- ∞-norm, defined as $\|\mathbf{x}\|_\infty = \max_{1 \leqslant i \leqslant n} |x_i|$, and
- ℓ_0-norm (or) 0-norm, which captures the sparsity, or number of non-zero elements in a the vector!

We will use the ℓ_0-, ℓ_1- and ℓ_2-norms in this chapter and the next, to set up suitable optimisation problems for studying metabolic networks.

Mathematically, pFBA involves minimising the sparsity of the flux vector, or the ℓ_0-norm, in addition to the regular flux balance constraints. Since the minimisation of the ℓ_0-norm of the flux vector involves the solution of a mixed–integer LP (MILP) problem, in practice, the ℓ_1-norm is often minimised. This optimisation also ensures that any unnecessary (stoichiometrically balanced) cycles are driven to zero. The formulation is as follows:

$$\min \|\mathbf{v}\|_0 \tag{8.15a}$$

$$subject\ to\ \sum_{j=1}^{r} s_{ij}v_j = 0 \qquad \forall i \in M \tag{8.15b}$$

$$v_{lb,j} \leqslant v_j \leqslant v_{ub,j} \qquad \forall j \in J \tag{8.15c}$$

$$v_{bio} = v_{bio,max} \tag{8.15d}$$

On the basis of pFBA, genes (or reactions) can be classified into six different categories [36]:

1. *essential*—indispensable for growth in the given medium),
2. *pFBA optima*—non-essential genes contributing to the optimal growth rate and minimum gene-associated flux,
3. *enzymatically less efficient (ELE)*—genes requiring more flux through reactions compared to alternative pathways that can meet the same growth rate (*e.g.* longer pathways),
4. *metabolically less efficient (MLE)*—genes resulting in a growth-rate reduction, if used (*e.g.* because the enzyme uses a sub-optimal co-factor),
5. *pFBA no-flux*—genes that are associated with reactions unable to carry flux in the present conditions, and
6. *blocked*— genes that are associated with reactions unable to carry flux in any condition (*i.e.* blocked reactions)

The pFBA function in COBRA toolbox readily enables the classification of genes (or reactions) as outlined above. pFBA also finds application in the identification of minimal metabolic networks under any condition [37].

8.8.6   ATP maintenance fluxes

Given the key aim of constraint-based modelling methods is to predict growth rates reliably, both in wild-type strains and in other conditions such as alternative environments or genetic perturbations, it becomes rather important to capture the energy requirements of a cell accurately in the metabolic model. Two forms of energy, in terms of ATP, are typically required: growth-associated maintenance (GAM) energy, and non-growth-associated maintenance (NGAM) energy.

GAM essentially represents the energy required by the cell to replicate itself, which includes the energy required for polymerisation of macromolecules, such as DNA, RNA, proteins, and glycogen, during growth. In models, it appears as the stoichiometric coefficient of ATP in the biomass reaction:

$$x\,ATP + x\,H_2O \rightarrow x\,ADP + x\,P_i + x\,H^+$$

where x is the number of phosphate bonds required. Note that this value is 75.55223 for the *i*ML1515 model (see Table 8.4).

Beyond GAM, the cell also has other requirements for ATP, *e.g.* for the maintenance of turgor pressure, cell repair, or motility, which must be satisfied. This ATP requirement is reflected in terms of a lower bound constraint on another ATP hydrolysis reaction, the NGAM reaction[5] ($ATP + H_2O \rightarrow ADP + P_i + H^+$). The value is normally accurately estimated from experiments; for the *E. coli* model *i*AF1260, this value was estimated as 8.39 mmol/gDW/h [38].

It is interesting to note that this constraint may at times lead to *infeasibility* of the LP problem—the usual flux balance constraints (bounds) always admit a (trivial) solution of $\mathbf{v} = \mathbf{0}$. Thus, the solver always returns a feasible solution for the growth rate (or any other objective), although it could be zero if the current conditions (constraints) do not support growth (*i.e.*, a *lethal* phenotype). On the other hand, a non-zero lower bound for NGAM can result in infeasibility, if the current conditions cannot support the NGAM flux—in such cases, a lethal phenotype can be concluded again, even in the absence of the solver returning a zero growth rate. This is also revisited in §10.8.1.

8.9   TROUBLESHOOTING

8.9.1   Zero growth rate

The most common problem one encounters while working with constraint-based models is a zero growth rate. That is, having optimised for maximum biomass formation, you encounter a solution where $\mathbf{v} = \mathbf{0}$. If this were a result of a perturbation, like the knock-out of a gene, the zero solution would signify a lethal phenotype. Otherwise, it is generally a consequence of a model(ling) error. How does one troubleshoot this scenario?

As a first model diagnostic, it is also advisable to study the blocked reactions and dead-end metabolites, as discussed in §8.8.1. Are there any essential reactions, *i.e.* reactions essential for growth/production of biomass components that are blocked? Are there important precursors that show up as dead-end metabolites in the model? Some of these may need to be provided in the medium as exchange fluxes. It is also possible that these issues arise because some of the irreversible reactions are in the wrong direction.

Another cause for the zero growth rate could arise from the inability of the model to satisfy the NGAM ATP requirements. Of course, this may not show up as a zero growth rate; rather, it will result in the solver reporting that the LP problem is infeasible, as discussed above (§8.8.6).

8.9.2   Objective values vs. flux values

It is common for an LP problem to have multiple optimal solutions, for the same value of the objective. In the context of FBA, this translates to multiple flux distributions that can satisfy the flux-balance equations but produce the same growth rate. Therefore, the predicted growth rate is more reliable than the values

---

[5]typically denoted "ATPM" in genome-scale reconstructions

predicted for the various internal fluxes—the multiple optima do indicate the different 'flux states' the cell can inhabit under the given conditions. Another way to study a metabolic network more carefully involves performing an FVA (§8.7), to understand the boundaries of each of the fluxes, for a given objective. This can also highlight the bottlenecks in the metabolic network and the constraints that actually have a strong effect on the objective itself.

As we saw in the example of the toy metabolic network of Figure 8.1, in Equation 8.12, even though the constraints on reactions can be quite 'loose', *e.g.* $\{-1000, 1000\}$, in practice, reactions such as R_5 can only carry a flux of 50 mmol/gDW/h. Also, always remember that FBA presents a view of the theoretical capabilities of the cell. The more the constraints, on fluxes such as metabolite uptakes, the smaller and more reliable the space of observed/allowable flux distributions. Internal flux measurements, using ^{13}C-MFA also come in very handy in such cases, as we will discuss later (§9.7).

8.10 SOFTWARE TOOLS

COBRA toolbox [39] (https://opencobra.github.io/) is one of the most widely used MATLAB toolboxes for performing COnstraint-Based Reconstruction and Analysis of metabolic networks. It is an open-source community-developed toolbox that supports a huge number of routines for performing various flavours of constraint-based analyses. It also has been ported to Python (COBRApy) and Julia (COBRA.jl).

CellNetAnalyzer [40] (https://www2.mpi-magdeburg.mpg.de/projects/cna/cna. html) is a MATLAB toolbox that provides various computational methods (including FBA, and elementary mode analysis (see §9.8)) for the analysis of metabolic networks.

RAVEN Toolbox [41] (https://github.com/SysBioChalmers/RAVEN) is a MATLAB toolbox for Reconstruction, Analysis, and Visualisation of Metabolic Networks.

All the above tools rely on at least one of the solvers for LP/QP/MILP, such as IBM CPLEX (https://www.ibm.com/analytics/cplex-optimizer), GLPK (https://www.gnu.org/software/glpk/), Gurobi (https://www.gurobi.com/), MOSEK (https://www.mosek.com/), PDCO (https://web.stanford.edu/group/SOL/software/pdco/), and Tomlab CPLEX (https://tomopt.com/tomlab/products/cplex/), or even MATLAB linprog. More details can be found in Appendix D.

EXERCISES

8.1 Write down the stoichiometric matrix for glycolysis (you can refer to the reactions in Table 2.3). What are the reactions you must *add* to this system, so that it can carry a flux at steady-state?

8.2 Consider a metabolic network characterised by the stoichiometric matrix given below. Is there a dead–end metabolite? Are there any blocked reactions?

$$\begin{bmatrix} 1 & 0 & 0 & -2 & 0 & 0 \\ 0 & 1 & 0 & 3 & -1 & 0 \\ 0 & 0 & 0 & 0 & 3 & -2 \\ 0 & 0 & 1 & 0 & 0 & 1 \end{bmatrix}$$

8.3 Consider a simple system of reactions given below:

b_1: →A	r_1: 2A→3B	r_3: 2C→D
b_2: →B	r_2: B→3C	b_3: D→

If the objective function for this system is the maximisation of b_3, predict the following:

a. If, at a particular steady–state, $b_3 = 12$ mM/s, what is the rate of flux r_2?

b. If the constraints on the exchanges are: $0 \leqslant b_1 \leqslant 1$, $0 \leqslant b_2 \leqslant 2$, $0 \leqslant b_3 \leqslant 10$, and the reactions r_1, r_2, and r_3 are unconstrained, what will be value of r_2 and the objective function?

[LAB EXERCISES]

8.4 The *E. coli* 'core model'[6] is a simplified model of *E. coli* covering reactions of glycolysis, pentose phosphate pathway, TCA cycle, glyoxylate cycle, gluconeogenesis, anaplerotic reactions, etc [42]. Compared to the larger models that contain thousands of reactions, the core model contains a mere 95 reactions and 72 metabolites, and is very useful for learning the ropes of constraint-based modelling.

a. Perform FBA and compute the optimal biomass flux for the core model. Note the default bounds, especially for oxygen and glucose.

b. Examine the lower bounds of the reactions and identify the ATP maintenance flux. What is its value?

c. Plot the change in growth rate as a function of change in glucose uptake rate (from 10–25 mmol/gDW/h), keeping the oxygen uptake rate fixed at 16 mmol/gDW/h.

d. Identify the flux variability of the 'fumarate reductase (FRD7)' reaction for growth on a glucose medium.

8.5 For the above model, change the growth medium from glucose to acetate to succinate (all 10 mmol/gDW/h) and predict the growth rate. What do you infer? How will you simulate growth under anaerobic conditions?

---

[6] Available from COBRA toolbox and also https://systemsbiology.ucsd.edu/Downloads/EcoliCore

8.6 For the above model with glucose as the sole carbon source, add a D-lactate uptake flux of 5 mmol/gDW/h and predict the growth rate. Comment on your observations. Perform your simulations using Escher FBA (https://sbrg.github.io/escher-fba/#/app).

8.7 Can you encode the toy metabolic network of Figure 8.1 in Excel and load it into MATLAB using the xls2model function in the COBRA Toolbox?

8.8 Consider the metabolic network shown in Figure 8.3. Can you perform an FVA to confirm that the blocked reactions are the same as those indicated in the text?

8.9 Obtain the *i*ML1515 metabolic model of *E. coli* from the BiGG database (http://bigg.ucsd.edu/models/iML1515; also see Appendix C). Can you identify the default medium for its growth? Can you also compute a 'minimal medium' for growth?

8.10 Simulate the *E. coli i*ML1515 model using optimizeCbModel from the COBRA toolbox with (a) minimisation of ℓ_1-norm and (b) minimisation of ℓ_0-norm. Can you see that the biomass flux remains the same? What are the fluxes that have changed? Interestingly, you might see different solutions for the same ℓ_0-norm minimisation problem, with different solvers.

8.11 Perform a pFBA on the *E. coli i*ML1515 model to identify the reactions belonging to the various classes. Now, incrementally pin the bounds of the following sets of reactions (based on the pFBA reaction classes identified above) to zero (use changeRxnBounds): (i) zero-flux, MLE, and ELE. Also compute the growth rate at each step. What do you understand from these simulations?

8.12 For the *i*ML1515 model above, compute the substrate graph adjacency matrix from the stoichiometric matrix. What is its degree distribution?

REFERENCES

[1] K. Smallbone, E. Simeonidis, N. Swainston, and P. Mendes, "Towards a genome-scale kinetic model of cellular metabolism", *BMC Systems Biology* 4:6+ (2010).

[2] M. R. Watson, "Metabolic maps for the Apple II", *Biochemical Society Transactions* 12:1093–1094 (1984).

[3] M. R. Watson, "A discrete model of bacterial metabolism", *Computer Applications in the Biosciences* 2:23–27 (1986).

[4] D. A. Fell and J. R. Small, "Fat synthesis in adipose tissue. An examination of stoichiometric constraints", *The Biochemical Journal* 238:781–786 (1986).

[5] A. Varma, B. W. Boesch, and B. Ø. Palsson, "Biochemical production capabilities of *Escherichia coli*", *Biotechnology and Bioengineering* 42:59–73 (1993).

[6]   A. Varma and B. Ø. Palsson, "Stoichiometric flux balance models quantitatively predict growth and metabolic by-product secretion in wild-type *Escherichia coli W3110*", *Applied and Environmental Microbiology* **60**:3724–3731 (1994).

[7]   J. B. S. Haldane, "On being the right size", *Harper's Magazine* (1926).

[8]   M. W. Covert, I. Famili, and B. Ø. Palsson, "Identifying constraints that govern cell behavior: a key to converting conceptual to computational models in biology?", *Biotechnology and Bioengineering* **84**:763–772 (2003).

[9]   J. S. Edwards, M. Covert, and B. Ø. Palsson, "Metabolic modelling of microbes: the flux-balance approach", *Environmental Microbiology* **4**:133–140 (2002).

[10] B. Niebel, S. Leupold, and M. Heinemann, "An upper limit on Gibbs energy dissipation governs cellular metabolism", *Nature Metabolism* **1**:125–132 (2019).

[11] N. D. Price, J. L. Reed, and B. Ø. Palsson, "Genome-scale models of microbial cells: evaluating the consequences of constraints", *Nature Reviews Microbiology* **2**:886–897 (2004).

[12] K. Zhuang, G. N. Vemuri, and R. Mahadevan, "Economics of membrane occupancy and respiro-fermentation", *Molecular Systems Biology* **7**:500 (2011).

[13] Z. A. King, A. Dräger, A. Ebrahim, N. Sonnenschein, N. E. Lewis, and B. Ø. Palsson, "Escher: a web application for building, sharing, and embedding Data-Rich visualizations of biological pathways", *PLoS Computational Biology* **11**:e1004321 (2015).

[14] S. A. Becker, N. D. Price, and B. Ø. Palsson, "Metabolite coupling in genome-scale metabolic networks", *BMC Bioinformatics* **7**:111–111 (2006).

[15] I. Famili and B. Ø. Palsson, "The convex basis of the left null space of the stoichiometric matrix leads to the definition of metabolically meaningful pools", *Biophysical Journal* **85**:16–26 (2003).

[16] R. Schuetz, L. Kuepfer, and U. Sauer, "Systematic evaluation of objective functions for predicting intracellular fluxes in *Escherichia coli*", *Molecular Systems Biology* **3**:119 (2007).

[17] A. M. Feist and B. Ø. Palsson, "The biomass objective function", *Current Opinion in Microbiology* **13**:344–349 (2010).

[18] H. Alper, Y.-S. S. Jin, J. F. Moxley, and G. Stephanopoulos, "Identifying gene targets for the metabolic engineering of lycopene biosynthesis in *Escherichia coli*", *Metabolic Engineering* **7**:155–164 (2005).

[19] J. M. Savinell and B. Ø. Palsson, "Network analysis of intermediary metabolism using linear optimization. i. development of mathematical formalism", *Journal of Theoretical Biology* **154**:421–454 (1992).

[20] J. S. Edwards and B. Ø. Palsson, "Systems properties of the *Haemophilus influenzae* rd metabolic genotype", *Journal of Biological Chemistry* **274**:17410–17416 (1999).

[21] J. S. Edwards, R. U. Ibarra, and B. Ø. Palsson, "*In silico* predictions of *Escherichia coli* metabolic capabilities are consistent with experimental data", *Nature Biotechnology* **19**:125–130 (2001).

[22] J. Pramanik and J. D. Keasling, "Stoichiometric model of *Escherichia coli* metabolism: incorporation of growth-rate dependent biomass composition and mechanistic energy requirements", *Biotechnology and Bioengineering* **56**:398–421 (1997).

[23] D. S. Ow, D. Y. Lee, M. G. Yap, and S. K. Oh, "Identification of cellular objective for elucidating the physiological state of plasmid-bearing *Escherichia coli* using genome-scale *in silico* analysis", *Biotechnology Progress* **25**:61–67 (2009).

[24] J. M. Monk, C. J. Lloyd, E. Brunk, et al., "*i*ML1515, a knowledgebase that computes *Escherichia coli* traits", *Nature Biotechnology* **35**:904–908 (2017).

[25] I. Thiele and B. Ø. Palsson, "A protocol for generating a high-quality genome-scale metabolic reconstruction", *Nature Protocols* **5**:93–121 (2010).

[26] M. Lakshmanan, S. Long, K. S. Ang, N. Lewis, and D.-Y. Lee, "On the impact of biomass composition in constraint-based flux analysis", *bioRxiv*:652040 (2019).

[27] S. Boyd and L. Vandenberghe, *Convex optimization* (Cambridge University Press, 2004).

[28] R. Mahadevan and C. H. Schilling, "The effects of alternate optimal solutions in constraint-based genome-scale metabolic models", *Metabolic Engineering* **5**:264–276 (2003).

[29] A. Bordbar, J. M. Monk, Z. A. King, and B. Ø. Palsson, "Constraint-based models predict metabolic and associated cellular functions", *Nature Reviews Genetics* **15**:107–120 (2014).

[30] J. Monk, J. Nogales, and B. Ø. Palsson, "Optimizing genome-scale network reconstructions", *Nature Biotechnology* **32**:447–452 (2014).

[31] J. D. Orth and B. Ø. Palsson, "Systematizing the generation of missing metabolic knowledge", *Biotechnology and Bioengineering* **107**:403–412 (2010).

[32] I. Thiele, N. Vlassis, and R. M. T. Fleming, "fastGapFill: efficient gap filling in metabolic networks", *Bioinformatics* **30**:2529–2531 (2014).

[33] J. L. Reed and B. Ø. Palsson, "Genome-scale *in silico* models of *E. coli* have multiple equivalent phenotypic states: assessment of correlated reaction subsets that comprise network states", *Genome Research* **14**:1797–1805 (2004).

[34] C. J. Fritzemeier, D. Hartleb, B. Szappanos, B. Papp, and M. J. Lercher, "Erroneous energy-generating cycles in published genome scale metabolic networks: identification and removal", *PLoS Computational Biology* **13**:e1005494+ (2017).

[35] N. D. Price, I. Famili, D. A. Beard, and B. Ø. Palsson, "Extreme pathways and kirchhoff's second law", *Biophysical Journal* **83**:2879–2882 (2002).

[36] N. E. Lewis, K. K. Hixson, T. M. Conrad, et al., "Omic data from evolved *E. coli* are consistent with computed optimal growth from genome-scale models", *Molecular Systems Biology* **6**:390+ (2010).

[37] G. Sambamoorthy and K. Raman, "MinReact: a systematic approach for identifying minimal metabolic networks", *Bioinformatics* **36**:4309–4315 (2020).

[38] A. M. Feist, C. S. Henry, J. L. Reed, et al., "A genome-scale metabolic reconstruction for *Escherichia coli* K-12 MG1655 that accounts for 1260 ORFs and thermodynamic information", *Molecular Systems Biology* **3**:121 (2007).

[39] L. Heirendt, S. Arreckx, T. Pfau, et al., "Creation and analysis of biochemical constraint-based models using the COBRA toolbox v.3.0", *Nature Protocols* **14**:639–702 (2019).

[40] A. von Kamp, S. Thiele, O. Hädicke, and S. Klamt, "Use of CellNetAnalyzer in biotechnology and metabolic engineering", *Journal of Biotechnology* **261**:221–228 (2017).

[41] H. Wang, S. Marcišauskas, B. J. Sánchez, et al., "RAVEN 2.0: a versatile toolbox for metabolic network reconstruction and a case study on *Streptomyces Coelicolor*", *PLoS Computational Biology* **14**:e1006541 (2018).

[42] J. D. Orth, R. M. T. Fleming, and B. Ø. Palsson, "Reconstruction and use of microbial metabolic networks: the core *Escherichia coli* metabolic model as an educational guide", *EcoSal Plus* **4** (2010).

FURTHER READING

L. Yang, J. T. Yurkovich, Z. A. King, and B. Ø. Palsson, "Modeling the multiscale mechanisms of macromolecular resource allocation", *Current Opinion in Microbiology* **45**:8–15 (2018).

B. Ø. Palsson, *Systems biology: constraint-based reconstruction and analysis* (Cambridge University Press, 2015).

B. Ø. Palsson, "Metabolic systems biology", *FEBS Letters* **583**:3900–3904 (2009).

J. D. Orth, I. Thiele, and B. Ø. Palsson, "What is flux balance analysis?", *Nature Biotechnology* **28**:245–248 (2010).

CHAPTER 9

Extending constraint-based approaches

CONTENTS

W HY do we need other constraint-based approaches? While we are very convinced with the constraints imposed on biological systems or metabolic networks, it is still difficult to second-guess the behaviour, or physiological objective, of a cell. FBA, which we have discussed at length in the previous chapter, is but one such technique, which seeks to predict the phenotype of the cell, utilising a linear objective function. It is easy to imagine that the linear objective function, and consequently, FBA, cannot conceivably cover all possible scenarios. As discussed in §8.4, it may be necessary to employ different optimisation criteria to predict the phenotype of a cell in various situations.

Apart from various linear objectives, such as maximisation of growth, minimisation of ATP or nutrient consumption, or maximisation of metabolite production, it is often necessary to employ non-linear objective functions. In this chapter, we will overview different constraint-based approaches, that go beyond the linear biomass objective, and attempt to predict the state of the cell in many other conditions.

For instance, if a cell is subject to multiple deletions, it is unreasonable to assume that it will maximise its biomass—the cell might be trying to minimise the damage, or readjust its metabolism to counter the perturbations. Further, while FBA does not account for regulatory effects, there are other techniques that capsulate the effect of regulation on the redistribution of cellular fluxes, in terms of the metabolic adjustment that the cell effects. In the following sections, we will discuss some techniques that adopt a different logic to the optimisation, compared to FBA. Later in the chapter, we will also discuss techniques to integrate omic data, particularly transcriptomic data, to explore the relationship between metabolism and regulation.

9.1 MINIMISATION OF METABOLIC ADJUSTMENT (MoMA)

Church and co-workers proposed MoMA in 2002 [1]. The key premise of MoMA is that an organism tries to minimise its metabolic adjustments on perturbation; in other words, an organism re-routes its fluxes such that it admits a flux configuration as *close* as possible to the *original* wild-type flux distribution, albeit satisfying any new constraints arising out of the perturbation(s).

Geometrically, this can be imagined as identifying a flux solution nearest in flux space to the original but obeying any additional constraints, such as the loss of one or more reactions. Figure 9.1 illustrates the concept of MoMA. Under a perturbation, typically, new constraints are introduced, *e.g.* the deletion of one or more reactions (see §10.1), which alter the feasible space for possible solutions, and consequently the 'corner points'—corresponding to the FBA optima. MoMA picks the solution closest to the original optimum but lying in the new feasible space. Note that the original optimum/solution need not have been identified by FBA—it could be an experimentally measured distribution too. MoMA has been shown to correct certain incorrect predictions made by FBA, following the deletion of genes such as *tpiA*, *fba*, and *pfkAB* in *E. coli* [1].

Mathematically, the MoMA optimisation problem can be written as:

$$\min \|\mathbf{v_w} - \mathbf{v_m}\|^2 \quad \textit{subject to } \mathbf{Sv_m} = 0 \tag{9.1}$$

where $\mathbf{v_w}$ is the wild-type flux, $\mathbf{v_m}$ is the flux corresponding to a mutant, and $\mathbf{S}$ is the stoichiometric matrix corresponding to the metabolic network. Other 'usual' FBA constraints, such as uptake rates, and lower/upper bounds are also part of the formulation (not shown). Mathematically, this translates to a quadratic programming (QP) problem. A canonical QP problem is written as follows:

$$\min \frac{1}{2}\mathbf{x}^\top \mathbf{Q}\mathbf{x} + \mathbf{c}^\top \mathbf{x} \quad \textit{subject to } \mathbf{Ax} \preceq \mathbf{b} \tag{9.2}$$

where $\mathbf{Q}$ is a real symmetric matrix, and the other terms are the same as what we saw in the LP of Equation 8.8. Simplifying Equation 9.1, we obtain

$$\min \mathbf{v_w}^\top \mathbf{v_w} + \mathbf{v_p}^\top \mathbf{v_p} - 2\mathbf{v_w}^\top \mathbf{v_p} \quad \textit{subject to } \mathbf{Sv_p} = 0 \tag{9.3a}$$

$$\Rightarrow \min \frac{1}{2}\mathbf{v_p}^\top \mathbf{I}\mathbf{v_p} + (-\mathbf{v_w})^\top \mathbf{v_p} \quad \textit{subject to } \mathbf{Sv_p} = 0 \tag{9.3b}$$

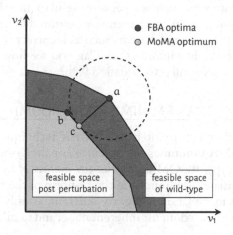

Figure 9.1 **Geometric interpretation of MoMA.** The different points indicated in the figure correspond to the FBA optimum corresponding to wild-type (a); the FBA optimum corresponding to a perturbation, *e.g.* knock-out (b) and the closest point to a in the new feasible space, which corresponds to the MoMA optimum corresponding to the perturbation (c). The dashed circle shows a region of solutions closest to the original optimum. c is where the circle intersects the new feasible space. Adapted with permission, from D. Segrè, D. Vitkup, and G. M. Church, "Analysis of optimality in natural and perturbed metabolic networks", *Proceedings of the National Academy of Sciences USA* **99**:15112–15117 (2002). © (2002) National Academy of Sciences, U.S.A.

where $\mathbf{I}$ is the identity matrix. Note that we have divided by two and discarded the strictly positive quantity $\mathbf{v_w}^\top \mathbf{v_w}$ from the minimisation problem, resulting in an equation that matches the canonical QP formulation of Equation 9.2.

The COBRA toolbox also implements a linear version of MoMA (linear-MOMA) that minimises the ℓ_1-norm of the difference between $\mathbf{v_w}$ and $\mathbf{v_p}$, rather than the ℓ_2-norm of the difference as in Equation 9.1 above.

9.1.1 Fitting experimentally measured fluxes

The MoMA formulation is also directly useful to fit experimentally measured fluxes; these are typically obtained through ^{13}C labelling (see §9.7). Of the n reactions in the system, perhaps only k could be measured/estimated accurately. Then, we would like to 'fit' the fluxes $\mathbf{v}$, employing a least-squares fit, as follows:

$$\sum_{i=1}^{k} (v_{w,i} - v_i)^2 \quad \textit{subject to} \ \mathbf{Sv} = 0 \tag{9.4}$$

Solving this optimisation problem is especially useful to (in)validate a model, *i.e.* to examine if a model can *support* experimentally measured fluxes. Notably, this can enable the identification of model errors such as incorrect reactions or reactions with incorrect directionality information. Can you see how this will transform into a QP problem, very similar to Equation 9.3b?

9.2 REGULATORY ON-OFF MINIMISATION (ROOM)

ROOM [2] is another constraint-based analysis technique, in the flavour of MoMA. While MoMA minimises the 'distance', or the ℓ_2-norm of the flux difference from wild-type, ROOM minimises the *number of significant flux changes*. This essentially translates to minimising the ℓ_0-norm of the flux difference. Precisely however, the formulation is slightly different, as only significant changes are counted, in order to account for inherent noise, and to minimise computation time.

Mathematically, this is formulated as a Mixed Integer Linear Programming (MILP) problem:

$$\min \|\mathbf{y}\|_0 \; i.e.$$

$$\min \sum_{i=1}^{n} y_i \tag{9.5a}$$

$$subject\ to\ \mathbf{Sv} = 0 \tag{9.5b}$$

$$\mathbf{v} - \mathbf{y}(\mathbf{v}_{\min} - \mathbf{v}_{\mathbf{w},l}) \succeq \mathbf{v}_{\mathbf{w},l}, \tag{9.5c}$$

$$\mathbf{v} - \mathbf{y}(\mathbf{v}_{\max} - \mathbf{v}_{\mathbf{w},u}) \preceq \mathbf{v}_{\mathbf{w},u}, \tag{9.5d}$$

$$v_j = 0, \; j \in D, \; y_i \in \{0, 1\} \tag{9.5e}$$

$$\mathbf{v}_{\mathbf{w},l} = \mathbf{v}_{\mathbf{w}} - \delta|\mathbf{v}_{\mathbf{w}}| - \epsilon \tag{9.5f}$$

$$\mathbf{v}_{\mathbf{w},u} = \mathbf{v}_{\mathbf{w}} + \delta|\mathbf{v}_{\mathbf{w}}| + \epsilon \tag{9.5g}$$

where D is the set of reactions associated with the deleted genes, and $y_i = 1$ for a significant flux change in v_i and $y_i = 0$ otherwise. The other symbols carry their usual meanings. Non-significant flux changes are defined using a range $[\mathbf{w}^l, \mathbf{w}^u]$ around the wild-type flux vector, $\mathbf{v}_{\mathbf{w}}$. For $y_i = 1$, *i.e.* a significant change in the flux of reaction i, the inequalities in Equations 9.5c and 9.5d do not impose any new constraints on v_i—they simplify to the original FBA constraints $\mathbf{v}_{\min} \preceq \mathbf{v} \preceq \mathbf{v}_{\max}$. On the other hand, $y_i = 0$ constrains v_i to the 'insignificant' ranges defined above. The values of δ and ϵ obviously affect the running time of the MILP solver, as also the accuracy of the predictions. Ruppin and co-workers used values of $\delta = 0.03$, $\epsilon = 0.001$ for flux predictions, and $\delta = 0.1$, $\epsilon = 0.01$ for lethality predictions [2].

9.2.1   ROOM vs. MoMA

ROOM was shown to outperform MoMA/FBA in certain scenarios; for instance, Ruppin and co-workers [2] showed that for an *E. coli* pyruvate kinase (*pyk*) knock-out, the predictions obtained by using ROOM showed higher correlations with experimentally observed values than those obtained from FBA/MoMA, particularly in a low-glucose condition. Also, in the case of *S. cerevisiae*, the authors showed that ROOM marginally outperforms FBA and significantly outperforms MoMA in lethality predictions, correctly predicting the non-lethality of as many as 50 genes falsely classified as lethal by MoMA. The authors attribute this difference to ROOM's ability to correctly identify alternate pathways for re-routing fluxes.

While methods such as MoMA and ROOM are nowhere as widely used as FBA, they present useful methodologies that provide a different perspective on studying/predicting the flux state of a cell. More importantly, they present other interesting physiologically plausible objectives, such as the minimisation of metabolic adjustment or the minimisation of the number of flux changes. These premises are plausible, as a cell might initially try to minimise its changes with respect to the wild-type, but finally settle down into a state that has relatively few reactions 're-regulated' compared to the wild-type. Indeed, MoMA has been shown to better predict the transient growth rates post-perturbation, while ROOM has been shown to better predict the final growth rate post-adaptation to the perturbation [2].

9.3   BI-LEVEL OPTIMISATION

All the constraint-based modelling techniques we have studied thus far tackle only a *single* optimisation problem, linear or otherwise. However, it is common to require satisfying more than one objective. For instance, in a metabolic engineering application (also see §10.6.1), one might want to optimise for both cellular production of a metabolite and cellular growth rate. These involve the formulation of bi-level optimisation problems. A more detailed account of bi-level optimisation frameworks and optimisation problems, in general, can be found in the excellent textbook by Maranas and Zomorrodi [3].

9.3.1   OptKnock

OptKnock is one of the classic strain design methods, which proposed a bi-level optimisation problem that maximises a given bio-engineering design objective, concurrently with the maximisation of biomass [4]. Bi-level optimisation problems are nested, with an *outer-level* problem corresponding to the desired bio-engineering objective (*e.g.* maximisation of the production of a vitamin) and an *inner* problem that optimises the usual cellular objective (*e.g.* maximisation of cellular growth, *i.e.* biomass objective function):

$$\max v_{\text{product}} \tag{9.6}$$

$$\text{subject to} \quad \begin{bmatrix} \max v_{\text{bio}} \\ \textit{subject to} \quad \mathbf{Sv} = 0 \\ v_{\text{EX}_{\text{glc}}} \geqslant v_{\text{glc,uptake}} \\ v_{\text{EX}_{O_2}} \geqslant v_{O_2,\text{uptake}} \\ v_{\text{bio}} \geqslant f v_{\text{bio}}, \qquad (0 < f \leqslant 1) \\ v_{\text{lb},j} y_j \leqslant v_j \leqslant v_{\text{ub},j} y_j, \quad \forall j \in J \end{bmatrix}$$

$$\sum_{j \in J} (1 - y_j) \leqslant K \qquad \text{(constrain number of deletions)}$$

$$y_j \in \{0, 1\}, \quad \forall j \in J$$

where y_j is a binary variable indicating whether or not a given reaction is knocked out of the model, and K constrains the maximum allowable gene deletions (*e.g.* K = 2 will identify double gene deletions leading to over-production of a given product). Note that the inner problem (bracketed optimisation problem) also places constraints on glucose and oxygen uptake ($v_{\text{EX}_{\text{glc}}}$ and $v_{\text{EX}_{O_2}}$, respectively, constrained to $v_{\text{glc,uptake}}$ and $v_{O_2,\text{uptake}}$), as well as on biomass, to be within a fraction f of v_{bio}, the wild-type biomass (*i.e.* growth rate).

The problem is actually a non-linear optimisation problem, but can be transformed into a single MILP, by computing the dual of the optimisation problem, as detailed elsewhere [3, 4]. Maranas and co-workers originally illustrated the application of OptKnock to identify combinations of gene knock-outs for the over-production of metabolites such as succinate, propanediol, and lactate in *E. coli* [4].

9.4   INTEGRATING REGULATORY INFORMATION

Although flux-balance models of metabolism are often effective in predicting metabolic phenotypes, they can go very wrong sometimes, as they do not consider regulation at all. Owing to this, for instance, in *E. coli*, FBA will predict a concurrent uptake of glucose and lactose, and consequently, a higher growth rate, as there are no *regulatory constraints* that we have imposed on the model. In reality, because of catabolite repression, glucose is first consumed, followed by a lag phase, and then lactose is consumed, with *E. coli* displaying diauxic growth.

9.4.1   Embedding regulatory logic: Regulatory FBA (rFBA)

The simplest method to integrate regulatory information into constraint-based models of metabolism is rFBA [5]. In rFBA, regulatory logic is embedded into flux balance models by means of Boolean regulatory rules. For instance,

$$g_1 = \text{IF } (v_2 > 0) \tag{9.7a}$$

$$g_2 = \text{IF NOT}(g_1 \text{ OR } g_3) \tag{9.7b}$$

$$g_3 = \text{IF}\ (Oxygen) \tag{9.7c}$$

Using Boolean rules predicated on the presence of other genes or transcription factors (TFs), or even values of reaction fluxes, the *states* of different genes are computed. These gene states, depending on the gene–protein–reaction (GPR) associations (see §10.1.1), translate to constraints on various reactions, notably whether they are on or off, *i.e.* present or absent. How does this work in practice? The following are the broad steps of an rFBA, starting with a given model and a carefully curated set of Boolean regulatory rules (on the lines of Equation 9.7):

1. Assume an initial *regulatory state* for the system of reactions, *i.e.* the on–off states for all of the genes in the system
2. Perform an FBA, computing a flux distribution, *i.e. metabolic state* (note that this flux distribution is but one of the possibly many allowable flux distributions under a given set of constraints; recall §8.8.3)
3. (Re-)Evaluate all regulatory rules, refreshing the constraints on the system
4. Repeat from Step 2 onwards

It is possible that the above sequence of steps might lead to a 'metabo-regulatory' steady-state, or even a limit cycle of sorts, where the cell cycles through different states. A limitation of rFBA is that we discretise the gene states to (fully) on or off—it is certainly likely that many genes/TFs may have only a partial effect on several reactions, rather than completely activating/inhibiting them, as captured by Boolean logic discussed herein. Another key limitation of this approach is that we arbitrarily pick a single metabolic state (*i.e.* flux distribution), from the set of possible solutions to the flux balance problem. Nevertheless, rFBA gives a much superior glimpse of how regulation affects metabolism, compared to FBA. SR-FBA [6] provides an alternative approach to characterise steady-state behaviours, integrating models of regulation, and metabolism to predict a combined metabolic–regulatory state. These metabolic–regulatory states are described by a pair of consistent metabolic and regulatory steady-states, which satisfy both the metabolic and regulatory constraints.

9.4.2 Informing metabolic models with omic data

Given the inherent difficulties in carefully curating the regulatory logic corresponding to a genome-scale metabolic model, it is far easier to study and understand regulation, by means of integrating omic data. Integrating transcriptomic data arising from microarray or RNAseq experiments is a relatively easier way to constrain reactions in metabolic models, based on the observed expression of corresponding genes. A host of techniques have been developed, which focus on the integration of omic data, notably gene expression data, into flux balance models of metabolism. Some of these techniques are discussed below.

E-Flux One of the first methods to elegantly integrate expression data into constraint-based models was E-Flux [7], by imposing suitable constraints on reaction fluxes. The premise of the method is that the level of mRNA can be used as an approximate upper bound for the maximum amount of enzyme available—and consequently, and upper bounds for the corresponding reaction(s). Therefore, E-Flux carefully specifies reaction constraints, based on the GPR associations (see §10.1.1) for any given reaction. Essentially, the constraint on the upper bound of every reaction is altered as follows:

$$v_{max,j} = f(\text{expression level of gene(s) associated with reaction } j)$$

where f captures the effect of the expression levels of the genes associated with a reaction as follows: if there is only one gene associated with a given reaction, then the upper bound is set to the expression value of that gene. If there are multiple genes involved in a reaction independently (*i.e.* isozymes), then the upper bound is computed as the sum of all the corresponding expression values. If the multiple genes are jointly required for the reaction (*e.g.* a multi-enzyme complex composed of many sub-units), the upper bound is taken as the minimum of all corresponding expression values. These values were finally normalised by dividing each of the bounds by the maximum value of $v_{max,j}$. Thus, the bounds were all constrained in the region $[-1, 1]$, for each of the reactions.

Probabilistic Regulation of Metabolism (PROM) PROM [8] uses a probabilistic framework that can be readily learnt from high-throughput datasets. PROM represents the regulatory interactions in a network as probabilities. The TF states are first determined from the environmental conditions. These TF states then dictate the on/off states of their corresponding target genes, based on gene expression data. The probabilities are used to constrain the reaction fluxes in the metabolic network. For example, if a given gene is expressed 8 out of 10 times (datasets) a TF is present, then a probability $p = 0.8$ is used to constrain the reaction flux as $v_{max,j} = 0.8v_{max,unregulated,j}$. Further, these constraints are encoded as *soft* constraints—constraints that can be violated to maximise the objective, but with a penalty. This is in contrast to the hard constraints that arise from stoichiometry/thermodynamics/environmental conditions.

While PROM requires no manual annotation of regulatory rules, unlike rFBA, it does require the structure of the regulatory network, in terms of TFs and their target genes. Therefore, it is possible to apply PROM only for organisms where the regulatory networks have been well-studied—such as E. coli, where the regulatory network data are available from RegulonDB [9] (also see Appendix C).

Other methods have also been proposed for the integration of gene expression data, *e.g.* Gene Inactivity Moderated by Metabolism and Expression (GIMME; [10]) and Gene Expression and Metabolism Integrated for Network Inference (GEMINI; [11]). For more comprehensive reviews, see [12, 13].

9.4.3 Tissue-specific models

With the development of human metabolic reconstructions, beginning with 'Recon1' [14], it became very important to also be able to adapt such models to capture tissue-specific metabolism. A number of algorithms have been developed to build such context-specific models, which can present a more accurate picture of metabolism in specific tissues. Shlomi and co-workers [15] developed one of the first algorithms to generate a tissue-specific model from the generic Recon1 model by integrating a variety of omic data—transcriptomic, proteomic and metabolomic—as well as literature-based knowledge and phenotypic data. The model expectedly outperformed the generic human model in predicting fluxes across a variety of hormonal and dietary conditions.

Another algorithm, metabolic Context-specificity Assessed by Deterministic Reaction Evaluation (mCADRE) was developed by Price and co-workers [16]. mCADRE was used to reconstruct draft genome-scale metabolic models for 126 human tissue and cell types, including 26 tumour tissues (paired with their normal counterparts), and 30 different brain tissues. mCADRE includes a high-confidence core set of reactions from an underlying generic genome-scale model by leveraging tissue-specific expression data, network structure and flux analyses. Additional non-core reactions are ranked and pruned based on gene expression, connectivity and flux values. mCADRE is fully automated, deterministic and is also computationally efficient, almost 1000-fold, compared to previous methods such as [15].

9.5   COMPARTMENTALISED MODELS

We have spent the last several pages discussing metabolic networks, the reactions that comprise them and how their fluxes can be predicted using various approaches. *Where* do these reactions actually occur, in a cell? In bacterial cells, most of the reactions occur in the cytosol, and are there are not many major compartments. In contrast, in eukaryotic cells, there exist a multitude of compartments, such as the mitochondrion, Golgi apparatus, endoplasmic reticulum, nucleus, vacuoles, and chloroplasts. It is essential to consider the compartmentalisation of cells in modelling, as has been well-illustrated in [17].

In the context of constraint-based models, even bacterial cells are typically modelled with three simple compartments—the periplasm, the cytoplasm and an extracellular space. Transport reactions carry metabolites to and fro between these various compartments. In the classic compartmentalised reconstruction of *S. cerevisiae* [18], Palsson and co-workers localised all the reactions to the cytosol or mitochondria. Using the information on compartmentalisation from resources such as the Comprehensive Yeast Genome Database and the Yeast Protein Database, reactions were localised to appropriate compartments, with any non-mitochondrial reaction being relegated to the cytosol, for simplicity. A link between cytosol and mitochondria was established through either known transport and shuttle systems or through inferred reactions to meet metabolic demands.

Metabolites in the mitochondria were specially tagged with an '*m*', to demarcate them from the default cytosolic metabolites. A major challenge in such compartmentalised reconstructions is the careful inclusion of exchange reactions, especially given the current state of knowledge on intracellular transport systems and their mechanisms [19]. Transport reactions may also lead to futile cycles (see §8.8.4) in metabolic networks, and must be given careful consideration.

A later reconstruction [14] considered many more compartments including the peroxisome, nucleus, Golgi apparatus, endoplasmic reticulum and vacuole, with as many as 80 reactions being added to represent the metabolite exchange for these five compartments. Different types of transport reactions were encoded in the model, including simple diffusion, symport and antiport reactions.

As noted by Klitgord and Segrè [17], compartmentalised models are able to more truly predict the fluxes in *S. cerevisiae*, and help understand energy limitations. On the other hand, the de-compartmentalised models likely over-estimate both the growth rate and individual fluxes. Interestingly, they also showed that better predictions could be obtained even from de-compartmentalised models, by constraining specific fluxes to their true (*i.e.* compartmentalised model) values.

For more complex organisms, it is important to consider both compartmentalisation and tissue-specific metabolism, during reconstruction. A good example of these is the reconstruction of *Arabidopsis thaliana* [20]—where the authors outline three stages of reconstruction, starting with the *global* reaction network, followed by a compartmentalised reconstruction, and finally incorporating tissue-specificity to obtain more accurate models.

Compartmentalised models also naturally serve as an inspiration to constructing community models of microbiomes/ecosystems, as we will see in §11.1.3.

9.6   DYNAMIC FLUX BALANCE ANALYSIS (dFBA)

Although FBA is predominantly used to perform steady-state analyses, FBA has also been extended to study dynamic processes such as microbial growth in batch reactors. Beginning with the classic study of Varma and Palsson (1994) [21], now called dynamic FBA (dFBA) [22], a set of approaches have been developed to study and understand the dynamics of microbial cultures using constraint-based methods. In contrast to FBA, these methods also take into account the concentrations of various substrates. dFBA can predict the time profiles of metabolites that are consumed or secreted in batch and fed-batch experiments.

Doyle and co-workers [22] proposed two methodologies to handle the dynamics using FBA:
(i) a dynamic optimisation algorithm, which is cast as a non-linear programming problem, and
(ii) a static optimisation algorithm that performs multiple iterative FBAs at different time-points based on a *quasi-steady-state assumption*. That is, the intracellular dynamics of cellular metabolism operates at a much faster time-scale than the time-scale for the dynamics of the extracellular environment—therefore, the metabolic network reaches an equilibrium for a given set of uptake fluxes.

Implementing dFBA

The COBRA toolbox implements dFBA in the function `dynamicFBA.m`. This function performs a dFBA using the static optimisation algorithm from [21]. A look into the code should give you a good idea of how the iterations discussed below are implemented in practice.

We will review the static optimisation algorithm here, which is a straightforward extension of FBA. The method involves iterative steps, as follows:

1. At every time-step, perform an FBA to predict growth rate (μ), nutrient uptake and product secretion rates (v_{S_i}'s)[1]
2. Use these rates to compute the biomass (X) as well as nutrient concentrations ($[S_i]$) at the end of the time-step
3. Based on these concentrations, compute the maximum uptake rate of nutrients for the next time-step
4. Repeat Step 1, having modified the constraints based on the above step

For the first time-step, the concentration of all substrates is the initial substrate concentration

$$[S_i] = [S_i]_0$$

For later steps, this is determined as the concentration predicted from the previous step. The substrate concentration is scaled to define the amount of substrate available per cell (unit biomass) per unit time, which is used as a constraint (lower bound), for performing an FBA:

$$\text{Substrate available} = \frac{[S_i]}{X \cdot \Delta t} = v_{\min,i}$$

where X (gDW/L) is the current cell density. Now, FBA can be used to compute the substrate uptake (v_{S_i}, mmol/L) and the growth rate (μ, h^{-1}). Note that

$$\mu = \sum w_i v_i = c^\top v$$

where μ is obtained as a weighted sum of the reaction fluxes corresponding to the synthesis of various growth precursors, and w_i capture the amounts of the growth precursors required per gram (DW) of biomass.

The values for substrate concentrations at each time-step are calculated from the following differential equations (integrating from 0 to Δt):

$$\frac{dX}{dt} = \mu X \Rightarrow X = X_0 e^{\mu \Delta t} \tag{9.8}$$

$$\frac{d[S_i]}{dt} = -[v_{S_i}] \cdot X \Rightarrow [S_i]_{\Delta t} = [S_i]_0 + \frac{v_{S_i}}{\mu} X_0 \left(1 - e^{\mu \Delta t}\right) \tag{9.9}$$

---

[1]The original papers use S_u to denote these uptakes.

dFBA has since been used in a number of applications, more recently even to study microbial communities (see §11.1.3.4). A comprehensive review of the applications of dFBA is available elsewhere [23].

9.7 ^{13}C-MFA

With our increasing ability to predict fluxes in genome-scale metabolic networks, it also becomes imperative for us to be able to validate these predictions. While growth rates are relatively easier to measure, it is much harder to quantify the various internal fluxes, viz. the partitioning of fluxes into various pathways/reactions happening within the cell. Further, while the concentrations of enzymes, metabolites or mRNA transcripts can be measured directly, the fluxes must necessarily be inferred from other measurable quantities, and must be resolved carefully, in the context of network structure.

In the last two decades, pivotal advances have been made towards understanding the internal flow of metabolites in various organisms. This has been made possible by the development of a suite of experimental and computational techniques—primarily network-wide stable isotope balancing methods involving isotopes such as ^{13}C/^{14}C. By growing organisms on ^{13}C-labelled substrates, it is possible to disentangle the internal pathways and infer flux partitioning based on the distribution of labels in the various metabolites. For instance, phosphoenolpyruvate molecules synthesised via glycolysis exhibit a different 'label signature' compared to those coming from the pentose phosphate pathway. Beginning with the choice of the isotopic tracers, experiments have to be carefully designed to estimate fluxes, as we discuss below.

Choosing the tracers   The choice of isotopic tracers is most crucial in designing an informative ^{13}C-MFA experiment that will yield precise estimates of fluxes. This is typically done using *in silico* simulations of the metabolic network in question—in addition to the usual stoichiometry, this network must also have atom transition data, to enable the selection of suitable labelled substrates. A combination of [1,2-^{13}C]glucose and [1,6-^{13}C]glucose tracers applied in parallel experiments is generally very informative, in the context of prokaryotic metabolic networks [24]. The tracers are very helpful in resolving the distribution of flux through alternate pathways.

Selecting experimental conditions   Two key steady-state assumptions underlie typical ^{13}C-MFA: (i) metabolic fluxes are constant during the labelling experiment, *i.e.* a *metabolic steady-state* has been reached, and (ii) the label has fully *propagated* throughout the metabolic network and an *isotopic steady-state* has been reached. These can be achieved in batch cultures, during the exponential growth phase, where the growth rate is constant, or in continuous cultivations, after 3–5 volume changes. Non-stationary ^{13}C-MFA [25, 26] or dynamic ^{13}C-MFA [27–29] can also be performed, but are obviously more demanding.

Analysing the samples For batch cultures, biomass and glucose concentration are measured at different time-points and can be used to determine the growth rate of cells, the substrate uptake rate, the biomass yields, as well as product secretion rates and yields. Metabolite labelling is measured using a variety of experimental techniques, viz. HPLC (high-performance liquid chromatography), GC–MS (gas chromatography–mass spectrometry), LC–MS (liquid chromatography–mass spectrometry), tandem mass spectrometry, or even NMR (nuclear magnetic resonance). Typical metabolites studied are biomass components, amino acids, fatty acids, and sugar constituents.

Estimating the fluxes Based on the metabolic model (with atom transitions), the external fluxes (*e.g.* substrate uptake and product secretion rates) and the metabolite labelling measurements, fluxes can be estimated. Many software tools such as Metran [30], INCA [31], and SUMOFLUX [32] use a variety of methods—from elementary metabolic unit (EMU) analysis [33] to flux ratio analysis and machine learning—to estimate the fluxes. These methods all use a least-squares approach (linear, non-linear, or weighted) to minimise the differences between the experimentally observed measurements and the *in silico* simulated measurements. A statistical analysis is essential to verify the goodness-of-fit and determine the confidence intervals for the fluxes.

Applications and limitations The key application of ^{13}C-MFA is in guiding metabolic engineering for the production of various metabolites [34]. In many under-characterised species, ^{13}C-MFA has helped unravel novel metabolic pathways [35, 36]. ^{13}C-MFA has also been used to understand the metabolism of cancer cells [37]. Applications of ^{13}C-MFA to the analysis of microbial co-cultures and consortia are also emerging [38].

Many challenges exist in applying ^{13}C-MFA effectively, to study intracellular fluxes in metabolic networks. Gaps in the model or errors in stoichiometric atom transition data can obviously adversely impact predictions. Another limitation stems from the fact that only central carbon metabolism, amino acid biosynthesis, and a few secondary metabolite pathways are typically investigated. ^{13}C-MFA studies also suffer from a lack of reproducibility and lack of data standards [39], as for many other models (see Epilogue, p. 316). Crown and Antoniewicz [39] also outline best practices for ^{13}C-MFA and a checklist for publishing.

9.8 ELEMENTARY FLUX MODES AND EXTREME PATHWAYS

We have previously discussed how the linear constraints imposed during FBA alter the space of allowable solutions for *permissible* flux distributions (Figure 8.2). Remember, any vector in a space can be written in terms of its *basis vectors*. Now, every point in the allowable space of solutions represents a *valid* flux configuration of the cell—one that does not violate the imposed thermodynamic/stoichiometric constraints. An interesting question now emerges: is it possible to characterise each of these allowable flux distributions by means of some fundamental pathways or *modes*, which define a basis for all these allowable vectors? Two parallel approaches have been developed—elementary flux modes (EFMs; [40, 41]) and extreme pathways (EPs; [42]).

An EFM, or just elementary mode, is a minimal set of enzymes that can operate in steady-state with all irreversible reactions proceeding in the correct direction. Reversible directions can obviously go either way, but irreversible reactions should all be proceeding in the thermodynamically feasible direction. Schuster *et al* (1999) [41] give the following operational definition of an EFM: beginning with a metabolic system, inhibit enzymes one by one, while ensuring that there is flux flowing through the system. As the process of blocking enzymes is continued, a stage is reached when the inhibition of any further enzyme in the system will lead to a cessation of any steady-state flux through the entire system— this *minimal* set of enzymes constitutes one EFM. Formally, every EFM (and EP) must satisfy the following three conditions:

1. *Steady-state*: The reactions in the EFM (or EP) must be able to operate at steady-state. That is, the corresponding fluxes must obey the mass balance equation (Equation 8.4).
2. *Feasibility*: If a reaction is irreversible, its flux must be ≥ 0. That is, all reactions must be proceeding in a thermodynamically feasible direction.
3. *Non-decomposability*: If any reaction is removed from the EFM (or EP), the EFM ceases to carry any flux. That is, there is no *subset* of the EFM (or EP) that is an EFM (or EP)! This is often also referred to as *genetic independence*, as this condition mandates that the enzymes in one EFM (or EP) are not (all) a subset of the enzymes from another EFM (or EP).

In addition to the above conditions, EPs must satisfy two more conditions:

1. *Network reconfiguration*: Every reaction in the network must be classified either as an exchange flux (through which metabolites enter or exit the system) or as an internal reaction. Then, all reversible internal reactions must be split into two—one in the forward direction and another in reverse. Consequently, no internal reaction can have a negative flux. Exchange fluxes, on the other hand, can be reversible, but there is only one exchange flux corresponding to each metabolite.
2. *Systemic independence*: The set of EPs for a network is *minimal*—that is, no EP can be represented as a non-negative linear combination of other EPs. Note that this condition does not apply to EFMs at all.

Figure 9.2 illustrates the EFMs and EPs for a simple system. Note that the set of all EFMs for a system, J_{EFM} is always a superset of the set of all EPs (J_{EP}), *i.e.* $J_{EP} \subseteq J_{EFM}$. Note that EFM5, EFM6, and EFM7 do not satisfy the systemic independence criterion listed above—as shown in the figure, they are all combinations of the EP denoted EFM1* with another EP, viz. EFM2*, EFM3*, or EFM4*, respectively. An excellent comparison of EFMs and EPs is available elsewhere [43].

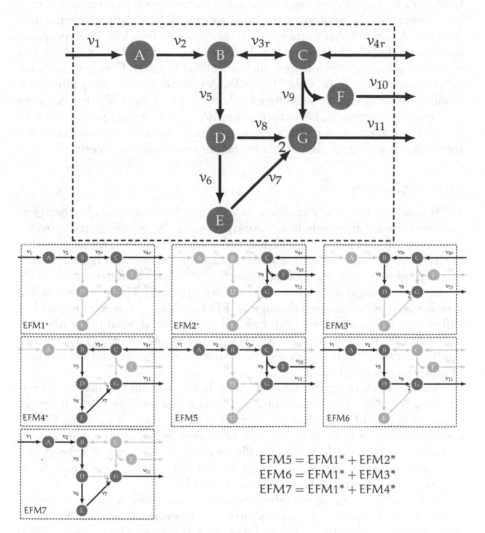

Figure 9.2 **An example to illustrate elementary flux modes.** A simple system with seven metabolites reacting is shown. The reactions represented by v_3 and v_4 are reversible. The reactions corresponding to v_7 and v_8 produce 2 moles of G, indicated at the terminus of the arrow. The system has seven EMs as shown in the seven panels below the main network, four of which are also EPs (EFM1*, EFM2*, EFM3*, and EFM4*). In each of the smaller panels, the reactions not part of the EFM are greyed out.

9.8.1 Computing EFMs and EPs

EFMs/EPs are basically derived from the stoichiometric matrix directly, but they are computationally very expensive to compute; in fact, their computation is recognised as an NP–hard problem [43]. Many algorithms have been proposed to compute these in the past, such as EFMtool [44] and METATOOL [45], and have also been integrated into tools such as CellNetAnalyzer. Owing to the combinatorial explosion of possible EFMs/EPs, despite advances in algorithms, it is challenging to exhaustively enumerate the EFMs/EPs, especially for genome-scale metabolic networks. Nevertheless, EFMs/EPs present an important advance in our understanding of the flux space, and have interesting consequences for metabolic network analysis, notably for guiding metabolic engineering.

9.8.2 Applications

EFMs (and EPs) have many applications, owing to the kind of insights they generate from even static metabolic networks, or essentially, stoichiometric matrices. First, they can shed light on which biochemical transformations are stoichiometrically and thermodynamically feasible. A classic problem in this regard pertains to the question *"Can sugars be produced from lipids?"*. We are all aware that our body converts excess sugar to fat—but is the converse possible? This is a question that is well–suited to be answered through an EFM analysis, as was demonstrated by Fell and co–workers in 2009 [46]. Fell and co–workers showed that while such pathways exist in plants, they do not exist in animals and humans. They showed that the enzymes of the glyoxylate shunt were critical in enabling the synthesis of glucose from acetyl CoA.

A very interesting application of EFM analysis is in the determination of maximal molar yields of various strains. For instance, from Figure 9.2, it can be observed that EFM3* produces two moles of G per mole of C, while EFM2* produces only one mole of G per mole of C. Furthermore, this computation is straightforward, as the EFMs are usually represented as vectors, and the yield is merely the ratio of v_{11} and v_{4r}. This approach is also useful in comparing various strains, *i.e.* their corresponding metabolic networks.

EFMs can also shed light on the robustness of metabolic networks to knock-outs, as well as the potential impact of enzyme deficiencies on a system. EFMs and EPs can also highlight futile cycles (also see §8.8.4) in metabolic networks. EFMs also find use in determining minimal media, or minimal reactomes, and also in gap–filling.

EXERCISES

9.1 You are trying to predict the growth rate of mutant yeast cells (with a single enzyme knocked out) using a metabolic model. Explain how you will make the best possible prediction for this *in silico* gene deletion, if

a. all fluxes ($v \in \mathbf{V}$) have been experimentally measured in the wild–type

b. only a subset $(V_m \subset V)$ has been experimentally measured in the wild-type

9.2 The synthesis of haeme in a cell takes place across the cytosol and mitochondria as shown in the table below. Reactions from the KEGG database are shown below. Construct the stoichiometric matrix of the system. Clearly state any assumptions that you make.

Cytosol
R01513: 3-Phospho-D-glycerate + NAD^+ ⇔ 3-Phosphonooxypyruvate + NADH + H^+
R04173: 3-Phosphonooxypyruvate + L-Glutamate ⇔ O-Phospho-L-serine + 2-Oxoglutarate
R00582: O-Phospho-L-serine + H_2O ⇔ L-Serine + Orthophosphate
R00036: 2 5-Aminolevulinate ⇔ Porphobilinogen + 2 H_2O
R00084: 4 Porphobilinogen + H_2O ⇔ Hydroxymethylbilane + 4 NH_3
R03165: Hydroxymethylbilane ⇔ Uroporphyrinogen III + H_2O
R03197: Uroporphyrinogen III ⇔ Coproporphyrinogen III + 4 CO_2
Mitochondria
R00945: Tetrahydrofolate + L-Serine ⇔ 5,10-Methylenetetrahydrofolate + Glycine + H_2O
R00830: Succinyl-CoA + Glycine ⇔ 5-Aminolevulinate + CoA + CO_2
R03220: Coproporphyrinogen III + Oxygen ⇔ Protoporphyrinogen IX + 2 CO_2 + 2 H_2O
R03222: Protoporphyrinogen IX + 3 Oxygen ⇔ Protoporphyrin + 3 Hydrogen peroxide
R00310: Protoporphyrin + Fe^{2+} ⇔ Haeme + 2 H^+

9.3 Consider the metabolic network below, where a small variation from Figure 9.2 has been introduced:

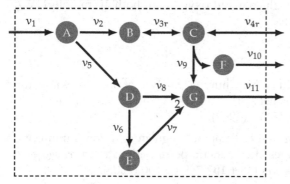

a. Can you compute all the EFMs and EPs?
b. How do the EFMs/EPs change if v_2 were also reversible?

9.4 Obtain an atom-mapped metabolic network for *E. coli* central carbon metabolism. If you feed the network with 1,2-^{13}C Glucose, which of the following statements will be true considering the network is at steady-state?

    a. Pyruvate would be labelled at the first and second positions if only the Entner–Doudoroff pathway were active

    b. Pyruvate would be labelled at the third position if only the pentose phosphate pathway were active

    c. Pyruvate would be labelled at the first position if only the pentose phosphate pathway were active

    d. Pyruvate would be labelled at the second and third positions if only glycolysis were active

9.5 Many studies have used constraint-based methods to predict over-expression targets for metabolic engineering of a variety of metabolites. Pick one such study from literature, and write a critical summary about its methodologies and main findings.

[LAB EXERCISES]

9.6 For the *E. coli* core model, use OptKnock[2] to identify a set of three reaction knock-outs that increase the production of lactate under anaerobic conditions, with a glucose uptake rate of 18 mmol/gDW/h. Constrain your search such that the growth rate of *E. coli* does not drop below 10% of the wild-type strain's growth rate under the same conditions. Is there a subset of reactions from the solution returned by OptKnock, which when knocked out, yields nearly the same lactate production, but at a higher growth rate?

9.7 Consider the *E. coli* core model. Can you predict how the concentrations of acetate, formate and glucose vary over the first 8h, for the following initial conditions: [Glucose] = 15 mM, initial biomass = 0.025 g/L. Assume that the maximum uptake rates for this strain for glucose and oxygen are both 10 mmol/gDW/h. Comment on your observations.

REFERENCES

[1] D. Segrè, D. Vitkup, and G. M. Church, "Analysis of optimality in natural and perturbed metabolic networks", *Proceedings of the National Academy of Sciences USA* **99**:15112–15117 (2002).

[2] T. Shlomi, O. Berkman, and E. Ruppin, "Regulatory on/off minimization of metabolic flux changes after genetic perturbations", *Proceedings of the National Academy of Sciences USA* **102**:7695–7700 (2005).

[3] C. D. Maranas and A. R. Zomorrodi, *Optimization methods in metabolic networks*, 1st ed. (Wiley, 2016).

---

[2]Note that OptKnock can be extremely slow for large models; much thought needs to be put into selecting the set of reactions that form the 'search space' for OptKnock.

[4] A. P. Burgard, P. Pharkya, and C. D. Maranas, "Optknock: a bilevel programming framework for identifying gene knockout strategies for microbial strain optimization", *Biotechnology and Bioengineering* **84**:647–657 (2003).

[5] M. W. Covert, C. H. Schilling, and B. Ø. Palsson, "Regulation of gene expression in flux balance models of metabolism", *Journal of Theoretical Biology* **213**:73–88 (2001).

[6] T. Shlomi, Y. Eisenberg, R. Sharan, and E. Ruppin, "A genome-scale computational study of the interplay between transcriptional regulation and metabolism", *Molecular Systems Biology* **3**:101 (2007).

[7] C. Colijn, A. Brandes, J. Zucker, et al., "Interpreting expression data with metabolic flux models: predicting *Mycobacterium tuberculosis* mycolic acid production", *PLoS Computational Biology* **5**:e1000489+ (2009).

[8] S. Chandrasekaran and N. D. Price, "Probabilistic integrative modeling of genome-scale metabolic and regulatory networks in *Escherichia coli* and *Mycobacterium tuberculosis*", *Proceedings of the National Academy of Sciences USA* **107**:17845–17850 (2010).

[9] A. Santos-Zavaleta, H. Salgado, S. Gama-Castro, et al., "RegulonDB v 10.5: tackling challenges to unify classic and high throughput knowledge of gene regulation in *E. coli* K-12", *Nucleic Acids Research* **47**:212–212 (2019).

[10] S. A. Becker and B. Ø. Palsson, "Context-specific metabolic networks are consistent with experiments", *PLoS Computational Biology* **4**:e1000082+ (2008).

[11] S. Chandrasekaran and N. D. Price, "Metabolic constraint-based refinement of transcriptional regulatory networks", *PLoS Computational Biology* **9**:e1003370 (2013).

[12] D. Machado and M. Herrgård, "Systematic evaluation of methods for integration of transcriptomic data into constraint-based models of metabolism", *PLoS Computational Biology* **10**:e1003580+ (2014).

[13] R. P. Vivek-Ananth and A. Samal, "Advances in the integration of transcriptional regulatory information into genome-scale metabolic models", *Biosystems* **147**:1–10 (2016).

[14] N. C. Duarte, M. J. Herrgård, and B. Ø. Palsson, "Reconstruction and validation of *Saccharomyces cerevisiae* iND750, a fully compartmentalized genome-scale metabolic model", *Genome Research* **14**:1298–1309 (2004).

[15] L. Jerby, T. Shlomi, and E. Ruppin, "Computational reconstruction of tissue-specific metabolic models: application to human liver metabolism", *Molecular Systems Biology* **6**:401 (2010).

[16] Y. Wang, J. Eddy, and N. Price, "Reconstruction of genome-scale metabolic models for 126 human tissues using mCADRE", *BMC Systems Biology* **6**:153+ (2012).

[17] N. Klitgord and D. Segrè, "The importance of compartmentalization in metabolic flux models: yeast as an ecosystem of organelles", *Genome Informatics* **22**:41–55 (2010).

[18] J. Förster, I. Famili, P. Fu, B. Ø. Palsson, and J. Nielsen, "Genome-scale reconstruction of the *Saccharomyces cerevisiae* metabolic network", *Genome Research* **13**:244–253 (2003).

[19] I. Thiele and B. Ø. Palsson, "A protocol for generating a high-quality genome-scale metabolic reconstruction", *Nature Protocols* **5**:93–121 (2010).

[20] S. Mintz-Oron, S. Meir, S. Malitsky, E. Ruppin, A. Aharoni, and T. Shlomi, "Reconstruction of *Arabidopsis* metabolic network models accounting for subcellular compartmentalization and tissue-specificity", *Proceedings of the National Academy of Sciences USA* **109**:339–344 (2012).

[21] A. Varma and B. Ø. Palsson, "Stoichiometric flux balance models quantitatively predict growth and metabolic by-product secretion in wild-type *Escherichia coli W3110*", *Applied and Environmental Microbiology* **60**:3724–3731 (1994).

[22] R. Mahadevan, J. S. Edwards, and F. J. Doyle, "Dynamic flux balance analysis of diauxic growth in *Escherichia coli*", *Biophysical Journal* **83**:1331–1340 (2002).

[23] M. Henson and T. J. Hanly, "Dynamic flux balance analysis for synthetic microbial communities", *Systems Biology, IET* **8**:214–229 (2014).

[24] C. P. Long and M. R. Antoniewicz, "High-resolution ^{13}C metabolic flux analysis", *Nature Protocols* **14**:2856–2877 (2019).

[25] K. Nöh, K. Grönke, B. Luo, R. Takors, M. Oldiges, and W. Wiechert, "Metabolic flux analysis at ultra short time scale: isotopically non-stationary ^{13}C labeling experiments", *Journal of Biotechnology* **129**:249–267 (2007).

[26] M. R. Antoniewicz, D. F. Kraynie, L. A. Laffend, J. González-Lergier, J. K. Kelleher, and G. Stephanopoulos, "Metabolic flux analysis in a non-stationary system: fed-batch fermentation of a high yielding strain of *E. coli* producing 1,3-propanediol", *Metabolic Engineering* **9**:277–292 (2007).

[27] R. W. Leighty and M. R. Antoniewicz, "Dynamic metabolic flux analysis (DMFA): a framework for determining fluxes at metabolic non-steady state", *Metabolic Engineering* **13**:745–755 (2011).

[28] M. R. Antoniewicz, "Dynamic metabolic flux analysis—tools for probing transient states of metabolic networks", *Current Opinion in Biotechnology*, Chemical Biotechnology • Pharmaceutical Biotechnology **24**:973–978 (2013).

[29] L.-E. Quek, J. R. Krycer, S. Ohno, et al., "Dynamic ^{13}C flux analysis captures the reorganization of adipocyte glucose metabolism in response to insulin", *iScience* **23**:100855 (2020).

[30] H. Yoo, M. R. Antoniewicz, G. Stephanopoulos, and J. K. Kelleher, "Quantifying reductive carboxylation flux of glutamine to lipid in a brown adipocyte cell line", *The Journal of Biological Chemistry* **283**:20621–20627 (2008).

[31] J. D. Young, "INCA: a computational platform for isotopically non-stationary metabolic flux analysis", *Bioinformatics* **30**:1333–1335 (2014).

[32] M. Kogadeeva and N. Zamboni, "SUMOFLUX: a generalized method for targeted ^{13}C metabolic flux ratio analysis", *PLoS Computational Biology* **12**:e1005109 (2016).

[33] M. R. Antoniewicz, J. K. Kelleher, and G. Stephanopoulos, "Elementary metabolite units (EMU): a novel framework for modeling isotopic distributions", *Metabolic Engineering* **9**:68–86 (2007).

[34] W. Guo, J. Sheng, and X. Feng, "Synergizing ^{13}C metabolic flux analysis and metabolic engineering for biochemical production", in *Synthetic Biology – Metabolic Engineering*, Advances in Biochemical Engineering/Biotechnology (Cham, 2018), pp. 265–299.

[35] Y. J. Tang, R. Chakraborty, H. G. Martín, J. Chu, T. C. Hazen, and J. D. Keasling, "Flux analysis of central metabolic pathways in *Geobacter metallireducens* during reduction of Soluble Fe(III)-Nitrilotriacetic Acid", *Applied and Environmental Microbiology* **73**:3859–3864 (2007).

[36] Y. Tang, F. Pingitore, A. Mukhopadhyay, R. Phan, T. C. Hazen, and J. D. Keasling, "Pathway confirmation and flux analysis of central metabolic pathways in *Desulfovibrio vulgaris* Hildenborough using Gas Chromatography-Mass Spectrometry and Fourier Transform-Ion Cyclotron Resonance Mass Spectrometry", *Journal of Bacteriology* **189**:940–949 (2007).

[37] L. Jiang, A. Boufersaoui, C. Yang, et al., "Quantitative metabolic flux analysis reveals an unconventional pathway of fatty acid synthesis in cancer cells deficient for the mitochondrial citrate transport protein", *Metabolic Engineering* **43**:198–207 (2017).

[38] N. A. Gebreselassie and M. R. Antoniewicz, "^{13}C-metabolic flux analysis of co-cultures: a novel approach", *Metabolic Engineering* **31**:132–139 (2015).

[39] S. B. Crown and M. R. Antoniewicz, "Publishing ^{13}C metabolic flux analysis studies: a review and future perspectives", *Metabolic Engineering* **20**:42–48 (2013).

[40] S. Schuster and C. Hilgetag, "On elementary flux modes in biochemical reaction systems at steady state", *Journal of Biological Systems* **2**:165–182 (1994).

[41] S. Schuster, T. Dandekar, and D. A. Fell, "Detection of elementary flux modes in biochemical networks: a promising tool for pathway analysis and metabolic engineering", *Trends in Biotechnology* **17**:53–60 (1999).

[42] C. H. Schilling, D. Letscher, and B. Ø. Palsson, "Theory for the systemic definition of metabolic pathways and their use in interpreting metabolic function from a pathway-oriented perspective", *Journal of Theoretical Biology* **203**:229–248 (2000).

[43] J. A. Papin, J. Stelling, N. D. Price, S. Klamt, S. Schuster, and B. Ø. Palsson, "Comparison of network-based pathway analysis methods", *Trends in Biotechnology* **22**:400–405 (2004).

[44] M. Terzer and J. Stelling, "Large-scale computation of elementary flux modes with bit pattern trees", *Bioinformatics* **24**:2229–2235 (2008).

[45] A. von Kamp and S. Schuster, "Metatool 5.0: fast and flexible elementary modes analysis", *Bioinformatics* **22**:1930–1931 (2006).

[46] L. F. de Figueiredo, S. Schuster, C. Kaleta, and D. A. Fell, "Can sugars be produced from fatty acids? a test case for pathway analysis tools", *Bioinformatics* **24**:2615–2621 (2008).

FURTHER READING

N. E. Lewis, H. Nagarajan, and B. Ø. Palsson, "Constraining the metabolic genotype-phenotype relationship using a phylogeny of *in silico* methods", *Nature Reviews Microbiology* **10**:291–305 (2012).

S. Imam, D. R. Noguera, and T. J. Donohue, "An integrated approach to reconstructing genome-scale transcriptional regulatory networks", *PLoS Computational Biology* **11**:e1004103+ (2015).

M. P. Pacheco, T. Pfau, and T. Sauter, "Benchmarking procedures for high-throughput context specific reconstruction algorithms", *Frontiers in Physiology* **6**:410–410 (2015).

J. Kim and J. L. Reed, "RELATCH: relative optimality in metabolic networks explains robust metabolic and regulatory responses to perturbations", *Genome Biology* **13**:R78+ (2012).

C. J. Lloyd, A. Ebrahim, L. Yang, et al., "COBRAme: a computational framework for genome-scale models of metabolism and gene expression", *PLoS Computational Biology* **14**:e1006302 (2018).

J. T. Yurkovich and B. Ø. Palsson, "Quantitative -omic data empowers bottom-up systems biology", *Current Opinion in Biotechnology* **51**:130–136 (2018).

N. Zamboni, S.-M. Fendt, M. Rühl, and U. Sauer, "^{13}C-based metabolic flux analysis", *Nature Protocols* **4**:878–892 (2009).

D. E. Ruckerbauer, C. Jungreuthmayer, and J. Zanghellini, "Predicting genetic engineering targets with elementary flux mode analysis: a review of four current methods", *New Biotechnology* **32**:534–546 (2015).

CHAPTER 10

Perturbations to metabolic networks

CONTENTS

A central philosophy of systems biology, and even biology in general, is to study and understand complex systems/networks through perturbations. A given metabolic network can be perturbed in multiple ways—genes or reactions can be deleted, either singly, or in combination. On the other hand, genes can be over-expressed, or even new reactions can be added to the system. How does the metabolic network behave, following such a perturbation? How do the fluxes or the growth rate change? In this chapter, we will study how constraint-based modelling techniques can be used to answer these questions.

Each perturbation can be viewed as a genotypic alteration, resulting in the loss or gain of one or more reactions. The ability to predict the phenotype for different genotypes, under various conditions, as well as exploring several *what-if* scenarios is a cornerstone of systems biology. The study of the different possible phenotypes that can arise on perturbation, from a given cell, finds direct application in diverse scenarios, from strain improvement for metabolic engineering, to drug target identification. For metabolic engineering, we are primarily interested in identifying genotypes that have improved phenotypes, such as higher growth, or the over-production of a vitamin. Whereas, for drug target identification, we want the opposite—to identify the gene deletions that are lethal, *i.e.* abrogate growth.

10.1   KNOCK-OUTS

A common way to deduce the function of a gene is by studying a knock-out mutant with the corresponding gene deleted. Such a gene deletion can be simulated *in silico*, by removing all reactions catalysed by the corresponding gene product (protein). Removing a reaction from a metabolic network is straightforward—all we have to do is to pin its flux to zero, by means of an additional constraint. This is undoubtedly easier than say, removing the corresponding column from the stoichiometric matrix.

How do we constrain a reaction to zero flux? Essentially, we need to add a constraint,

$$v_i = 0 \qquad (10.1)$$

where i denotes the reaction being deleted. In practice, this is enforced by setting both the upper and lower bounds of the reaction to zero:

$$v_{lb,i} = v_{ub,i} = 0 \qquad (10.2)$$

Following the addition of these constraints, the metabolic model can be simulated using FBA, MoMA, or ROOM (or other constraint-based modelling approaches), as discussed in the previous chapter. FBA has been proven to accurately predict phenotypes following various genetic perturbations [1, 2]. The FBA formulation for simulating the effect of gene or reaction deletions is as follows:

$$\max v_{bio} = \sum_{j=1}^{r} c_j v_j \qquad (10.3a)$$

$$subject\ to\ \sum_{j=1}^{r} s_{ij} v_j = 0 \qquad \forall i \in M \qquad (10.3b)$$

$$v_{lb,j} \leqslant v_j \leqslant v_{ub,j} \qquad \forall j \in J \qquad (10.3c)$$

$$v_d = 0 \qquad \forall d \in D \subset J \qquad (10.3d)$$

where all terms carry their usual meanings (similar to Equation 8.6). v_{bio} represents the flux through the biomass reaction, J and M represent the sets of all reactions and metabolites, respectively, and D is the set of reactions that are to be deleted, corresponding to the gene or reaction knock-out(s). Solving the above LP to identify the maximum obtainable biomass flux (v_{bio}) under the deletion constraints, we classify it as lethal/non-lethal based on a predetermined 'cut-off value', v_{co}, typically 1% or 5% of the wild-type growth rate ($v_{bio,WT}$). In practice, the simulations can throw up different types of results, which need to be interpreted carefully, as discussed in §10.8.1.

Many studies have performed single gene deletion simulations and predicted the phenotype (growth/no growth) for a variety of organisms, such as *Mycobacterium tuberculosis* [3], *Pseudomonas aeruginosa* [4], and *Staphylococcus aureus* [5] and *Toxoplasma gondii* [6].

Other methods to study knock-outs, for particular objectives, such as for metabolic engineering, include methods such as OptKnock (see §9.3.1), which employs a bi-level optimisation framework. Another tool, OptFlux [7], is user-friendly and versatile, and can simulate knock-outs using FBA/MoMA/ROOM. OptFlux can also search for *optimal* strains for metabolic engineering using direct search algorithms such as simulated annealing or evolutionary algorithms.

10.1.1 Gene deletions vs. reaction deletions

The biological *unit of deletion* is a gene—in laboratory experiments, genes can be silenced, knocked out or their corresponding enzyme inhibited by some chemical agent. There is no way to selectively remove a reaction from a cell *per se*. On the other hand, in constraint-based models, the unit of deletion is a reaction. That is, individual reactions can be removed from the metabolic network. In this section, we will understand how to reconcile the two approaches.

Note that the relationship between genes, proteins, and reaction is often not one-to-one. Every metabolic reaction in the cell can be carried out by one or more enzymes, each of which can be comprised of one or more gene products. For example, multiple isozymes can catalyse a single reaction. Many enzymes also exhibit substrate promiscuity, *i.e.* they may catalyse multiple reactions that utilise different substrates. This complex many-to-many relationship is carefully encoded in metabolic models as gene–protein–reaction (GPR) associations. The GPR relationships are conceivably critical in mapping gene deletions to reaction deletions. Owing to the many-to-many nature of GPRs, the removal of a reaction may require the deletion of multiple genes and may even be accompanied by the removal of additional reactions, which can certainly alter the potential metabolic solution space. Consider the following example,

R_1 : g_1 AND g_2 *e.g.* enzyme complex

R_2 : g_2 OR g_3 *e.g.* isozymes

R_3 : g_2 OR g_3 OR g_4 OR g_5 *e.g.* many isozymes

In this example, the deletion of gene g_1 (or g_2) will turn off reaction R_1. The simultaneous deletion of genes g_2 and g_3 is necessary for turning off reaction R_2—but it will in parallel turn off reaction R_1 as well, since g_2 has been deleted. Thus, while it is possible to simulate the removal of reactions R_1 or R_2 from the model, the reaction R_2 alone can never be removed from the cell, without also removing R_1. Also, it can be seen that it is very difficult to turn off reaction R_3—it would require the simultaneous deletion of four different genes g_2, g_3, g_4 and g_5. Indeed, such an example exists in nature—*E. coli* has at least four different serine deaminases [8]. For more details, see Appendix B.

10.2 SYNTHETIC LETHALS

Beyond single gene/reaction deletions, it is also possible to perform deletions of more than one gene/reaction. Among such combinatorial deletions, of special

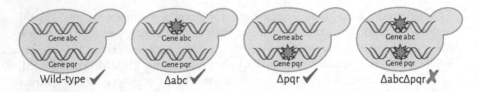

Figure 10.1 **An example to illustrate synthetic lethals.** A cell with two genes *abc* and *pqr* are shown. The mutants with only one gene deletion, viz. Δ*abc* and Δ*pqr* are viable (✓), as is the wild–type. The mutant Δ*abc*Δ*pqr* is not viable (✗); *abc* and *pqr* form a synthetic lethal pair.

interest are *synthetic lethals*, which are sets of genes (or reactions), where only the simultaneous removal of all genes in the set abrogates growth. Figure 10.1 shows a schematic to illustrate a synthetic lethal pair. In a synthetic lethal triplet, only the simultaneous removal of all three genes in the triplet will result in lethality; the deletion of any other subset of two or fewer genes from the triplet will not affect growth. Synthetic lethal sets are also known as *minimal cut sets* [9].

The study of synthetic lethals has many exciting applications. A key fundamental application is in understanding gene function and functional associations [10]. The study of synthetic lethality can shed light on what genes can buffer for one another, and even how different pathways can compensate for one another. Synthetic lethals also find applications in predicting putative *combinatorial* drug targets against pathogenic organisms [11] as well as in developing potential novel therapeutic strategies to combat cancer [12, 13].

Yeast synthetic lethals have been identified experimentally using yeast synthetic genetic arrays [14, 15]. Many computational methods have been developed to identify synthetic lethals, which have built on the framework of constraint-based analysis. FBA has been used to reliably predict synthetic lethal genes in metabolic networks of organisms such as yeast [16]. Broadly, there are three major approaches to identify synthetic lethals, as we will see in the rest of this section.

10.2.1 Exhaustive enumeration

A naïve method to identify synthetic lethals involves a simple exhaustive enumeration—take every pair of possible reactions (or genes), constrain them to zero, and simulate using FBA. However, for an organism with 1500 genes, solving an FBA LP for every pair will amount to $\binom{1500}{2} > 10^6$ LPs to be solved[1]! However, since the simulations are all independent of one another, they can be easily parallelised on a high–performance computing cluster. Still, for larger sets, the combinatorial explosion of possibilities makes it nearly impossible to resort to exhaustive enumeration.

---

[1]Essential genes will be eliminated, as they cannot form synthetic lethal sets, by definition. Therefore, the number will be somewhat lower, e.g. $\binom{1200}{2} > 7 \times 10^5$, if there are 300 essential genes.

Ruppin and co-workers identified synthetic lethal sets up to a size of four in a yeast metabolic network by exhaustive enumeration, parallelised on a computer cluster [17]. They also identified a few larger sets through a sampling approach. However, this model was fairly small, with only 484 genes considered for performing deletions. The parallel enumeration approach has also been used to identify synthetic lethal reaction sets up to the order of four, for the *Bacillus subtilis* model *i*Bsu1103 [18]. Using a BlueGene/P supercomputer with 65,536 processors, they evaluated > 18.2 billion combinations in a mere 162 minutes.

10.2.2   Bi-level optimisation

Exhaustive enumeration, while easy to implement and also parallelisable, is often impractical and infeasible for large metabolic networks, demanding the development of alternative methods. In this section, we will discuss two bi-level optimisation methods built on the framework of MILP, namely, SL Finder and the more recent MCSEnumerator, which exhibits much superior performance.

10.2.2.1   SL finder

SL Finder [19] poses the synthetic lethal identification problem elegantly as a bi-level MILP. The formulation involves a nested optimisation problem involving both real-valued variables (v_j's) and integer variables (y_j's). The integer variables are essentially binary in nature, and are encoded as below:

$$y_j = \begin{cases} 0, & \text{if reaction } j \text{ is eliminated} \\ 1, & \text{if reaction } j \text{ is active} \end{cases}, \quad \forall j \in J$$

By introducing this binary variable, the SL Finder formulation solves a min-max bi-level optimisation problem as shown below:

$$\min_{y_j} v_{bio} \qquad \text{[Outer]} \qquad (10.4a)$$

$$\text{subject to} \max_{v_j} v_{bio} \qquad \text{[Inner]} \qquad (10.4b)$$

$$\text{subject to} \quad \sum_{j=1}^{r} s_{ij} v_j = 0 \qquad \forall i \in M \qquad (10.4c)$$

$$v_j \leqslant v_{ub,j} \cdot y_j \qquad \forall j \in J \qquad (10.4d)$$

$$v_j \geqslant 0 \qquad \forall j \in J \qquad (10.4e)$$

$$\sum_j (1 - y_j) \leqslant k \qquad (10.4f)$$

$$y_j \in \{0, 1\} \qquad \forall j \in J \qquad (10.4g)$$

where y_j represents the binary variable encoding the presence/absence of the jth reaction, $v_{ub,j}$ represents the upper bound on the flux through the jth reaction and k is the number of simultaneous deletions *i.e.* the size of the synthetic lethal reaction set. Other variables have their usual meanings. Equation 10.4e suggests that all reactions have a lower bound of zero; therefore, the stoichiometric matrix must be transformed to contain only irreversible reactions. This is easily achieved, by replacing each reversible reaction with two reactions—one reaction in the forward direction and another in the backward direction.

The algorithm works as follows:

- When the SL Finder (Outer) problem is solved for the first time, $k = 1$ (Equation 10.4f), the solver identifies one single lethal reaction. This reaction corresponds to the variable y_j with a value of 0 in the solution.
- Now, to identify the remaining single lethal reactions, a new constraint is added—an *integer cut*—corresponding to the binary variable y_j identified in the previous step. For example, if reaction a is identified to be a lethal deletion, the new constraint is simply:

$$y_a = 1$$

which effectively excludes $y_a = 1$ as a solution. Subsequently, the SL Finder problem is solved repeatedly, until all remaining single lethal reactions are found.

- Next, to identify synthetic lethals of order two, *i.e.* synthetic lethal reaction pairs, we set $k = 2$ in Equation 10.4f. Similar to the above idea for single lethal reactions, one lethal pair is first identified, the corresponding additional integer cut constraint is added, to another lethal reaction pair, and the process is repeated. For example, if reactions b and c form a synthetic lethal pair, then the new constraint would be:

$$y_b + y_c \geqslant 1$$

This constraint ensures that at least one of the reactions, b or c, is active at a given time when searching for other lethal pairs—effectively eliminating the previously obtained solution $\{b, c\}$.

- Higher order synthetic lethals can be identified in a similar fashion: the steps mentioned above are repeated by changing the value of k and incrementally including integer cut constraints corresponding to the identified lethal sets.

Synthetic lethal double and triple reaction deletions have been reported for *E. coli*. However, the MILP problems become incrementally difficult to solve, owing to the imposition of the additional integer cut constraints described above. The time taken, on a workstation, was ≈ 6.75 days, to identify synthetic lethal reaction sets of order three in the *E. coli* iAF1260 model with ≈ 2400 reactions.

10.2.2.2 *MCSEnumerator*

A more recent and a much more efficient method, MCSEnumerator, was proposed by von Kamp and Klamt in 2014 [20]. MCSEnumerator builds on a previous study [21], which illustrated and exploited the dual relationship between synthetic lethal sets, or minimal cut sets (MCSs), in a metabolic network and elementary modes (recall §9.8). Essentially, the MCSs of a given metabolic network can be computed as certain EFMs of a dual network, derived by a simple transformation of the metabolic network. The method essentially builds on the shortest EFMs in the network obtained by (a modified version of) the algorithm in [22]. Since the EFMs in the dual network correspond to the smallest MCSs in the primal, a synthetic lethal set (MCS) of the desired size can be readily identified. Finally, an exclusion constraint is added after obtaining each solution similar to the integer cuts in SL Finder.

This method is reported to outperform previous methods using an advanced MILP solver [20]. However, MILP problems are not easy to parallelise—although there is instruction-level parallelism in that solving a single large MILP problem can be parallelised on a multi-core computer, MCSEnumerator does not have the data-level parallelism of exhaustive enumeration. In the following section, we will study another formulation, Fast-SL, which adopts a very different approach, essentially an intelligent enumeration, by exploiting the structure of the search space.

10.2.3 Fast-SL: Massively pruning the search space

The key idea behind the Fast-SL algorithm [23], developed earlier in our laboratory is that exhaustive enumeration solves far too many LPs than actually required to determine synthetic lethals—many pairs (or sets) of reactions can be eliminated easily, as they are guaranteed to not form synthetic lethals. We achieved this by performing a bunch of intelligent optimisations, to heavily prune the search space for lethals, and then *exhaustively* iterating through the remaining (much fewer) combinations.

10.2.3.1 *Most reactions do not carry a flux in a given condition*

An interesting aspect of nearly all metabolic networks is that not all reactions carry flux under all conditions. Indeed, some reactions are blocked, as we have discussed earlier (§8.8.1). However, even beyond blocked reactions, there are many reactions that may carry a flux only in a given environment, or only when an alternative reaction is blocked, and so on. A very interesting consequence of this fact is that these reactions, which do not carry a flux in a given environmental condition, can of course not be essential reactions for growth in that environment. Consequently, while performing FBA to identify singly essential reactions in an environment, we can first perform an FBA to identify a solution v^*. Next, instead of looping through all the reactions in J, we need only loop through the reactions carrying a non-zero flux in v^*(J_{nz}), constrain them to zero as in Equation 10.3 and identify the lethal deletions. This set of singly lethal reactions, J_{sl}, is contained entirely in J_{nz}, as shown in Figure 10.2a.

Figure 10.2b represents a matrix of size $|J| \times |J|$. If a pair of reactions i, j carry zero flux in the FBA solution v^* (i, j $\notin J_{nz}$), they cannot be a synthetic lethal pair

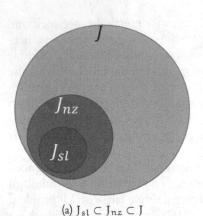

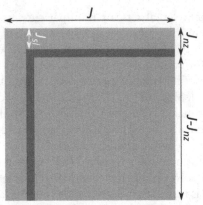

(a) $J_{sl} \subset J_{nz} \subset J$

(b) Eliminating large portions of the search space

Figure 10.2 **A schematic illustration of the search space for Fast-SL.** The Venn diagram in (a) indicates the set of all reactions, J, the set of reactions carrying a non-zero flux in a flux solution (J_{nz}), and J_{sl}, the set of single lethal reactions. The matrix in (b) illustrates the different portions of the space eliminated (light colour) and the space in which the Fast-SL actually searches for solutions (dark colour). Note that the matrix is symmetric and is shown fully for ease of representation. Adapted with permission from Springer Nature: K. Raman, A. Pratapa, O. Mohite, and S. Balachandran, "Computational prediction of synthetic lethals in genome-scale metabolic models using Fast-SL", *Methods in Molecular Biology* **1716**:315–336 (2018). © Springer Science+Business Media, LLC (2018).

(by extending the argument for single lethals). All synthetic lethal pairs lie in the narrow dark region (roughly $|J_{nz} - J_{sl}| \times |J - J_{sl}|$), drawn to scale for the *E. coli* *i*AF1260 model. Even in the narrow dark region of Figure 10.2b, further gains are made as we discuss below. Thus, Fast-SL ends up solving a tiny fraction of the LPs solved during exhaustive enumeration, to obtain the same result, by the elimination of many non-lethal sets.

10.2.3.2 *Choosing a sparse* **v***: The minimal norm solution*

We have already seen at length that any FBA LP problem has many solutions that can produce the same growth rate (§8.8.3). Which of these enables the best performance of Fast-SL, *i.e.* which of these gives the largest reduction in the search space? Clearly, the smaller the set of non-zero reactions, J_{nz}, the lesser the number of LPs to be solved for identifying lethal sets. Therefore, we use FBA to compute a flux distribution, corresponding to maximum growth rate, *while minimising the sum of absolute values of the fluxes*, *i.e.* the ℓ_1-norm of the flux vector, $\|\mathbf{v}\|_1$ — the *'minimal norm'* solution of the FBA LP problem:

$$\min \|\mathbf{v}\|_1 = \Sigma_j |v_j| \tag{10.5a}$$

TABLE 10.1 Extent of search space reduction in Fast-SL.

Order	Exhaustive LPs	LPs solved after eliminating non–lethal sets	Reduction in search space
Single	2.05×10^3	379	$\approx$ 5-fold
Double	1.57×10^6	6,084	$\approx$ 250-fold
Triple	9.27×10^8	223,469	$\approx$ 4,150-fold
Quadruple	4.10×10^{11}	1.43×10^7	$\approx$ 28,000-fold

$$subject\ to\ \sum_{j=1}^{r} s_{ij} v_j = 0 \qquad \forall i \in M \qquad (10.5b)$$

$$v_{lb,j} \leqslant v_j \leqslant v_{ub,j} \qquad \forall j \in J \qquad (10.5c)$$

$$v_{bio} = v_{bio,max} \qquad (10.5d)$$

Ideally, a pFBA-like solution, which maximises the sparsity of the flux vector should be even more advantageous—essentially it minimises the cardinality of J_{nz}, making it as sparse as possible. This translates to minimising the ℓ_0-norm of the flux vector (see Equation 8.15, which is identical to the above set of equations, save for the fact that it minimises the ℓ_0 norm), $\|\mathbf{v}\|_0$, and can afford further improvement, but we observe that the improvement is marginal in the context of the additional effort invested in solving the ℓ_0-norm minimisation problem, which is an MILP [23]. Table 10.1 illustrates the extent of reduction in search space achieved by Fast-SL for the *E. coli* iAF1260 model.

Importantly, Fast-SL can be readily parallelised, leading to further speed-ups, and it achieved $\approx 4\times$ speed-up over MCSEnumerator for *E. coli* iAF1260 model, on a system with only six cores. We also extended Fast-SL to identify lethal gene sets by incorporating gene–reaction rules in COBRA models, which capture the GPR associations. Fast-SL formulation identified 75 new gene triplets in *E. coli* that were not identified previously by SL Finder (see §10.2.2.1). Further, we identified up to synthetic lethal gene and reaction quadruplets for pathogenic organisms such as *S*. Typhimurium, *Mycobacterium tuberculosis*, *Staphylococcus aureus*, and *Neisseria meningitidis*.

In this section, we surveyed a range of methods to surmount the computational challenges posed by the combinatorial explosion of possibilities, while simulating multiple reaction/gene knock-outs in a metabolic network. Beginning with naïve exhaustive enumeration, which can be parallelised for efficiency, to more complex MILP-based formulations, and finally Fast-SL, which massively prunes the search space for synthetic lethals, a wide range of methods are available, to predict synthetic lethals in metabolic networks. Studying synthetic lethals can shed light on compensatory mechanisms in metabolism, as well as novel genetic interactions.

10.3   OVER-EXPRESSION

Another commonly employed perturbation, notably for strain improvement in metabolic engineering, is the amplification of one or more reaction fluxes, to mimic the over-expression of one or more genes in an organism. As discussed in the previous chapters, notably in §9.4, gene expression can be manifest as constraints on reaction fluxes, in constraint-based models. How does one impose the 'over-expression constraints'? Similar to the deletion constraints (Equations 10.1 and 10.2), we can amplify a reaction as follows:

$$v_i = 2v_{i,WT} \tag{10.6}$$

where the ith reaction is amplified to twice its wild-type flux. In practice, this can be achieved by altering both lower and upper bounds (or even just the lower bound) of the reaction in question. Following the imposition of the over-expression constraints, the metabolic network can be simulated using FBA/-MoMA/ROOM to predict its phenotype, notably the growth rate and product formation rates. The resulting flux distributions can be examined to identify the over-expression strategies, which have the best impact on the phenotype, as we discuss later in this chapter (see §10.5). Beyond mere FBA, it is always very insightful to perform an FVA (§8.7), to understand if the *allowable flux ranges* have been altered by the perturbation.

In practice, it may so happen that the amplification is infeasible—this would mean that the amplification is beyond the capability of the metabolic network, under the given conditions (environment and constraints).

10.3.1   Flux Scanning based on Enforced Objective Flux (FSEOF)

FSEOF [25] is an interesting and widely used approach to predict over-expression targets for strain improvement. Many of the knock-out approaches discussed above predict targets by randomly/exhaustively knocking out genes. In contrast, FSEOF evaluates a bunch of flux distributions, including the wild-type and those corresponding to incrementally higher fluxes of product formation. By analysing these flux distributions, FSEOF predicts the genes that could potentially be over-expressed in order to improve the flux of product formation.

The key steps of the FSEOF algorithm are as follows:

1. Identify the maximum biomass flux permissible through the system, by maximising v_{bio}

2. Identify the maximum product flux possible through the system, by maximising v_{prod}. Since there is often an inverse relationship between biomass formation and product flux, the biomass flux drops to zero as product flux reaches its theoretical maximum, $v_{prod,max}$.

3. Perform additional simulations, *enforcing* the product flux, to increasing fractions, f, of $v_{prod,max}$, for say, f = 0.1, 0.2, ..., 0.9, with the objective function being the maximisation of biomass. We will likely observe

that the v_{bio} keeps decreasing with an increase in v_{prod}—particularly for non-growth-associated metabolites and secondary metabolites.

4. Examine the flux distributions obtained in the above step, and classify reactions into three categories, based on those that:

    (a) monotonically admit higher fluxes, with an increase in v_{prod},

    (b) monotonically admit lower fluxes, with an increase in v_{prod}, and

    (c) reactions that remain unaffected (or change arbitrarily), with an increase in v_{prod}

FSEOF posits that those reactions that monotonically admit higher fluxes, with an increase in product formation flux, are the *roads that need to be widened*, to increase traffic through the product formation reaction. FSEOF has been used to identify gene targets for over-expression thus, in *Escherichia coli* for improved production of lycopene [25] and taxadiene [26], as well as in other organisms [27–30]. Other improvements have since been incorporated into FSEOF, bringing in the concepts of FVA and reaction grouping constraints based on similar flux values from omics datasets [31].

10.4 OTHER PERTURBATIONS

We have thus far discussed the key perturbations to metabolic networks, namely, deletions—single and multiple, and over-expression of reactions. Another common perturbation is the addition of reactions, a common scenario in metabolic engineering, when one or more reactions are heterologously expressed in a cell [32, 33]. This obviously involves altering the metabolic network itself, rather than merely adding constraints. Many tools/frameworks employ constraint-based modelling techniques to predict the changes in cellular phenotype following the incorporation of heterologous reactions/pathways [34–36].

10.5 EVALUATING AND RANKING PERTURBATIONS

We have seen how metabolic networks can be perturbed in multiple different ways—all of these approaches find application in metabolic engineering, or in identifying novel drug targets. While the modelling paradigms for either application are practically the same, the goals are starkly different. On the one hand, for metabolic engineering, we are interested in identifying knock-outs or over-expressions such that the corresponding phenotypes exhibit a much higher product yield (and at the same time sufficient growth). On the other hand, for drug target identification, we are interested in identifying synthetic lethal deletions or essential reactions, *i.e.* zero-growth phenotypes. For evaluating lethality, as mentioned earlier, we only need to predict the growth rate following the perturbation, and look at the ratio of the biomass reaction flux (v_{bio}) in the wild-type (WT) and mutant (Δ):

$$\frac{v_{bio,\Delta}}{v_{bio,WT}}$$

If this ratio is below a certain threshold (*e.g.* 0.05, or 0.01), then, the deletion is considered lethal. Similarly, a threshold of, say, 0.8 could be used to classify phenotypes as *sick*.

An important practical consideration here is that we may occasionally not be able to solve the LP, and the solver may report an infeasible solution. This happens in cases where $v = 0$ is not an admissible solution to the flux balance equation—typically because of a non-zero lower bound for one (or more) of the fluxes, such as ATP maintenance (also see §8.8.6 and §8.3). In such cases, infeasibility should be used to conclude that no growth is possible, and that the perturbation is indeed lethal.

Pfeifer and co-workers [37] defined a useful parameter, f_{ph}, to quantify the effect of any given perturbation, on a particular reaction (*e.g.* product biosynthesis reaction) as follows:

$$f_{ph} = \frac{v_{bio,KO}}{v_{bio,WT}} \cdot \frac{v_{p,KO}}{v_{p,WT}} \tag{10.7}$$

where v_{bio} and v_p refer to the fluxes of biomass and product-forming reactions, respectively, and the subscripts WT and KO denote the wild-type and knock-out strains, respectively. The parameter f_{ph}, which they called *phenotype fraction*, can be similarly used in the case of an over-expression or a reaction knock-in as well.

10.6 APPLICATIONS OF CONSTRAINT-BASED MODELS

Constraint-based models have been very popular for the systems-levels modelling of metabolism, in a wide variety of organisms. While genome-scale metabolic network reconstruction itself is an arduous task (see Appendix B), constraint-based models provide excellent insights into the organisation of metabolism, and a systems-level understanding that can be leveraged for strain improvement in metabolic engineering or identifying essential genes for drug target identification. Many review papers have already discussed the broad spectrum of applications of constraint-based modelling [38–40].

A vast majority of applications of genome-scale metabolic models stem from their ability to predict cellular phenotypes in various scenarios, including different environments (change in nutrient media *e.g.* carbon sources, aerobic vs. anaerobic conditions, etc.) and genetic perturbations (removal/inhibition of one or more genes/enzymes). Predicting cellular phenotypes in such conditions also enables us to pinpoint model inadequacies or missing biological knowledge (such as metabolic gaps), enabling us to improve the annotation of genes, and so on, ultimately enabling new biological discoveries. Further, as described above, the study of synthetic lethals can be instrumental in uncovering novel genetic interactions. Many studies have also studied the interaction between organisms, either host and pathogen, or in a microbiome, as we will study in greater detail in Chapter 11.

In this section, we will discuss a few studies, focussing on metabolic engineering and drug target identification.

10.6.1   Metabolic engineering

A very popular application of constraint-based modelling is in metabolic engineering, to predict strategies for strain improvement, with a number of excellent reviews written on the subject [41–44]. Constraint-based models enable a rational predictive approach to identifying key targets for knock-out or over-expression, to improve the yield of a particular metabolite. While traditional metabolic engineering has often focussed on a few product-forming pathways and *nearby* enzymes, systems-level modelling, by taking a more holistic perspective, is able to come up with strategies for strain improvement that are non-intuitive, yet novel and valuable.

A classic study by Stephanopoulos and co-workers [45] identified strategies to over-produce lycopene in an already engineered strain of *E. coli*. Using FBA and MoMA to predict phenotypes following the deletion of 2–3 genes, Stephanopoulos and co-workers identified a triple gene knock-out ($\Delta gdhA\Delta aceE\Delta fdhF$) that afforded a 37% increase in lycopene yield compared to the parent over-producing strain. The authors emphasised the combinatorial complexity that arises while looking for multiple deletions; they evaluated single and double deletions exhaustively, and subjected the double gene deletions with the best predicted phenotypes to every possible additional third deletion. It is worth noting that, while the model predictions did not have a 100% match with the experimentally observed yields, the predictions still showed a similar trend to the experiments, and ultimately enabled the identification of the triple gene deletion discussed above. Other tools such as OptKnock (see §9.3.1), Genetic Design through Local Search [46], and RobustKnock [47] have also been developed, which have been very popular for metabolic engineering applications.

Methods such as FSEOF (§10.3.1) have also been useful in reliably predicting strategies for strain improvement. Choi *et al* [25] used a combination of over-expression using FSEOF and knock-outs modelled using MoMA to identify an *E. coli* strain that remarkably over-produced lycopene (23.97 mg/L) compared to the parent strain (4.95 mg/L). Xu *at al* [36] used combinatorial gene knock-out and up-regulation using OptForce [48] to obtain a 474 mg/L yield of naringenin in *E. coli*, over 5.5-fold higher than the initial yield of 85 mg/L. A number of such case studies are illustrated elsewhere [49].

Overall, the approaches are straightforward: we seek to identify a phenotype with superior characteristics, such as product yield, by performing a combination of perturbations on a given metabolic network. It is interesting to note that these approaches have yielded many strategies for metabolic engineering, in a variety of organisms. Notably, a number of these studies have also been validated experimentally, underlining the utility of constraint-based approaches, even when experimental data, such as ^{13}C-based flux measurements, or transcriptomic data, have not been used to build the model.

10.6.2 Drug target identification

One of the early applications of constraint-based modelling to pathogenic metabolic networks was the reconstruction of the *S. aureus* genome-scale metabolic model [5]. Genome-scale metabolic models of several other pathogens have been reconstructed since [3, 50–52], and many of these studies have also performed experimental validations [3, 52, 53]. An excellent review on the application of constraint-based methods to drug target identification was published by Papin and co-workers [54].

An illustrative study, highlighting the potential of constraint-based modelling for modelling *Mycobacterium tuberculosis* metabolism was carried out by Raman and Chandra [55]. Modelling the mycolic acid pathway in *M. tuberculosis*, they illustrated many potential drug targets in the pathway. They further filtered the targets based on similarity to human proteins. This is important, as drug targets chosen in the pathogen should ideally be unique, without homologues in the human. Otherwise, the drugs that bind the pathogenic protein may also have unintended binding to human proteins, and consequently cause adverse drug reactions. This study was further extended to the complete organism [56], predicting more than 450 high-confidence targets in *M. tuberculosis*. This study also considered expression data to refine the targets, and considered the possibility of adverse drug reactions arising out of possible drugs binding off-target to gut flora proteins or 'anti-targets' in the human host.

10.6.2.1 Human metabolic models

Beginning with Recon 1 [57], till the more recent Recon 3D [58], a number of increasingly detailed models of human metabolism have been reconstructed. A large number of studies have used the various available human metabolic models, to derive interesting insights into both health and disease [59, 60]. Nielsen and co-workers built a consensus hepatocyte metabolic model *i*Hepatocytes2322 [61], based on the Human Metabolic Reaction database (HMR version 2.0) and proteomics data in Human Protein Atlas (https://www.proteinatlas.org; [62]). Integrating clinical data from non-alcoholic fatty liver disease patients into the hepatocyte model, they identified key disease-specific biomarkers.

The most recent work in this direction, by Thiele *et al* [63], presents very detailed gender-specific reconstructions of human metabolic networks—"Harvey" and "Harvetta" models representing male and female physiologies, respectively—accounting for as many as 26 different organs and six blood cell types. In all, the reconstruction comprised over 80,000 reactions resolved across the organs. The model is able to provide insights into inter-organ metabolic cycles and predict biomarkers for inherited metabolic diseases. Notably, the model also provided the first stepping stone towards the analysis of host–microbiome co-metabolism.

Other studies have studied cancer-specific metabolic models, comparing them with those of healthy cells, to identify critical metabolic differences, or cancer-specific metabolic signatures, and consequently, biomarkers. Nielsen and co-workers [64] identified key reporter metabolites, which were preferentially enriched only in cancer cells, suggesting unique reactions that could be targeted in anti-cancer therapies. Sahoo and co-workers [65] have studied retinoblastoma, a childhood eye cancer, and identified essential reactions as well as synthetic lethals unique to retinoblastoma.

Beyond merely studying individual genes/metabolites, these studies enable the comparison of healthy and diseased cells at a systems-level, presenting a more holistic picture of the changes taking place during disease. Ultimately, such models serve to better understand disease aetiology, and consequently, identify novel therapeutic strategies.

10.7 LIMITATIONS OF CONSTRAINT-BASED APPROACHES

Like any computational/modelling approach, constraint-based approaches also have their limitations. These limitations stem from the various *components* of a constraint-based model—the metabolic network itself, annotations and GPR associations, the extent of curation, the constraints themselves, knowledge of thermodynamics, the objective function used, the accuracy of environment/medium composition, the biomass composition, ATP maintenance fluxes, and so on! Fundamentally, we mostly assume a steady-state system, with no room for metabolite accumulation. Also, it is essential to remember the scope of constraint-based models—they predict the theoretical capabilities of a metabolic network, with little information on the kinetics, or other complexities arising from the other *layers* of regulation or signalling.

While constraint-based models have often performed admirably in both predicting steady-state growth rates and gene essentiality or even synthetic lethality, they may fall short where regulatory effects predominate.

For instance, as discussed in §9.4, FBA will predict a concurrent uptake of lactose and glucose, as the regulatory effects of the *lac* operon are often not modelled. Nevertheless, constraint-based models can be augmented with regulatory information (§9.4) or even integrated with signalling networks, although such efforts have been few and far between. Also, there is no explicit representation of metabolite concentrations in constraint-based models. A major scope for future improvement of constraint-based modelling lies in the integration of other kinds of networks, such as regulatory or signalling networks, alongside metabolism.

A detailed discussion of the strengths and limitations of constraint-based models is available elsewhere [38]. We will now focus on some of the reasons why incorrect predictions are made by constraint-based models.

10.7.1 Incorrect predictions

An important aspect of discussion is the prediction of growth rates, or simpler still, the prediction of metabolic network *viability*, in a given medium (environment), or following gene deletions/perturbations. Two types of erroneous prediction scenarios commonly arise [38]:

G/NG where the model predicts growth (G), but no growth (NG) is experimentally observed (*false positive*), and

NG/G where the model predicts no growth, but growth is observed experimentally (*false negative*).

G/NG false positives arise from an over-estimation of the organism's metabolic capabilities. Possible causes include:

1. The biomass equation may be missing one or more metabolites essential for cell growth
2. The gene deleted may have a non-metabolic essential function
3. The gene deletion/perturbation may cause the accumulation of some toxic by-products that ultimately prove lethal to the cell, which is not captured in the model
4. Regulatory effects that cause certain reactions to not function under the given conditions
5. ATP maintenance flux is incorrectly calibrated (*i.e.*, insufficient)
6. Isozymes are often not created equal—one of them may be more dominant; this is not captured in current constraint-based models. Even when the dominant isozyme is deleted, the model would predict growth, whereas *in vivo*, the minor isozyme may not be able to rescue the strain [66].

NG/G false negatives arise from an under-estimation of the organism's metabolic capabilities. Possible causes include:

1. The biomass equation may include one or more metabolites not essential for cellular growth
2. There exist alternate pathways that are hitherto unknown, or unknown isozymes—knowledge gaps, as also regulatory effects that activate alternate pathways

Notably, the study of these false negatives can aid in the design of new targeted experiments that ultimately uncover new biology; such studies have been carried out in organisms ranging from *E. coli*, to yeast, to human, as detailed in [39].

Other limitations that we have discussed elaborately in previous chapters include the existence of multiple optima, limiting our ability to make accurate inferences about the exact physiological state of the cell. The objective function is often a point of contention, both in terms of its nature (linear, quadratic, etc.) as well as composition, as it significantly impacts predictions. By and large, these incorrect model predictions stem from knowledge gaps and model incompleteness, rather than fundamental flaws in simulation techniques.

All these limitations notwithstanding, constraint-based models provide excellent insights at a systems-level, into complex networks. Many other paradigms such as dynamic modelling, while can give accurate predictions of regulatory effects and the like, are very limited in terms of the size of the systems they can handle. Thus, constraint-based modelling methods provide a first glimpse of emergent behaviours in complex systems, and also provide an excellent framework for the integration of omics data and coupling with regulatory/signalling networks within a cell.

10.8 TROUBLESHOOTING

10.8.1 Interpreting gene deletion simulations

As discussed in §10.1, we can simulate the deletion of genes by constraining the flux through the corresponding reactions to zero (similar to Equation 10.2). We then solve the modified LP problem, to obtain a solution. An obvious scenario is that the LP terminates successfully, and an optimum is found. Let the flux distribution corresponding to this optimum be $\mathbf{x}$. If $\mathbf{c}$ is the objective, then $\mathbf{c}^T\mathbf{v} = v_{bio}$ will yield the objective function value. If this is zero, then, it means that we have a lethal phenotype. That is, the optimal flux through the system that admits all the input constraints is $\mathbf{v} = \mathbf{0}$. Another possibility is that v_{bio} is non-zero, but *very low*. As discussed in §10.1, deletion phenotypes can also be determined by comparing the objective flux with a cut-off value, typically 1% or 5% of the wild-type objective flux ($v_{bio,WT}$).

A third possibility exists, where the LP problem is *infeasible*. That is, it is not possible to solve the LP, under the given set of constraints. Why would this occur? At first, glance, $\mathbf{v} = \mathbf{0}$ seems to be a feasible solution to any FBA problem, as it satisfies $\mathbf{Sv} = \mathbf{0}$. The FBA formulation in Equation 10.3 also seems to admit $\mathbf{v} = \mathbf{0}$, when *typical* lower and upper bounds are used, where 0 is either an admissible lower bound (in case of irreversible reactions), or is an admissible value, in case of irreversible reactions ($-\infty \leqslant v_i \leqslant \infty$). The problem comes when there are certain constraints that specify non-zero lower bounds for one or more fluxes. Typically, this happens with the fluxes such as NGAM (non-growth associated ATP maintenance flux). In the case of *i*AF1260, this value is 8.39 mmol/gDW/h, which may not be satisfied under certain deletion constraints, leading to infeasible LPs. Thus, infeasible LPs can also be regarded as indicators of lethal phenotypes, or essentially, zero objective function values.

10.9 SOFTWARE TOOLS

Fast-SL [23, 24] implements the Fast-SL algorithm for efficiently identifying synthetic lethals up to the order of four.

MCSEnumerator [20] is available as part of the CellNetAnalyzer (see p. 195).

OptKnock is available as part of the COBRA toolbox itself (OptKnock.m).

OptFlux [7] is available from http://www.optflux.org/.

EXERCISES

10.1 In the genome-scale metabolic network *i*NJ661, corresponding to *Mycobacterium tuberculosis*, the reactions corresponding to CTP synthases, CTPS1 and CTPS2 form a synthetic lethal reaction pair. If you performed an FVA, what will be the minimal permissible flux through each of these reactions, for a growth rate that is at least 5% of wild-type (WT) growth rate? You can indicate the values as a fraction of WT fluxes. Verify your answers by actually performing an FVA.

10.2 Consider a network with the following reactions and gene–protein–reaction relationships as indicated:

R1:	$A + B \rightarrow C$	g_1 or g_2
R2:	$C \rightarrow D + E$	g_1 and g_3
R3:	$B \rightarrow E$	g_4
R4:	$E \rightarrow F$	g_5

Which of the following statements is/are true in a medium containing A and B, if the metabolite F is required for cell survival?

a. g_1 is a single lethal

b. g_1 and g_5 are double lethals

c. g_1 and g_4 are double lethals

d. g_5 is a single lethal

[LAB EXERCISES]

10.3 The *E. coli* core model has two transketolases, 'TKT1' and 'TKT2'. Delete each of the reactions individually and compute the growth rate. Also delete both reactions simultaneously from the model and predict the growth rate. What do you infer?

10.4 In the above problem, what is the difference between the 'flux states' of the cell when TKT1 is active, versus when TKT2 is active. How will you identify the minimal set of reactions that need to be altered as the cell switches from using TKT1 to TKT2?

10.5 Perform FSEOF on *E. coli* to identify possible targets for over-expression, for the over-production of taxadiene, as discussed in reference [26]. Compute f_{ph} for various over-expression targets. Also try to identify if there are possible deletion targets, using OptKnock.

10.6 Obtain the model for *Mycobacterium tuberculosis* *i*NJ661 from the BiGG database. What is the number of single gene deletions that lead to lethal phenotypes? Consider deletions where growth rate falls to less than 5% of the wild-type growth rate to be lethal.

10.7 For the same model, identify all the synthetic double lethals using Fast-SL, using the same cut-off for lethality. These lethals represent genetic interactions [67]—how many of these gene pairs already have edges between them in the STRING database?

10.8 Identify a 'minimal reactome' for the iML1515 model. A minimal reactome must be such that the removal of *any* reaction from the network results in a lethal phenotype. Can you see that the minimal reactome is not merely a collection of all singly essential reactions from the metabolic network? Further, can you think of efficient ways to compute this?

10.9 The Recon2 model of human metabolism [68] defines 354 metabolic *tasks* ranging from the production of various amino acids, haeme synthesis to biomass production. Identify essential genes in the human metabolic network corresponding to each of these tasks. Comment on any interesting observations.

10.10 Many studies have used constraint-based methods to study the metabolic networks of pathogenic organisms, including strain-specific reconstructions (see Epilogue, p. 311). Pick one such study from literature, and write a critical summary describing its methodologies and main findings.

REFERENCES

[1] J. S. Edwards and B. Ø. Palsson, "Metabolic flux balance analysis and the *in silico* analysis of *Escherichia coli* K-12 gene deletions", *BMC Bioinformatics* **1**:1 (2000).

[2] I. Famili, J. Forster, J. Nielsen, and B. Ø. Palsson, "*Saccharomyces cerevisiae* phenotypes can be predicted by using constraint-based analysis of a genome-scale reconstructed metabolic network", *Proceedings of the National Academy of Sciences USA* **100**:13134–13139 (2003).

[3] D. J. Beste, T. Hooper, G. Stewart, et al., "GSMN-TB: a web-based genome-scale network model of *Mycobacterium tuberculosis* metabolism", *Genome Biology* **8**:R89 (2007).

[4] M. A. Oberhardt, J. Puchałka, K. E. Fryer, V. A. Martins dos Santos, and J. A. Papin, "Genome-scale metabolic network analysis of the opportunistic pathogen *Pseudomonas aeruginosa* PAO1", *Journal of Bacteriology* **190**:2790–2803 (2008).

[5] S. A. Becker and B. Ø. Palsson, "Genome-scale reconstruction of the metabolic network in *Staphylococcus aureus* N315: an initial draft to the two-dimensional annotation", *BMC Microbiology* **5**:8 (2005).

[6] S. Tymoshenko, R. D. Oppenheim, R. Agren, J. Nielsen, D. Soldati-Favre, and V. Hatzimanikatis, "Metabolic needs and capabilities of *Toxoplasma gondii* through combined computational and experimental analysis", *PLoS Computational Biology* **11**:e1004261+ (2015).

[7] I. Rocha, P. Maia, P. Evangelista, et al., "OptFlux: an open-source software platform for *in silico* metabolic engineering", *BMC Systems Biology* **4**:45+ (2010).

[8] J. Kim and J. L. Reed, "OptORF: optimal metabolic and regulatory perturbations for metabolic engineering of microbial strains", *BMC Systems Biology* **4**:53+ (2010).

[9] S. Klamt and E. D. Gilles, "Minimal cut sets in biochemical reaction networks", *Bioinformatics* **20**:226–234 (2004).

[10] S. L. L. Ooi, X. Pan, B. D. Peyser, et al., "Global synthetic-lethality analysis and yeast functional profiling", *Trends in Genetics* **22**:56–63 (2006).

[11] K.-C. Hsu, W.-C. Cheng, Y.-F. Chen, W.-C. Wang, and J.-M. Yang, "Pathway-based screening strategy for multitarget inhibitors of diverse proteins in metabolic pathways", *PLoS Computational Biology* **9**:e1003127+ (2013).

[12] W. G. Kaelin Jr, "The concept of synthetic lethality in the context of anticancer therapy", *Nature Reviews Cancer* **5**:689–698 (2005).

[13] F. L. Muller, E. A. Aquilanti, and R. A. DePinho, "Collateral lethality: a new therapeutic strategy in oncology", *Trends in Cancer* **1**:161–173 (2015).

[14] A. H. Y. Tong, M. Evangelista, A. B. Parsons, et al., "Systematic genetic analysis with ordered arrays of yeast deletion mutants", *Science* **294**:2364–2368 (2001).

[15] A. H. Y. Tong, G. Lesage, G. D. Bader, et al., "Global mapping of the yeast genetic interaction network", *Science* **303**:808–813 (2004).

[16] R. Harrison, B. Papp, C. Pál, S. G. Oliver, and D. Delneri, "Plasticity of genetic interactions in metabolic networks of yeast", *Proceedings of the National Academy of Sciences USA* **104**:2307–2312 (2007).

[17] D. Deutscher, I. Meilijson, M. Kupiec, and E. Ruppin, "Multiple knockout analysis of genetic robustness in the yeast metabolic network", *Nature Genetics* **38**:993–8 (2006).

[18] C. S. Henry, F. Xia, and R. Stevens, "Application of high-performance computing to the reconstruction, analysis, and optimization of genome-scale metabolic models", *Journal of Physics: Conference Series* **180**:012025 (2009).

[19] P. F. Suthers, A. Zomorrodi, and C. D. Maranas, "Genome-scale gene/reaction essentiality and synthetic lethality analysis", *Molecular Systems Biology* **5**:301 (2009).

[20] A. von Kamp and S. Klamt, "Enumeration of smallest intervention strategies in genome-scale metabolic networks", *PLoS Computational Biology* **10**:e1003378 (2014).

[21] K. Ballerstein, A. von Kamp, S. Klamt, and U. U. Haus, "Minimal cut sets in a metabolic network are elementary modes in a dual network", *Bioinformatics* **28**:381–387 (2012).

[22] L. F. de Figueiredo, A. Podhorski, A. Rubio, et al., "Computing the shortest elementary flux modes in genome-scale metabolic networks", *Bioinformatics* **25**:3158–3165 (2009).

[23] A. Pratapa, S. Balachandran, and K. Raman, "Fast-SL: an efficient algorithm to identify synthetic lethal sets in metabolic networks", *Bioinformatics* **31**:3299–3305 (2015).

[24] K. Raman, A. Pratapa, O. Mohite, and S. Balachandran, "Computational prediction of synthetic lethals in genome-scale metabolic models using Fast-SL", *Methods in Molecular Biology* **1716**:315–336 (2018).

[25] H. S. Choi, S. Y. Lee, T. Y. Kim, and H. M. Woo, "*In silico* identification of gene amplification targets for improvement of lycopene production", *Applied and Environmental Microbiology* **76**:3097–3105 (2010).

[26] B. A. Boghigian, J. Armando, D. Salas, and B. A. Pfeifer, "Computational identification of gene over-expression targets for metabolic engineering of taxadiene production", *Applied Microbiology and Biotechnology* **93**:2063–2073 (2012).

[27] J. Nocon, M. G. Steiger, M. Pfeffer, et al., "Model based engineering of *Pichia pastoris* central metabolism enhances recombinant protein production", *Metabolic Engineering* **24**:129–138 (2014).

[28] V. Razmilic, J. F. Castro, B. Andrews, and J. A. Asenjo, "Analysis of metabolic networks of *Streptomyces leeuwenhoekii* C34 by means of a genome scale model: prediction of modifications that enhance the production of specialized metabolites", *Biotechnology and Bioengineering* **115**:1815–1828 (2018).

[29] A. Badri, K. Raman, and G. Jayaraman, "Uncovering novel pathways for enhancing hyaluronan synthesis in recombinant *Lactococcus lactis*: genome-scale metabolic modeling and experimental validation", *Processes* **7**:343 (2019).

[30] A. Srinivasan, V. S, K. Raman, and S. Srivastava, "Rational metabolic engineering for enhanced alpha-tocopherol production in *Helianthus annuus* cell culture", *Biochemical Engineering Journal* **151**:107256 (2019).

[31] J. Park, H. Park, W. Kim, H. Kim, T. Kim, and S. Lee, "Flux variability scanning based on enforced objective flux for identifying gene amplification targets", *BMC Systems Biology* **6**:106+ (2012).

[32] A. R. Brochado, C. Matos, B. L. Moller, J. Hansen, U. H. Mortensen, and K. R. Patil, "Improved vanillin production in baker's yeast through *in silico* design", *Microbial Cell Factories* **9**:84–84 (2010).

[33] Z. Sun, H. Meng, J. Li, et al., "Identification of novel knockout targets for improving terpenoids biosynthesis in *Saccharomyces cerevisiae*", *PLoS ONE* **9**:e112615 (2014).

[34] S. Chatsurachai, C. Furusawa, and H. Shimizu, "An *in silico* platform for the design of heterologous pathways in nonnative metabolite production", *BMC Bioinformatics* **13**:93 (2012).

[35] P. Carbonell, P. Parutto, J. Herisson, S. B. Pandit, and J.-L. L. Faulon, "XTMS: pathway design in an eXTended metabolic space", *Nucleic Acids Research* **42**:W389–W394 (2014).

[36] P. Xu, S. Ranganathan, Z. L. Fowler, C. D. Maranas, and M. A. Koffas, "Genome-scale metabolic network modeling results in minimal interventions that cooperatively force carbon flux towards malonyl-CoA", *Metabolic Engineering* **13**:578–587 (2011).

[37] B. A. Boghigian, K. Lee, and B. A. Pfeifer, "Computational analysis of phenotypic space in heterologous polyketide biosynthesis—Applications to *Escherichia coli*, *Bacillus subtilis*, and *Saccharomyces cerevisiae*", *Journal of Theoretical Biology* **262**:197–207 (2010).

[38] D. McCloskey, B. Ø. Palsson, and A. M. Feist, "Basic and applied uses of genome-scale metabolic network reconstructions of *Escherichia coli*", *Molecular Systems Biology* **9**:661 (2013).

[39] E. J. O'Brien, J. M. Monk, and B. Ø. Palsson, "Using genome-scale models to predict biological capabilities", *Cell* **161**:971–987 (2015).

[40] A. Bordbar, J. M. Monk, Z. A. King, and B. Ø. Palsson, "Constraint-based models predict metabolic and associated cellular functions", *Nature Reviews Genetics* **15**:107–120 (2014).

[41] M. R. Long, W. K. K. Ong, and J. L. Reed, "Computational methods in metabolic engineering for strain design", *Current Opinion in Biotechnology* **34**:135–141 (2015).

[42] B. Kim, W. Kim, D. Kim, and S. Lee, "Applications of genome-scale metabolic network model in metabolic engineering", *Journal of Industrial Microbiology & Biotechnology* **42**:339–348 (2015).

[43] E. Simeonidis and N. D. Price, "Genome-scale modeling for metabolic engineering", *Journal of Industrial Microbiology & Biotechnology* **42**:327–338 (2015).

[44] J. Yen, I. Tanniche, A. Fisher, G. Gillaspy, D. Bevan, and R. Senger, "Designing metabolic engineering strategies with genome-scale metabolic flux modeling", *Advances in Genomics and Genetics*:93–105 (2015).

[45] H. Alper, Y.-S. S. Jin, J. F. Moxley, and G. Stephanopoulos, "Identifying gene targets for the metabolic engineering of lycopene biosynthesis in *Escherichia coli*", *Metabolic Engineering* **7**:155–164 (2005).

[46] D. S. Lun, G. Rockwell, N. J. Guido, et al., "Large-scale identification of genetic design strategies using local search", *Molecular Systems Biology* 5:296 (2009).

[47] N. Tepper and T. Shlomi, "Predicting metabolic engineering knockout strategies for chemical production: accounting for competing pathways", *Bioinformatics* 26:536–543 (2010).

[48] S. Ranganathan, P. F. Suthers, and C. D. Maranas, "OptForce: an optimization procedure for identifying all genetic manipulations leading to targeted overproductions", *PLoS Computational Biology* 6:e1000744+ (2010).

[49] A. Badri, A. Srinivasan, and K. Raman, "*In Silico* approaches to metabolic engineering", in *Current Developments in Biotechnology and Bioengineering* (2016).

[50] Y. Seif, E. Kavvas, J.-C. Lachance, et al., "Genome-scale metabolic reconstructions of multiple *Salmonella* strains reveal serovar-specific metabolic traits", *Nature Communications* 9:3771 (2018).

[51] A. Raghunathan, J. Reed, S. Shin, B. Ø. Palsson, and S. Daefler, "Constraint-based analysis of metabolic capacity of *Salmonella typhimurium* during host-pathogen interaction", *BMC Systems Biology* 3:38+ (2009).

[52] G. Plata, T.-L. Hsiao, K. L. Olszewski, L. Manuel, and D. Vitkup, "Reconstruction and flux-balance analysis of the *Plasmodium falciparum* metabolic network", *Molecular Systems Biology* 6:408 (2010).

[53] G. J. E. Baart, B. Zomer, A. de Haan, et al., "Modeling *Neisseria meningitidis* metabolism: from genome to metabolic fluxes", *Genome Biology* 8:R136 (2007).

[54] A. K. Chavali, K. M. D'Auria, E. L. Hewlett, R. D. Pearson, and J. A. Papin, "A metabolic network approach for the identification and prioritization of antimicrobial drug targets", *Trends in Microbiology* 20:113–123 (2012).

[55] K. Raman, P. Rajagopalan, and N. Chandra, "Flux balance analysis of mycolic acid pathway: targets for Anti-Tubercular drugs", *PLoS Computational Biology* 1:e46+ (2005).

[56] K. Raman, K. Yeturu, and N. Chandra, "targetTB: a target identification pipeline for *Mycobacterium tuberculosis* through an interactome, reactome and genome-scale structural analysis", *BMC Systems Biology* 2:109+ (2008).

[57] N. C. Duarte, S. A. Becker, N. Jamshidi, et al., "Global reconstruction of the human metabolic network based on genomic and bibliomic data", *Proceedings of the National Academy of Sciences USA* 104:1777–1782 (2007).

[58] E. Brunk, S. Sahoo, D. C. Zielinski, et al., "Recon3D enables a three-dimensional view of gene variation in human metabolism", *Nature Biotechnology* 36:272– (2018).

[59] J. Geng and J. Nielsen, "*In silico* analysis of human metabolism: reconstruction, contextualization and application of genome-scale models", *Current Opinion in Systems Biology* **2**:28–37 (2017).

[60] A. Mardinoglu and J. Nielsen, "New paradigms for metabolic modeling of human cells", *Current Opinion in Biotechnology* **34**:91–97 (2015).

[61] A. Mardinoglu, R. Agren, C. Kampf, A. Asplund, M. Uhlen, and J. Nielsen, "Genome-scale metabolic modelling of hepatocytes reveals serine deficiency in patients with non-alcoholic fatty liver disease", *Nature Communications* **5**:3083 (2014).

[62] M. Uhlén, L. Fagerberg, B. M. Hallström, et al., "Tissue-based map of the human proteome", *Science* **347**:1260419 (2015).

[63] I. Thiele, S. Sahoo, A. Heinken, et al., "Personalized whole-body models integrate metabolism, physiology, and the gut microbiome", *Molecular Systems Biology* **16**:e8982 (2020).

[64] R. Agren, S. Bordel, A. Mardinoglu, N. Pornputtapong, I. Nookaew, and J. Nielsen, "Reconstruction of genome-scale active metabolic networks for 69 human cell types and 16 cancer types using INIT", *PLoS Computational Biology* **8**:e1002518+ (2012).

[65] S. Sahoo, R. K. Ravi Kumar, B. Nicolay, et al., "Metabolite systems profiling identifies exploitable weaknesses in retinoblastoma", *FEBS Letters* **593**:23–41 (2019).

[66] A. R. Joyce and B. Ø. Palsson, "Predicting gene essentiality using genome-scale *in silico* models", *Methods in Molecular Biology* **416**:433–457 (2008).

[67] R. Mani, R. P. St.Onge, J. L. Hartman, G. Giaever, and F. P. Roth, "Defining genetic interaction", *Proceedings of the National Academy of Sciences USA* **105**:3461–3466 (2008).

[68] I. Thiele, N. Swainston, R. M. T. Fleming, et al., "A community-driven global reconstruction of human metabolism", *Nature biotechnology* **31**:419–425 (2013).

FURTHER READING

P. A. Saa, M. P. Cortés, J. López, D. Bustos, A. Maass, and E. Agosin, "Expanding metabolic capabilities using novel pathway designs: computational tools and case studies", *Biotechnology Journal*:1800734 (2019).

X. Fang, C. J. Lloyd, and B. Ø. Palsson, "Reconstructing organisms *in silico*: genome-scale models and their emerging applications", *Nature Reviews Microbiology*:731–743 (2020).

IV

Advanced Topics

Modelling cellular interactions

CONTENTS

W<span></span>E have thus far studied several paradigms to model various cellular networks. Typically, these paradigms have focussed on the networks of a single *type of* cell, *i.e.* a single species. However, microbes seldom exist in isolation. In practically every ecosystem, microbes thrive in complex and diverse communities—known as microbiomes. A variety of microbiomes have been studied and characterised, such as the soil microbiome [1], or the ocean microbiome [2], and in the human body, the gut microbiome [3], or the skin microbiome [4]. These microbiomes comprise a variety of micro-organisms, ranging from viruses, archaea, and bacteria, to fungi, protists, and other eukaryotes.

How do we extend the approaches that we have studied thus far, to modelling not a single cell, but multiple microbes interacting in natural/synthetic communities? In this chapter, we will overview some of the methods for modelling microbial communities. Once again, we will pay special attention to metabolic networks, as metabolic exchanges and interactions predominate and drive most microbial communities.

11.1 MICROBIAL COMMUNITIES

Microbial communities have been modelled for two broad purposes: (i) to understand microbial interactions in various natural communities (*e.g.* the gut microbiome), and (ii) to engineer microbial interactions for a particular purpose, such as metabolic engineering. A rich repertoire of mathematical and computational tools have been developed over the last decade or so, for modelling various aspects of microbial interactions in communities.

These methods can be broadly classified into four paradigms: (i) network-based, (ii) population-based, (iii) individual-based, and (iv) constraint-based modelling techniques, based on the level of abstraction, the nature of problem formulation and also the applications. Each of these modelling techniques considers a different set of assumptions, and focusses on modelling specific aspects of a community. Together, the methods seek to understand the abundances of different species, their abilities to produce various metabolites in one or more environmental niches, the competition between species for resources, the ability of the species to cross-feed, microbial associations, and so on. Figure 11.1 presents an overview of these key paradigms for modelling microbial communities. Table 11.1 samples some of the many useful modelling approaches that have been developed in recent years, to understand and characterise microbial interactions in communities.

11.1.1 Network-based approaches

Network-based or graph-theoretic approaches have been employed in two complementary strategies to understand model communities. One strategy more broadly deals with characterisation, through the construction of microbial association networks, primarily from metagenomics datasets. The other strategy involves the construction of the more familiar metabolic networks, but built to capture metabolite exchanges in communities.

11.1.1.1 Microbial association networks

The recent years have seen a rise in high-throughput sequencing of *metagenomes*, *i.e. all* microbial genomes in a sample, beyond only those that are culturable. Shotgun metagenomics[1] is one such technology that can be used to profile the taxonomic composition of complex microbial communities. With a wealth of metagenomic data emerging from various high-throughput sequencing studies, it is possible to construct microbial association networks based on the relative abundances of the organisms and their co-occurrences. Typically, the raw sequencing data are processed in a pipeline such as QIIME [25]. The sequences are clustered into operational taxonomic units (OTUs)[2] using a suitable threshold, typically 97%, or lower, *e.g.* 90% [26]. Following this, a variety of statistical and network analyses are carried out.

von Mering and co-workers reported a very interesting study analysing the co-existence of microbes across metagenomic datasets obtained from different environments [27]. After identifying statistically significant co-occurrences, they performed a network inference followed by Markov clustering, to identify clusters of co-occurring lineages. Several statistical approaches are possible to analyse the co-occurrences, and also to perform network inference, as nicely reviewed in [28].

---

[1]Other approaches such as high-throughput 16S rRNA gene sequencing do not target the entire genomic content of a sample, and truly do not generate 'metagenomic' data [24].

[2]Essentially a proxy for species in large metagenomic datasets with many unknown organisms.

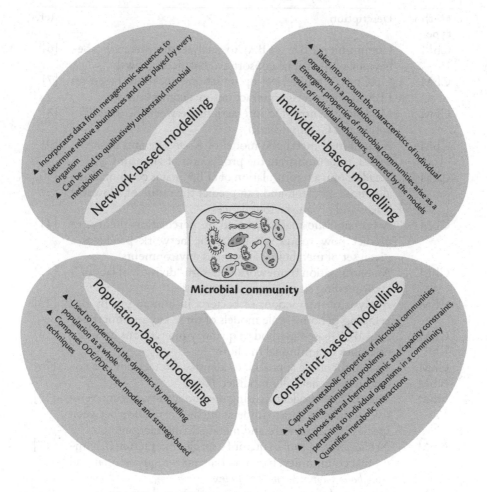

Figure 11.1 **Overview of techniques for modelling microbial communities.** Microbial communities can be broadly modelled using four different types of techniques. Network-based modelling techniques seek to understand the association and qualitatively decipher the interactions between the microbes in a community. Population-based and individual-based models capture the population and spatial dynamics at a collective and single organism level, respectively. Constraint-based modelling techniques have also been used to predict the metabolic phenotype of microbial communities. Adapted by permission from [5]. © (2018) CRC Press.

TABLE 11.1 Some recent advances in modelling microbial communities. Adapted by permission from [5]. © (2018) CRC Press.

Method type	Description	Ref.
CbM	Community FBA (cFBA) to analyse the metabolic behaviour of microbial consortia at balanced growth	[6]
CbM	OptCom, a multi-level optimisation procedure based on FBA to characterise interactions in a community	[7]
CbM	Modelling syntrophic interactions between *G. metallireducens* and *G. sulfurreducens*	[8]
CbM	A combined metabolic model of *K. vulgare* and *B. megaterium*, to demonstrate the production of vitamin C	[9]
CbM	SteadyCom, a reformulation of cFBA that is computationally more efficient, and can predict the range of allowable fluxes and organism abundances	[10]
CbM	A probabilistic method, based on percolation theory, to predict how robustly a metabolic network produces a given set of metabolites in various environments	[11]
CbM	An optimisation framework to design "division of labour"/cross-feeding strategies in communities	[12]
CbM	Analysis of interactions of bacteria in human gut by developing genome-scale models of three key gut microbes	[13]
CbM	An MILP-based method to quantify interactions between microbes in a minimal medium	[14]
CbM, NbM	Graph-theoretic approach to identify interactions in a co-culture of genetically modified *E. coli*	[15]
NbM	Microbiome modelling using metabolic network and relative abundance data to understand the correlation between species co-occurrence and metabolic interactions	[16]
NbM	*NetCooperate*, a tool to quantify species interaction in a microbial community, based on *Biosynthetic Support Score* and *Metabolic Complementarity Index*	[17]
NbM	Analysis of co-occurrence networks of soil microbes	[18]
NbM	Host–pathogen interaction studied using a graph-based algorithm that uses the data from metabolic network	[19]
NbM	*MetQuest*, a dynamic programming-based algorithm to enumerate biosynthetic pathways in communities and *Metabolic Support Index*, to quantify microbial interactions	[20] [21]
IbM	*BacSim* to model the behaviour of microbes using individual properties	[22]
PbM	Understanding co-operations during range expansions using partial differential equations (PDEs)	[23]

CbM: Constraint-based methods, NbM: Network-based methods, IbM: Individual-based methods, PbM: Population-based methods

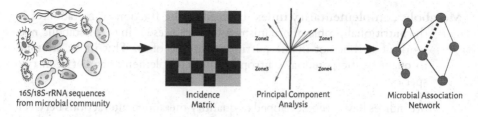

16S/18S-rRNA sequences
from microbial community

Incidence
Matrix

Principal Component
Analysis

Microbial Association
Network

Figure 11.2 Inference of microbial association networks from metagenomic data. Beginning with 16S/18S rRNA sequences from microbial communities (across space/time), an incidence matrix is constructed. Various statistical procedures are typically employed to construct a microbial association network, with different types of edges (positive and negative relationships, represented by continuous and dotted lines, respectively) and edge weights that denote the strength of either type of relationship. Adapted by permission from [5]. © (2018) CRC Press.

Figure 11.2 illustrates a typical workflow to construct microbial association networks from metagenomic data. The networks thus constructed can be analysed in various ways, including the methodologies/parameters detailed in Chapter 3.

11.1.1.2 Community metabolic networks

In a complementary approach to the construction of microbial association networks, it is possible to study the metabolic underpinnings of microbial interactions in communities through models of 'community metabolic networks'. These are essentially a set union of individual organism metabolic networks, besides accounting for the exchanges of metabolites between various organisms through an extracellular medium. These networks are typically studied by examining how a pair of organisms can compete with or complement one another in a community, through the exchange of metabolites.

Metabolic competition and complementarity Ruppin and co-workers have developed a reverse ecology framework [29] to identify a set of seed compounds that an organism must exogenously acquire from its environment in order to grow. This set of compounds essentially represents the nutritional profile of a given organism; building on this concept, Levy and Borenstein [16] developed a pair of indices to quantify the interactions between microbes in the human microbiome:

Metabolic competition index Defined as the fraction of metabolites in the nutritional profile of organism A, overlapping with those in the nutritional profile of organism B. This index seeks to measure the potential level of competition between two species.

Metabolic complementarity index Defined as the fraction of metabolites in the nutritional profile of organism A that are present in the metabolic network of B, but not in the nutritional profile of B. This index attempts to quantify the potential syntrophy[3] and complementarity between two species.

More indices have been developed to quantify metabolic interactions between organisms, *e.g. effective metabolic overlap* [30] and *biosynthetic support score* [17], which seek to capture the level of competition and the extent of nutritional support between two species, respectively.

Enumerating metabolic pathways across organisms MetQuest [20] is an algorithm that enumerates all possible pathways up to a given size, between a set of source and target metabolites in a given metabolic network. MetQuest takes as input the metabolic network corresponding to an organism, a set of seed metabolites[4], a set of target metabolites that need to be produced (*e.g.* biomass components) and an integer that bounds the size of the pathways enumerated. The algorithm consists of two phases:

(i) a *guided* breadth-first-search to identify the *scope* of the metabolic network, *i.e.* the set of all metabolites that can be produced from a given 'seed set' of metabolites.

(ii) a dynamic programming-based approach to assemble reactions into pathways of various sizes producing the target metabolite(s) of interest.

MetQuest uses a bipartite graph representation for metabolic networks (recall §2.5.5.3). It also efficiently handles cycles and branched pathways. MetQuest scales well to large metabolic networks, and was demonstrated on a community metabolic network corresponding to a previously studied community [31] comprising three organisms—*Clostridium cellulolyticum, Desulfovibrio vulgaris,* and *Geobacter sulfurreducens.* The resulting bipartite graph contained 14,265 nodes and 29,073 edges, from which pathways of size 20 were computed.

On the basis of the paths generated from MetQuest, Ravikrishnan *et al* defined an index, known as the *Metabolic Support Index* (MSI) [21] to quantify the metabolic interactions taking place between a pair of organisms. The index essentially computes the fraction of reactions in a metabolic network that are newly functional in the presence of another organism, as follows:

$$\text{MSI}(A|A \cup B) = 1 - \frac{n_{\text{stuck},A|A \cup B}}{n_{\text{stuck},A|A}} \tag{11.1}$$

---

[3]Syntrophy, or 'cross-feeding', denotes one species living off of metabolites produced by another.
[4]Note that the seed metabolites described here are distinct from the seed set/nutritional profile described in the previous section. Here, they account for the carbon source and other co-factors, akin to the medium of growth for an organism/community.

where $A \cup B$ denotes the metabolic network corresponding to the two organisms A and B in a community, A denotes the (individual) metabolic network of organism A, $n_{\text{stuck},A|A}$ denotes the number of *stuck*[5] reactions in A, and $n_{\text{stuck},A|A \cup B}$ denotes the number of stuck reactions of A in the community $A \cup B$. MSI thus quantifies the fraction of reactions stuck in the original metabolic network, but relieved in the presence of another organism in a community. Note that this definition is *asymmetric*; MSI$(A|A \cup B)$ is distinct from MSI$(B|A \cup B)$. An MSI of 0 means that there is no support afforded by the community to the individual organism; while an MSI of 1 indicates that all reactions that were originally stuck in the individual organism can be active when present in the community. MetQuest and MSI have been used to understand microbial interactions in the gut microbiome associated with recovery post-antibiotic treatment [32].

11.1.2 Population-based and agent-based approaches

Population-based approaches to model microbial communities capture the spatio-temporal dynamics of communities using ODEs or PDEs. The classic Lotka–Volterra model (see §1.5.1) can be extended to model microbial interactions in communities.

11.1.2.1 Generalised Lotka–Volterra (gLV) model

The gLV model is described by the following equations:

$$\frac{dx_i(t)}{dt} = x_i(t) \left(b_i + \sum_{j=1}^{n} a_{ij}x_j(t) \right) \qquad i = 1, \ldots, n \qquad (11.2)$$

where n is the number of organisms in the community, x_i is the abundance of organism i, b_i is the intrinsic growth rate of organism i, and a_{ij} capture the pairwise interaction coefficients, with a_{ii} accounting for intra-species interactions[6].

The gLV model has been used to study microbial dynamics in the intestine [33] and suggest how groups of microbes in the gut offer protection against *Clostridium difficile*, and how antibiotic perturbations can actually make the host more susceptible to infection. Other studies have also studied bacterial communities that can offer protection against *C. difficile* [34], as well as overall community dynamics in gut, oral and skin microbiomes [35].

Spatial dynamics of microbial communities have been captured using PDE models, as detailed elsewhere [36–39]. PDEs have further been combined with constraint-based modelling techniques such as COMETS (see §11.1.3.4). Other models, based on evolutionary game theory, have also been applied to study microbial interactions in communities [40, 41].

---

[5]reactions that cannot proceed due to the unavailability of one or more metabolites
[6]These terms typically take on negative values to account for intra-species competition.

11.1.3 Constraint-based approaches

The constraint-based approaches we studied in Part III are also widely used to study microbial interactions in communities. Again, constraint-based methods can present insights into fluxes, and abundances, in terms of the relative biomasses of the different species, over and above associations and dependencies indicated by graph-theoretic approaches. Of course, these methods also demand better-curated models with accurate stoichiometries and biomass equations. The modelling of microbiomes has been enabled by the development of excellent large-scale metabolic reconstruction resources such as AGORA [42] and a variety of algorithms. Powerful automated tools such as CarveMe [43] are rapidly enabling the reconstruction of single-species and community metabolic models in a fast and scalable way.

Various methods have been developed, building on the framework of FBA (see §8.3), which vary in their approach to capture the community-level objective function, as well as their approach to characterising the dynamics.

11.1.3.1 Compartmentalised FBA

As a natural extension of compartmentalised models (see §9.5), it is possible to think of each individual organism's genome-scale model as an individual compartment inside a larger 'community model', connected via a common extracellular medium. Such an approach (also referred to as "Joint FBA" [10]) was first employed by Stahl and co-workers [44], where they modelled a mutualistic community comprising *Desulfovibrio vulgaris* and *Methanococcus maripaludis* using a multispecies stoichiometric metabolic model. The objective function used was simply a weighted biomass for both organisms, *e.g.* $v_{bio,Dvu} + 0.1v_{bio,Mma}$, with the function heavily weighted towards the growth of *D. vulgaris*. The community was also cultured experimentally on lactate, in the absence of sulphate; the model could accurately predict various interesting aspects of the community, such as the ratio of *D. vulgaris* to *M. maripaludis*. Klitgord and Segrè [45] extended this approach further, by introducing a fictitious compartment for the extracellular environment shared by both species, in addition to the extracellular spaces for individual species. This method is implemented in the `createMultipleSpeciesModel` method in the Microbiome Modelling Toolbox [46]. Another study, by Freilich, Ruppin and co-workers [47], also adopts a compartmentalised FBA approach to predict the potential for competition/co-operation amongst nearly 7,000 pairs of bacteria.

11.1.3.2 Community FBA (cFBA)

cFBA [6] models communities at metabolic steady-state, *i.e.* the entire consortium of organisms is at steady-state, with the rate of change of every metabolite being zero (steady-state mass balance). This scenario can occur when the net growth of all organisms is zero, or more interestingly, when the organisms all grow at the same *specific growth rate*[7], with a concomitant increase in the exchange rates with the environment, maintaining the mass balance. In such communities, if the organisms do not grow *equally fast*, metabolites that are being exchanged would accumulate (or deplete), violating the steady-state conditions.

11.1.3.3 Multi-level optimisation

Multi-level optimisation approaches (on the lines of the bi-level optimisations described in §9.3) have also been developed, to study microbial communities.

OptCom Maranas and co-workers [7] proposed a method based on multi-level multi-objective optimisation to model microbial communities. The method, OptCom, maximises the fitness of a microbial community, at the same time not compromising on the fitness of each individual organism in the community. Opt-Com circumvents the approach of a single community objective of compartmentalised ("Joint FBA") approaches, which inherently cannot reconcile the trade-offs between individual organism objectives and community-level objectives. This is achieved by means of a multi-level objective function: at the inner level, separate biomass maximisation problems are defined for each of the individual species. At the outer level, there is a community-level objective function, that is an integration of the inner problems, and additionally, the flow of metabolites between organisms. Using this formulation, OptCom predicts a community scenario such that the combined biomass of the community and the individual organisms are maximised. The mass balance equations are identical to compartmentalised FBA described above.

d-OptCom [48] is an extension of OptCom, that also accounts for temporal dynamics of community biomass as well as the kinetics of substrate uptake and the concentrations of shared metabolites.

Community And Systems-level INteractive Optimization (CASINO) Nielsen and co-workers [49] proposed a two-stage multi-level optimisation algorithm to simulate microbial communities in the gut. In the first initialisation stage, species are iteratively *activated*, based on their ability to grow—while *primary* species grow based on system-level inputs, *non-primary* species rely on the metabolites produced by the primary species. An appropriate community topology is constructed, defining the community constraints matrix and the community objective function. The second stage involves community optimisation, through an iterative procedure. In each step, the community biomass is optimised. Relative carbohydrate uptake rates for the systems-level optimum condition are fed to the individual species models to simulate for organism-level optimality. The metabolite secretion rates obtained from the simulation of organism-level models are plugged back to find a new systems-level optimum. This process is iterated until the system converges to an optimum.

---

[7]This is the rate of increase of the biomass of a cell population, *per unit biomass*.

This method is implemented as part of a toolbox called CASINO. Two microbial communities comprising four organisms each were designed and simulated using the CASINO toolbox, and the results were compared with data generated from *in vitro* experiments growing the same communities. The model simulations were in agreement with faecal metabolomics data, and also correctly captured the serum levels of 10 amino acids as well as acetate. Most interestingly, the simulations could point towards the dominant species in the communities studied, and the relative contributions of different species towards the amino acid production in the community.

11.1.3.4 Dynamic approaches

dFBA (§9.6) can be directly applied to microbial communities, as demonstrated by Mahadevan and co-workers [50], who modelled the competition between *Rhodoferax ferrireducens* and *Geobacter sulfurreducens*. Their method, the dynamic multi-species metabolic modelling (DMMM) framework, accounts for the growth of different organisms in the community at different rates, as well as for the production and consumption of various metabolites, influenced by the presence (and abundance) of different microbes in the community. Henson and co-workers [51, 52] have also employed dFBA-like methods to model microbial communities.

Another method, Computation Of Microbial Ecosystems in Time and Space (COMETS), implements a dFBA algorithm on a spatially structured 2D lattice [53]. Each box in this lattice (an $N \times M$ grid of boxes) spatially resolves different interacting metabolic subsystems (one or more microbial species and extracellular metabolites). The COMETS algorithm comprises two key alternating steps:

(i) *cellular growth*, modelled using a dFBA-like algorithm, capturing the increase (or decrease) in biomass at different spatial locations, based on nutrient availability and each organism's metabolic network (and user-defined initial conditions)

(ii) *diffusion* of various metabolites into the environment, described by a simplified finite differences approximation to the standard two-dimensional diffusion equation on a 2D lattice

which are used to update all the values in every box, corresponding to biomass and metabolites concentrations.

FLYCOP (FLexible sYnthetic Consortium OPtimization) [54] is a framework that extends COMETS, for the design and optimisation of microbial consortia.

11.1.3.5 SMETANA

Species METabolic interaction ANAlysis (SMETANA) [14] is a constraint-based approach that predicts the interaction potential between organisms in a community based on three separate scores that capture (i) species coupling, (ii) metabolite uptake, and (iii) metabolite production. These scores are computed based on solving MILP/LP optimisation problems, and combined together to obtain a "SMETANA score" that essentially quantifies the possible metabolic coupling between organisms in a community based on all possible metabolite exchanges.

11.1.3.6 SteadyCom

SteadyCom [10] is a scalable constraint-based modelling framework that can predict community compositions in microbial communities. SteadyCom can be thought of as an extension of Joint FBA, and a generalisation of basic FBA (§8.3) itself, which carefully considers the community growth rate as well as the relative abundances of the organisms:

$$\max v_{\text{comm}} \tag{11.3}$$

$$\text{subject to} \quad
\left[
\begin{array}{ll}
\sum_{j \in J^{(k)}} s_{ij}^{(k)} V_j^{(k)} = 0, & \forall i \in I^{(k)} \\[2ex]
\alpha_j^{(k)} X^{(k)} \leqslant V_j^{(k)} \leqslant \beta_j^{(k)} X^{(k)}, & \forall j \in J^{(k)} \\[2ex]
V_{\text{bio}}^{(k)} = X^{(k)} v_{\text{comm}} \\[2ex]
X^{(k)} \geqslant 0
\end{array}
\right] \quad \forall k \in K$$

$$u_i^c - e_i^c + \sum_{k \in K} V_{\text{ex}(i)}^{(k)} = 0, \quad \forall i \in I^{\text{com}}$$

$$\sum_{k \in K} X^{(k)} = X_0$$

$$V_{\text{bio}}^{(k)} = X^{(k)} v_j^{(k)}, \quad \forall j \in J^{(k)}, k \in K$$

$$v_{\text{comm}}, \quad e_i^c \geqslant 0, \quad \forall i \in I^{\text{com}}$$

where v_{comm} is the community growth rate, $s_{ij}^{(k)}$ represents entries from the stoichiometric matrix of the kth organism in the community K, $V_j^{(k)}$ represents the aggregate flux of the jth reaction in the entire population of the kth organism[8] (with units of mmol/h), $I^{(k)}$ is the set of metabolites in organism k, $\alpha_j^{(k)}$ and $\beta_j^{(k)}$ represent the lower and upper bounds of the jth reaction in organism k ($v_j^{(k)}$, with units mmol/gDW/h), $V_{\text{bio}}^{(k)}$ is the aggregate biomass production, $J^{(k)}$ is the set of

---

[8]This differs from $v_j^{(k)}$, the reaction flux normalised by the biomass of organism k.

reactions in organism k, $X^{(k)}$ is the biomass (in gDW) for organism k, X_0 is a non-zero total biomass, defined to prevent trivial zero solutions to the problem[9], u_i^c and e_i^c represent the community uptake and export rates for the ith metabolite, respectively, I^{com} is the set of shared community metabolites, and $V_{ex(i)}^{(k)}$ represents the transport reaction fluxes from the extracellular space into the kth organism in the community also taking into account its abundance.

The k sets of constraints, for each organism k in the community **K** are basically a generalisation of FBA. Setting $X^{(1)} = 1$ and $X^{(k)} = 0$ for $k > 1$ reduces the above SteadyCom formulation to the basic FBA formulation, with the aggregate biomass flux coinciding with the specific growth rate v_{comm}.

Importantly, SteadyCom is compatible with other constraint-based methods such as FVA (see §8.7), which in this case enables it to estimate the range of allowable fluxes for each organism as well as the variability in the biomass of a given organism in a community. SteadyCom is also able to account for important constraints such as ATP maintenance (see §8.8.6).

The SteadyCom approach was demonstrated on two different communities: (a) a consortium involving four *E. coli* mutants, each auxotrophic for two amino acids and missing an exporter for one amino acid, and (b) a community model of the gut, comprising nine bacteria spanning four major phyla. SteadyCom was able to accurately predict the compositions of the gut microbiota, as determined by previous experiments.

11.2   HOST–PATHOGEN INTERACTIONS (HPIs)

Another key application of modelling cellular interactions lies in the understanding of HPIs. The occurrence of any infection is contingent upon a complex interplay of various factors: the virulence mechanisms of an invading pathogen, the defence mechanisms of the host immune system, and the counter-defences of either organism. Nearly all of the paradigms discussed in the previous chapters have been brought to bear on modelling HPIs. These include network models of host-pathogen PPIs and metabolism, constraint-based models of HPIs, as well as dynamic models of immune system dynamics, captured using ODEs or Boolean networks. All of these models adopt different levels of abstraction to capture various cellular components and processes that ultimately dictate the outcome of infection.

11.2.1   Network models

As in community modelling, network models have been widely used to study HPI networks. In particular, PPI networks have been studied across the host and pathogen.

---

[9]X_0 is set to 1 gDW, resulting in $X^{(k)}$ becoming equal to the relative abundance of organism k.

11.2.1.1 Host–pathogen PPIs

Host–pathogen PPI networks are essentially very large graphs, where the nodes include proteins from both the host and the pathogen, and edges represent interactions within host proteins, within pathogen proteins and between host and pathogen proteins. A very interesting study by Asensio *et al* (2017) [55] illustrated the applicability of the classic centrality–lethality hypothesis (see §4.1) to host–pathogen networks. They studied the interaction network of the pathogen *Yersinia pestis* with its human host and showed that the centrality–lethality hypothesis holds for pathogen proteins during infection, but only in the context of the host–pathogen interactome. They further illustrated that the importance of pathogen proteins during infection was directly related to their number of interactions with human proteins. The study is an excellent exemplar of how simplistic models, can lead to valuable insights into network function, and in this case, the aetiology of *Y. pestis* infection.

Another extensive study [56] integrated experimentally verified human–pathogen PPIs for 190 pathogen strains from multiple public databases, viz. MINT, IntAct, DIP, Reactome, MIPS, and HPRD (see Appendix C). They showed that both viral and bacterial pathogens exhibited a higher tendency to interact with hubs (proteins with high degree in the PPI network) and bottlenecks (proteins with high betweenness centrality) in the human PPI network.

Memišević *et al* (2015) [57] performed a network analysis of human–*Burkholderia mallei* protein interactions and once again reaffirmed the importance of hub proteins, showing that known and putative *B. mallei* virulence factors tended to target host proteins with greater degree. They also developed an HPI Alignment (HPIA) algorithm to identify common sets of HPIs across different pathogen–human PPI networks. These aligned sets of interactions were useful in predicting the functional roles of *B. mallei* proteins, based on the roles of their aligned counterparts in *Salmonella enterica* and *Y. pestis*. An interesting study of the human coronavirus—host protein interaction network revealed possible drug targets and drug re-purposings, as we discuss in the Epilogue, p. 311.

A nice review on experimental approaches to elucidate host–pathogen PPIs has been published by Collins and co-workers [58]. Computational approaches have also been developed to predict host–pathogen PPIs, *e.g.* [59].

11.2.1.2 Response networks

A term attributed by Forst [60] to Boris Magasanik, *response networks* involve the identification of groups of genes, *i.e.* a sub-network, that shows significant changes in expression, over certain conditions [61, 62]. Integrating databases of yeast PPIs and protein–DNA interactions with gene expression data, Siegel and co-workers [61] identified significant sub-networks that corresponded well with known regulatory mechanisms. Forst and co-workers also carried out a *differential network expression analysis* in *M. tuberculosis*, studying gene expression in response to drugs such as isoniazid. Their methodology illustrated the differences between a generic stress response and a drug-specific response.

Forst has also constructed a host–pathogen metabolic network capturing the metabolic linkages between *C. psittaci* and human, particularly between trypto-phan biosynthesis in the bacterium and tryptophan catabolism in humans [63]. Their model illustrated how *C. psittaci* hijacks kynurenine from the host to drive its own metabolism.

11.2.2 Dynamic models

Beyond the static models discussed in Part I of this book, the dynamic modelling paradigms discussed in Part II have also been used extensively to study HPIs. Dynamic models based on both ODEs and Boolean networks have been used to study the interactions between different pathogens and the human immune system.

11.2.2.1 Boolean network modelling

Boolean network modelling has been a useful paradigm to model HPIs. The nodes in the Boolean network capture the various components of the pathogen and the host, including various immune system components such as T cells, B cells, dendritic cells, and macrophages. Transfer functions, as described in §7.2 capture the relationships between the various components using Boolean opera-tors such as OR, AND, and AND NOT. Thakar *et al* [64] modelled the host response to infection for two bacteria from the *Bordetella* species, which cause different dis-eases in their respective hosts. Their studies revealed three distinct phases of infec-tion corresponding to different immune processes. They also simulated the effect of knocking out immune system components, which serves to underline the im-portance of various processes in the overall immune response. A similar approach for modelling the complex interaction between *M. tuberculosis* and the human immune system was reported by Chandra and co-workers [65]. This study built a relatively large Boolean model comprising 75 nodes that spanned the host and pathogen cells, and also included various cellular states, processes, and signalling molecules. Importantly, these models can serve as an excellent framework for the development of more detailed mechanistic models.

Boolean networks have also been modelled as logical interaction hypergraphs (see §7.3.2) [66]. Constructed carefully from previously published experiments in literature, the model could illustrate the differences and similarities of the network response to c-Met signalling induced by hepatocyte growth factor and *Helicobac-ter pylori*. The model also pointed towards targets for therapeutic interventions.

11.2.2.2 Immune system dynamics

The dynamics of the host immune response to any infection is a critical aspect of HPIs. A detailed introduction to mathematical modelling of the immune system has been presented by Perelson and Weisbuch [67]. Another excellent review by Perelson [68] details the mathematical modelling of viral and immune system dynamics, using ODEs. For a more recent review of viral dynamics, see reference [69]. Kirschner and co-workers have built a number of mathematical models that capture the interaction of *M. tuberculosis* with the human immune system, as also recently reviewed in reference [70]. As illustrated nicely in reference [71], computational models can be extremely valuable to enhance our understanding of various immune processes. In this study, Cilfone *et al* describe *GranSim*, a hybrid agent-based model (ABM) of *M. tuberculosis* infection. GranSim captures the dynamics of key cytokines such as IL-10 and TNF. Ultimately, the study highlights several roles for the IL-10 in tubercular infection and how it could be exploited to design adjunctive therapies for TB.

11.2.3 Constraint-based models

Constraint-based models have also been helpful to study HPIs. While the host and pathogen metabolisms can be individually studied using methods such as FBA described in Chapter 8, it is even more interesting to study integrated models of host and pathogen metabolism, to provide insights into the key metabolic reactions, and possible points of intervention. The use of constraint-based models to model pathogens and HPIs has been reviewed elsewhere [72].

The classic study of HPIs using constraint-based modelling was performed by Raghunathan *et al* [73]. They built a metabolic model of the pathogen *Salmonella* Typhimurium LT2 but constrained the model using gene expression data obtained from *S.* Typhimurium isolated from macrophage cell lines to predict the pathogen's metabolism during infection. The first study to integrate host and pathogen metabolic networks was by Bordbar *et al* in 2010 [74]. The model, identified as "iAB-AMØ-1410-Mt-661", integrated models of the human alveolar macrophage (based on Recon1) as well as *M. tuberculosis*, to study the metabolic states of the host and pathogen in three different types of TB infection. Others have since used metabolic modelling to study *M. tuberculosis*–host interactions [75], or the human parasite *P. falciparum*, and its interactions with erythrocyte/hepatocyte cells [76, 77].

Integrating multi-omic data with metabolic modelling, Palsson and co-workers [78] studied the murine leukaemic macrophage cell line RAW 264.7, to shed light on the various factors modulating the metabolic activation of macrophages. Such studies are crucial to decode the relationship between metabolism and the immune system.

A very recent study explores the metabolic network of human alveolar macrophages and SARS-CoV-2 [79]. Building on the alveolar macrophage model described above (iAB-AMØ-1410), Dräger and co-workers studied the metabolism of both infected and uninfected host cells, and identified the critical metabolic differences between the two states. These alterations in metabolism were also used to suggest possible antiviral targets, such as guanylate kinase (GK1), whose knock-out was lethal for the virus in *in silico* simulations.

Metabolic models have also been constructed for plant pathogens such as *Phytophthora infestans* and integrated with genome-scale models of its host, tomato [80]. The simulations elucidated how the pathogen scavenges more metabolites from the host, itself reducing *de novo* synthesis, during various stages of the infection cycle.

11.2.3.1 Microbiome–pathogen interactions

As we build better models for modelling microbial interactions, it is now possible to study the interactions of pathogens with the gut microflora. By building on extensive resources such as AGORA (Assembly of Gut Organisms through Reconstruction and Analysis)[10], and available reconstructions of pathogenic organisms (*e.g. C. difficile* [81], *Enterococcus faecalis* [82]), it will be possible to simulate the effect of gut pathogens on bacterial dynamics in the gut. An interesting study by Nagarajan and co-workers has illustrated the importance of microbial interactions in the gut, for gut recovery post-antibiotic treatment [32].

11.2.3.2 Host–microbiome models

A natural extension of the human metabolic models (§10.6.2.1) and models of the gut microbiome is the development of host–microbiome models that capture the interaction between the human host and the rich microbiomes that reside in it. A number of exciting studies have been pioneered by Ines Thiele. As an interesting proof-of-concept, Thiele and co-workers constructed a host–microbiome model comprising (a) *Bacteroides thetaiotaomicron*, a prominent gut microbe and (b) a mouse metabolic model. Extending the mouse model with reactions for intestinal transport and absorption, the two models were linked via the lumen as a joint external compartment through which metabolites could be exchanged.

In another classic study, Heinken and Thiele integrated a human metabolic model with 11 gut microbe models and studied the metabolism under four dietary conditions [83], showing the important contributions of the gut microflora to human metabolism. More recently, Thiele and co-workers constructed personalised gut microbiome models using metagenomics data obtained from individuals suffering from inflammatory bowel disease (IBD) as well as healthy controls, and predicted each individual's potential to synthesise bile acids [84]. Bile acid biosynthesis is an important example of a co-operative task in the gut microbiome; no single strain is known to harbour the complete biosynthetic pathways. This study hence serves as an important example to understand the complex web of metabolic interactions in the gut, the differences between the metabolic profiles of healthy and IBD patients, as well as critical bottlenecks for the synthesis of bile acids.

With the advances in modelling of human metabolism, and the availability of large numbers of gut microbe models [85], the stage is set to deeply examine

---

[10]presently contains reconstructions of over 800 gut microbes, available from https://github.com/VirtualMetabolicHuman/AGORA

the interplay between diet, the gut microbiome, and the host metabolism, by building integrated constraint-based models that can readily capture all of these aspects. Nevertheless, many challenges remain to be surmounted, particularly in terms of identifying suitable trade-offs and capturing them mathematically using appropriate objective functions.

11.3 SUMMARY

Systems-level modelling has come a long way from the modelling of single pathways in cells, to larger genome-scale models of metabolism or regulation, all the way up to models that capture the interplay between one or more cells in a community or microbiome, or interactions between a pathogen, and the host. Depending upon the key questions that need to be answered, a variety of models have been employed, each able to provide some valuable insights into such complex systems.

The coming years will likely see more complex models, which are able to model larger communities of micro-organisms and make deeper inferences about their behaviour. Multi-scale models, which are able to capture varied facets of cellular interactions, spanning across metabolism, regulation, and signalling, may provide further insights to unravel the complexity of cellular interactions in microbiomes or in infection.

It is likely that microbiome modelling will become even more widespread over the coming years, with the continued increase in metagenomic studies of different microbiomes. Many new applications are likely to emerge, such as microbiome engineering for enabling recovery from gut dysbiosis, bioremediation of harmful chemicals, or soil microbiome engineering for improved agricultural productivity.

HPI models have evolved extensively over the last two decades, and have provided major insights into the underlying complexity. Modelling HPIs shares some commonalities with modelling microbial communities, particularly in terms of the methodologies employed, but comes with its own set of challenges, notably because of the added complexity of host cells and the immune system machinery. Systems-level models of HPIs are able to disentangle various mechanisms and effects, shedding further light on the key interactions and immune processes. HPI modelling holds great promise for improving our understanding of the molecular basis of diseases, and ultimately, also advance the development of novel therapeutic intervention strategies.

11.4 SOFTWARE TOOLS

CASINO toolbox [49] implements an iterative multi-level optimisation procedure to predict fluxes in a communities such as the gut microbiome.

NetCooperate [17] is a web-based tool and software package, that computes two indices—the biosynthetic support score and metabolic complementarity index—given a pair of metabolic networks. It is useful to assess microbe–microbe or host–microbe co-operative potential.

SteadyCom [10] is a constraint-based modelling framework that can predict community compositions in microbial communities. SteadyCom is now integrated into the COBRA Toolbox.

MetQuest [20] enumerates metabolic pathways of a specified size from metabolic networks comprising one or more organisms.

Microbiome Modelling Toolbox [46] is a versatile toolbox for MATLAB, which integrates a large number of tools for modelling microbe–microbe, host–microbe interactions as well as microbial community modelling.

Virtual Metabolic Human [85] is an excellent multi-faceted resource for metabolic modelling of humans, their gut microbiomes, and their interactions. It mainly includes metabolic models of human and 800+ gut microbes, and tools for visualisation. It even includes a large number of dietary components, mapped to metabolites in the database, to enable modelling the interplay between diet and the host/microbiome.

EXERCISES

11.1 The gLV model can also be used to model trophic food webs. Consider a simple ecosystem consisting of grass, grains, mice, snakes, rabbits, and hawks. For such a system, what would the 'community' matrix of the gLV model look like?

11.2 Consider a community comprising *Lactobacillus acidophilus* ATCC 4796 (*La*) and *Lactobacillus casei* ATCC 334 (*Lc*). The metabolic network of *La* comprises 739 metabolites participating in 776 reactions, including 121 exchange reactions. The metabolic network of *Lc* comprises 947 metabolites participating in 1020 reactions, including 139 exchange reactions. Amongst the exchange reactions, 8 are unique to *La*, and 26 are unique to *Lc*. If the two models are combined to build a compartmentalised FBA model, what will be the corresponding numbers (reactions, metabolites, and exchange fluxes) for the community metabolic network? Use `createMultipleSpeciesModel` to verify your calculations.

[LAB EXERCISES]

11.3 The first step of computing Metabolic Support Index [21] involves identifying the reactions that are *stuck* in a given reaction network. Use the `forward_pass` function from the *MetQuest* Python package to identify the reactions that are stuck in the metabolic network of the *E. coli iAF1260* model grown in a minimal glucose medium. The components of glucose minimal medium can be obtained from https://mediadb.systemsbiology.net/

defined_media/media/115/. Once you have done this, repeat the same for the *i*MM904 model of *S. cerevisiae*, and then compute the MSI for each of the organisms in the community. How does the *scope* of each metabolic network change, when in a community?

11.4 Determine the relative abundances of the gut bacteria *Bacteroides thetaiotaomicron* and *Faecalibacterium prausnitzii* (models available from the Virtual Metabolic Human resource) in the community using SteadyCom. Use different diet constraints such as the "high-fibre diet" or the "Western diet", as mentioned in reference [86], and observe the changes in relative abundances and growth rates. Comment on any correlations that you observe between the abundances and growth rates.

Can you also determine the type of interaction, viz. commensalism, amensalism, mutualism, parasitism, competition, or neutralism (See Table 1 in [86]) in both diets.

11.5 Using functions such as `createMultipleSpeciesModel` from the Microbiome Modelling Toolbox, design a two-species community with predominant the gut bacteria *B. thetaiotaomicron* and *F. prausnitzii* (available from the Virtual Metabolic Human resource). Compute and compare growth rates for each species individually and when present in a community. Alter oxygen availability and observe how this affects the growth rates. Refer to the study by Heinken and Thiele [86] for additional insights.

REFERENCES

[1] N. Fierer, "Embracing the unknown: disentangling the complexities of the soil microbiome", *Nature Reviews Microbiology* **15**:579–590 (2017).

[2] S. Sunagawa, L. P. Coelho, S. Chaffron, et al., "Structure and function of the global ocean microbiome", *Science* **348**:1261359 (2015).

[3] S. R. Gill, M. Pop, R. T. DeBoy, et al., "Metagenomic analysis of the human distal gut microbiome", *Science* **312**:1355–1359 (2006).

[4] E. A. Grice, H. H. Kong, S. Conlan, et al., "Topographical and temporal diversity of the human skin microbiome", *Science* **324**:1190–1192 (2009).

[5] A. Ravikrishnan and K. Raman, *Systems-level modelling of microbial communities: theory and practice* (CRC Press, 2018).

[6] R. A. Khandelwal, B. G. Olivier, W. F. M. Röling, B. Teusink, and F. J. Bruggeman, "Community flux balance analysis for microbial consortia at balanced growth", *PLoS ONE* **8**:e64567 (2013).

[7] A. R. Zomorrodi and C. D. Maranas, "OptCom: A multi-level optimization framework for the metabolic modeling and analysis of microbial communities", *PLoS Computational Biology* **8**:e1002363 (2012).

[8] H. Nagarajan, M. Embree, A.-E. Rotaru, et al., "Characterization and modelling of interspecies electron transfer mechanisms and microbial community dynamics of a syntrophic association", *Nature Communications* **4**:2809 (2013).

[9] C. Ye, W. Zou, N. Xu, and L. Liu, "Metabolic model reconstruction and analysis of an artificial microbial ecosystem for vitamin C production", *Journal of Biotechnology* **182–183**:61–67 (2014).

[10] S. H. J. Chan, M. N. Simons, and C. D. Maranas, "SteadyCom: predicting microbial abundances while ensuring community stability", *PLoS Computational Biology* **13**:e1005539 (2017).

[11] D. B. Bernstein, F. E. Dewhirst, and D. Segrè, "Metabolic network percolation quantifies biosynthetic capabilities across the human oral microbiome", *Elife* **8**:e39733 (2019).

[12] M. Thommes, T. Wang, Q. Zhao, I. C. Paschalidis, and D. Segrè, "Designing metabolic division of labor in microbial communities", *mSystems* **4**:e00263–18 (2019).

[13] S. Shoaie, F. Karlsson, A. Mardinoglu, I. Nookaew, S. Bordel, and J. Nielsen, "Understanding the interactions between bacteria in the human gut through metabolic modeling", *Scientific Reports* **3**:2532 (2013).

[14] A. Zelezniak, S. Andrejev, O. Ponomarova, D. R. Mende, P. Bork, and K. R. Patil, "Metabolic dependencies drive species co-occurrence in diverse microbial communities", *Proceedings of the National Academy of Sciences USA* **112**:6449–6454 (2015).

[15] E. Tzamali, P. Poirazi, I. G. Tollis, and M. Reczko, "A computational exploration of bacterial metabolic diversity identifying metabolic interactions and growth-efficient strain communities", *BMC Systems Biology* **5**:167 (2011).

[16] R. Levy and E. Borenstein, "Metabolic modeling of species interaction in the human microbiome elucidates community-level assembly rules", *Proceedings of the National Academy of Sciences USA* **110**:12804–12809 (2013).

[17] R. Levy, R. Carr, A. Kreimer, S. Freilich, and E. Borenstein, "NetCooperate: a network-based tool for inferring host-microbe and microbe-microbe cooperation", *BMC Bioinformatics* **16**:164 (2015).

[18] K. Faust, G. Lima-Mendez, J.-S. Lerat, et al., "Cross-biome comparison of microbial association networks", *Frontiers in Microbiology* **6**:1200 (2015).

[19] E. Borenstein and M. W. Feldman, "Topological signatures of species interactions in metabolic networks", *Journal of Computational Biology* **16**:191–200 (2009).

[20] A. Ravikrishnan, M. Nasre, and K. Raman, "Enumerating all possible biosynthetic pathways in metabolic networks", *Scientific Reports* **8**:9932 (2018).

[21]   A. Ravikrishnan, L. M. Blank, S. Srivastava, and K. Raman, "Investigating metabolic interactions in a microbial co-culture through integrated modelling and experiments", *Computational and Structural Biotechnology Journal* **18**:1249–1258 (2020).

[22]   A. Matyjaszkiewicz, G. Fiore, F. Annunziata, et al., "BSim 2.0: An advanced agent-based cell simulator", *ACS Synthetic Biology* **6**:1969–1972 (2017).

[23]   K. S. Korolev, "The fate of cooperation during range expansions", *PLoS Computational Biology* **9**:e1002994 (2013).

[24]   C. Quince, A. W. Walker, J. T. Simpson, N. J. Loman, and N. Segata, "Shotgun metagenomics, from sampling to analysis", *Nature Biotechnology* **35**:833–844 (2017).

[25]   J. G. Caporaso, J. Kuczynski, J. Stombaugh, et al., "QIIME allows analysis of high-throughput community sequencing data", *Nature Methods* **7**:335–336 (2010).

[26]   A. Barberán, S. T. Bates, E. O. Casamayor, and N. Fierer, "Using network analysis to explore co-occurrence patterns in soil microbial communities", *The ISME Journal* **6**:343–351 (2012).

[27]   S. Chaffron, H. Rehrauer, J. Pernthaler, and C. von Mering, "A global network of coexisting microbes from environmental and whole-genome sequence data", *Genome Research* **20**:947–959 (2010).

[28]   K. Faust and J. Raes, "Microbial interactions: from networks to models", *Nature Reviews Microbiology* **10**:538–550 (2012).

[29]   E. Borenstein, M. Kupiec, M. W. Feldman, and E. Ruppin, "Large-scale reconstruction and phylogenetic analysis of metabolic environments", *Proceedings of the National Academy of Sciences USA* **105**:14482–14487 (2008).

[30]   S. Freilich, A. Kreimer, I. Meilijson, U. Gophna, R. Sharan, and E. Ruppin, "The large-scale organization of the bacterial network of ecological co-occurrence interactions", *Nucleic Acids Research* **38**:3857–3868 (2010).

[31]   L. D. Miller, J. J. Mosher, A. Venkateswaran, et al., "Establishment and metabolic analysis of a model microbial community for understanding trophic and electron accepting interactions of subsurface anaerobic environments", *BMC Microbiology* **10**:149 (2010).

[32]   K. R. Chng, T. S. Ghosh, Y. H. Tan, et al., "Metagenome-wide association analysis identifies microbial determinants of post-antibiotic ecological recovery in the gut", *Nature Ecology & Evolution* **4**:1256–1267 (2020).

[33]   R. R. Stein, V. Bucci, N. C. Toussaint, et al., "Ecological modeling from time-series inference: insight into dynamics and stability of intestinal microbiota", *PLoS Computational Biology* **9**:e1003388 (2013).

[34] C. G. Buffie, V. Bucci, R. R. Stein, et al., "Precision microbiome reconstitution restores bile acid mediated resistance to *Clostridium difficile*", *Nature* 517:205–208 (2015).

[35] A. Bashan, T. E. Gibson, J. Friedman, et al., "Universality of human microbial dynamics", *Nature* 534:259–262 (2016).

[36] R. S. Cantrell and C. Cosner, *Spatial ecology via reaction-diffusion equations* (John Wiley & Sons Ltd, 2003), p. 411.

[37] E. E. Holmes, M. A. Lewis, J. E. Banks, and R. R. Veit, "Partial differential equations in ecology: spatial interactions and population dynamics", *Ecology* 75:17–29 (1994).

[38] M. S. Datta, K. S. Korolev, I. Cvijovic, C. Dudley, and J. Gore, "Range expansion promotes cooperation in an experimental microbial metapopulation", *Proceedings of the National Academy of Sciences USA* 110:7354–7359 (2013).

[39] M. J. I. Müller, B. I. Neugeboren, D. R. Nelson, and A. W. Murray, "Genetic drift opposes mutualism during spatial population expansion", *Proceedings of the National Academy of Sciences USA* 111:1037–1042 (2014).

[40] S. M. Stump, E. C. Johnson, and C. A. Klausmeier, "Local interactions and self-organized spatial patterns stabilize microbial cross-feeding against cheaters", *Journal of the Royal Society Interface* 15:20170822 (2018).

[41] J. L. Mark Welch, Y. Hasegawa, N. P. McNulty, J. I. Gordon, and G. G. Borisy, "Spatial organization of a model 15-member human gut microbiota established in gnotobiotic mice", *Proceedings of the National Academy of Sciences USA* 114:E9105–E9114 (2017).

[42] S. Magnúsdóttir, A. Heinken, L. Kutt, et al., "Generation of genome-scale metabolic reconstructions for 773 members of the human gut microbiota", *Nature Biotechnology* 35:81–89 (2017).

[43] D. Machado, S. Andrejev, M. Tramontano, and K. R. Patil, "Fast automated reconstruction of genome-scale metabolic models for microbial species and communities", *Nucleic Acids Research* 46:7542–7553 (2018).

[44] S. Stolyar, S. Van Dien, K. L. Hillesland, et al., "Metabolic modeling of a mutualistic microbial community", *Molecular Systems Biology* 3:92 (2007).

[45] N. Klitgord and D. Segrè, "Environments that induce synthetic microbial ecosystems", *PLoS Computational Biology* 6:e1001002 (2010).

[46] F. Baldini, A. Heinken, L. Heirendt, S. Magnusdottir, R. M. T. Fleming, and I. Thiele, "The Microbiome Modeling Toolbox: from microbial interactions to personalized microbial communities", *Bioinformatics* 35:2332–2334 (2019).

[47] S. Freilich, R. Zarecki, O. Eilam, et al., "Competitive and cooperative metabolic interactions in bacterial communities", *Nature Communications* 2:589 (2011).

[48] A. R. Zomorrodi, M. M. Islam, and C. D. Maranas, "d-OptCom: Dynamic multi-level and multi-objective metabolic modeling of microbial communities", *ACS Synthetic Biology* 3:247–257 (2014).

[49] S. Shoaie, P. Ghaffari, P. Kovatcheva-Datchary, et al., "Quantifying diet-induced metabolic changes of the human gut microbiome", *Cell Metabolism* 22:320–331 (2015).

[50] K. Zhuang, M. Izallalen, P. Mouser, et al., "Genome-scale dynamic modeling of the competition between *Rhodoferax* and *Geobacter* in anoxic subsurface environments", *The ISME Journal* 5:305–316 (2011).

[51] M. Henson and T. J. Hanly, "Dynamic flux balance analysis for synthetic microbial communities", *Systems Biology, IET* 8:214–229 (2014).

[52] T. J. Hanly, M. Urello, and M. A. Henson, "Dynamic flux balance modeling of *S. cerevisiae* and *E. coli* co-cultures for efficient consumption of glucose/xylose mixtures", *Applied Microbiology and Biotechnology* 93:2529–2541 (2012).

[53] W. R. Harcombe, W. J. Riehl, I. Dukovski, et al., "Metabolic resource allocation in individual microbes determines ecosystem interactions and spatial dynamics", *Cell Reports* 7:1104 (2014).

[54] B. García-Jiménez, J. L. García, and J. Nogales, "FLYCOP: metabolic modeling-based analysis and engineering microbial communities", *Bioinformatics* 34:i954 (2018).

[55] N. C. Asensio, E. M. Giner, N. S. de Groot, and M. T. Burgas, "Centrality in the host–pathogen interactome is associated with pathogen fitness during infection", *Nature Communications* 8:1–6 (2017).

[56] M. D. Dyer, T. M. Murali, and B. W. Sobral, "The landscape of human proteins interacting with viruses and other pathogens", *PLoS Pathogens* 4:e32 (2008).

[57] V. Memišević, N. Zavaljevski, S. V. Rajagopala, et al., "Mining host-pathogen protein interactions to characterize *Burkholderia mallei* infectivity mechanisms", *PLoS Computational Biology* 11:e1004088 (2015).

[58] C. Nicod, A. Banaei-Esfahani, and B. C. Collins, "Elucidation of host–pathogen protein–protein interactions to uncover mechanisms of host cell rewiring", *Current Opinion in Microbiology* 39:7–15 (2017).

[59] M. D. Dyer, T. M. Murali, and B. W. Sobral, "Computational prediction of host-pathogen protein–protein interactions", *Bioinformatics* 23:i159–i166 (2007).

[60] C. V. Forst, "Host–pathogen systems biology", *Drug Discovery Today* 11:220–227 (2006).

[61] T. Ideker, O. Ozier, B. Schwikowski, and A. F. Siegel, "Discovering regulatory and signalling circuits in molecular interaction networks", *Bioinformatics* 18:S233–S240 (2002).

[62] L. Cabusora, E. Sutton, A. Fulmer, and C. V. Forst, "Differential network expression during drug and stress response", *Bioinformatics* 21:2898–2905 (2005).

[63] C. V. Forst, "Network genomics – A novel approach for the analysis of biological systems in the post-genomic era", *Molecular Biology Reports* 29:265–280 (2002).

[64] J. Thakar, M. Pilione, G. Kirimanjeswara, E. T. Harvill, and R. Albert, "Modeling systems-level regulation of host immune responses", *PLoS Computational Biology* 3:e109 (2007).

[65] K. Raman, A. G. Bhat, and N. Chandra, "A systems perspective of host–pathogen interactions: predicting disease outcome in tuberculosis", *Molecular BioSystems* 6:516–530 (2010).

[66] R. Franke, M. Müller, N. Wundrack, et al., "Host-pathogen systems biology: logical modelling of hepatocyte growth factor and *Helicobacter pylori* induced c-Met signal transduction", *BMC Systems Biology* 2:4 (2008).

[67] A. S. Perelson and G. Weisbuch, "Immunology for physicists", *Reviews of Modern Physics* 69:1219–1267 (1997).

[68] A. S. Perelson, "Modelling viral and immune system dynamics", *Nature Reviews Immunology* 2:28–36 (2002).

[69] P. Padmanabhan and N. M. Dixit, "Models of viral population dynamics", in *Quasispecies: From Theory to Experimental Systems*, Current Topics in Microbiology and Immunology (Cham, 2016), pp. 277–302.

[70] S. Marino, J. J. Linderman, and D. E. Kirschner, "A multifaceted approach to modeling the immune response in tuberculosis", *WIREs Systems Biology and Medicine* 3:479–489 (2011).

[71] N. A. Cilfone, C. B. Ford, S. Marino, et al., "Computational modeling predicts IL-10 control of lesion sterilization by balancing early host immunity–mediated antimicrobial responses with caseation during *Mycobacterium tuberculosis* infection", *The Journal of Immunology* 194:664–677 (2015).

[72] L. J. Dunphy and J. A. Papin, "Biomedical applications of genome-scale metabolic network reconstructions of human pathogens", *Current Opinion in Biotechnology* 51:70–79 (2018).

[73] A. Raghunathan, J. Reed, S. Shin, B. Ø. Palsson, and S. Daefler, "Constraint-based analysis of metabolic capacity of *Salmonella typhimurium* during host-pathogen interaction", *BMC Systems Biology* 3:38+ (2009).

[74] A. Bordbar, N. E. Lewis, J. Schellenberger, B. Ø. Palsson, and N. Jamshidi, "Insight into human alveolar macrophage and *M. tuberculosis* interactions via metabolic reconstructions", *Molecular Systems Biology* 6:422 (2010).

[75] R. A. Rienksma, P. J. Schaap, V. A. P. Martins dos Santos, and M. Suarez-Diez, "Modeling host-pathogen interaction to elucidate the metabolic drug response of intracellular *Mycobacterium tuberculosis*", *Frontiers in Cellular and Infection Microbiology* **9**:144 (2019).

[76] C. Huthmacher, A. Hoppe, S. Bulik, and H.-G. Holzhütter, "Antimalarial drug targets in *Plasmodium falciparum* predicted by stage-specific metabolic network analysis", *BMC Systems Biology* **4**:120 (2010).

[77] S. Bazzani, A. Hoppe, and H.-G. Holzhütter, "Network-based assessment of the selectivity of metabolic drug targets in *Plasmodium falciparum* with respect to human liver metabolism", *BMC Systems Biology* **6**:118 (2012).

[78] A. Bordbar, M. L. Mo, E. S. Nakayasu, et al., "Model-driven multi-omic data analysis elucidates metabolic immunomodulators of macrophage activation", *Molecular Systems Biology* **8**:558 (2012).

[79] A. Renz, L. Widerspick, and A. Dräger, *FBA reveals guanylate kinase as a potential target for antiviral therapies against SARS-CoV-2*, tech. rep. (2020).

[80] S. Y. A. Rodenburg, M. F. Seidl, H. S. Judelson, A. L. Vu, F. Govers, and D. de Ridder, "Metabolic model of the *Phytophthora infestans*–tomato interaction reveals metabolic switches during host colonization", *mBio* **10**:e00454–19 (2019).

[81] M. Larocque, T. Chénard, and R. Najmanovich, "A curated *C. difficile* strain 630 metabolic network: prediction of essential targets and inhibitors", *BMC Systems Biology* **8**:117 (2014).

[82] N. Veith, M. Solheim, K. W. A. van Grinsven, et al., "Using a genome-scale metabolic model of *Enterococcus faecalis* V583 to assess amino acid uptake and its impact on central metabolism", *Applied and Environmental Microbiology* **81**:1622–1633 (2015).

[83] A. Heinken and I. Thiele, "Systematic prediction of health-relevant human-microbial co-metabolism through a computational framework", *Gut Microbes* **6**:120–130 (2015).

[84] A. Heinken, D. A. Ravcheev, F. Baldini, L. Heirendt, R. M. T. Fleming, and I. Thiele, "Systematic assessment of secondary bile acid metabolism in gut microbes reveals distinct metabolic capabilities in inflammatory bowel disease", *Microbiome* **7**:75 (2019).

[85] A. Noronha, J. Modamio, Y. Jarosz, et al., "The Virtual Metabolic Human database: integrating human and gut microbiome metabolism with nutrition and disease", *Nucleic Acids Research* **47**:D614–D624 (2019).

[86] A. Heinken and I. Thiele, "Anoxic conditions promote species-specific mutualism between gut microbes *in silico*", *Applied and Environmental Microbiology* **81**:4049–4061 (2015).

FURTHER READING

M. Kumar, B. Ji, K. Zengler, and J. Nielsen, "Modelling approaches for studying the microbiome", *Nature Microbiology* **4**:1253–1267 (2019).

A. Eng and E. Borenstein, "Microbial community design: methods, applications, and opportunities", *Current Opinion in Biotechnology* **58**:117–128 (2019).

A. Succurro and O. Ebenhöh, "Review and perspective on mathematical modeling of microbial ecosystems", *Biochemical Society Transactions* **46**:403–412 (2018).

W. Gottstein, B. G. Olivier, F. J. Bruggeman, and B. Teusink, "Constraint-based stoichiometric modelling from single organisms to microbial communities", *Journal of the Royal Society Interface* **13**:20160627 (2016).

M. B. Biggs, G. L. Medlock, G. L. Kolling, and J. A. Papin, "Metabolic network modeling of microbial communities", *WIREs Systems Biology and Medicine* **7**:317–334 (2015).

K. S. Ang, M. Lakshmanan, N.-R. Lee, and D.-Y. Lee, "Metabolic modeling of microbial community interactions for health, environmental and biotechnological applications", *Current Genomics* **19**:712–722 (2018).

A. R. Zomorrodi and D. Segrè, "Synthetic ecology of microbes: mathematical models and applications", *Journal of Molecular Biology* **428**:837–861 (2016).

E. Bauer and I. Thiele, "From network analysis to functional metabolic modeling of the human gut microbiota", *mSystems* **3**:e00209–17 (2018).

K. C. H. van der Ark, R. G. A. van Heck, V. A. P. Martins Dos Santos, C. Belzer, and W. M. de Vos, "More than just a gut feeling: constraint-based genome-scale metabolic models for predicting functions of human intestinal microbes", *Microbiome* **5**:78 (2017).

S. Mukherjee, A. Sambarey, K. Prashanthi, and N. Chandra, "Current trends in modeling host–pathogen interactions", *WIREs Data Mining and Knowledge Discovery* **3**:109–128 (2013).

S. Durmuş, T. Çakır, A. Özgür, and R. Guthke, "A review on computational systems biology of pathogen–host interactions", *Frontiers in Microbiology* **6**:235 (2015).

Designing biological circuits

CONTENTS

S YNTHETIC BIOLOGY has generated a tremendous amount of buzz over the last two decades or so. Driven primarily by the desire to make biology more *engineerable*, synthetic biology has become a meeting point for multiple disciplines, with the ultimate goal of standardising the design of biological systems, in a predictive fashion. Engineering, as particularly exemplified by electronics, involves wiring various components in a specific fashion, to create circuits, or devices, with desired functionality. Synthetic biology, among many other things, represents the initiative to bring such predictive engineering to biology, to build systems—ideally from standardised parts—that can show a desired behaviour. Indeed, this construction of complex systems from simpler parts in a reliable fashion, is a hallmark of modern engineering and technology. To predictively engineer biological systems, though, we need to be able to model the system well enough. In this chapter, we will briefly overview some of the methodologies, and some interesting use cases for systems-level modelling to inform circuit design for synthetic biology.

12.1   WHAT IS SYNTHETIC BIOLOGY?

A quote often seen in synthetic biology talks is one from Richard Feynman, "*What I cannot create, I do not understand*". This quote reminds us that synthetic biology is not merely an effort to engineer biology for novel applications, but has a loftier goal, in terms of unravelling and understanding the extraordinary complexity underlying every living cell, by dissecting its circuitry. Synthetic biology is also a departure from the more conventional genetic engineering, which typically focusses on altering a few genes within an organism. Synthetic biology seeks to engineer more wholesale changes to existing cells, heralding the construction of more elaborate systems from the scratch. These new designs need not necessarily mimic extant biology—they can be radically different, such as using D–amino acids or unnatural amino acids to build truly novel proteins.

An ideal outcome of synthetic biology experiments is a deeper understanding of the underlying cellular machinery—which may or may not lend itself to convenient manipulation! Can the engineering efforts of synthetic biology shed more light on the *design principles* of biological circuitry? How does one design an oscillator? Or a switch? Or a circuit that can adapt to changes in input? Is it really possible to tame the seemingly interminable complexity of even a simple *E. coli* cell and engineer it for our purposes? These exciting questions are being answered by synthetic biologists, as they embark on the design of biological circuits.

The international genetically engineered machine (iGEM) competition

The iGEM competition (often just referred to as "iGEM"), organised by the iGEM foundation, is a unique annual competition in synthetic biology, where multidisciplinary teams from all over the world design, build, test, and characterise a system of their own design. These systems, or *circuits*, are built using interchangeable "biological parts" and standard molecular biology techniques.

Beginning with a mere five teams in 2004, recent editions have featured hundreds of teams from all over the world. The finals, or the annual giant jamboree, typically happened at the Massachusetts Institute of Technology (MIT), Cambridge, USA, although since 2014, the ever-increasing attendance pushed the meeting to larger convention centres outside of MIT! Originally targeted at undergraduate students, the iGEM now includes graduate students and even high-schoolers. The website https://igem.org/ presents a wealth of details about past competitions and projects.

The iGEM has routinely featured very novel projects, some of which have gone on to become full-scale laboratory projects and resulting in highly cited publications as well. Beyond all of this, the iGEM provides an enriching experience to young researchers world over, and remains a wonderful model for other fields to emulate. Kelvick *et al* [1] provide a nice account of the iGEM.

12.1.1   From LEGO bricks to biobricks

An excellent 'toy analogy' is that of LEGO bricks[1], which can be used to build incredibly complex objects. Can we make similar *bricks* out of biological parts,

---

[1]Indeed, the LEGO metaphor is enshrined in the iGEM biobrick trophy!

which can be re-used to build incrementally complex biological circuits? The iGEM registry (http://parts.igem.org/) has 20,000+ documented 'parts' or Bio-Bricks [2]. The BioBricks Foundation (http://biobricks.org/) was founded in 2006, with the goal of standardising biological parts. There is also the BioBrick assembly standard, which was proposed as an approach to combine parts to form larger composites, essentially to enable wider re-use of specific parts. Many other assembly standards have since been introduced [3].

Endy and co-workers [4] defined a standard biological part as "a genetically encoded object that performs a biological function and that has been engineered to meet specified design or performance requirements, especially *reliable physical and functional composition*". The reliable composition mentioned above becomes quite a challenge in most cases for synthetic biology.

Arkin and co-workers specify five desiderata for biological parts: independence, reliability, tunability, orthogonality, and composability [5].

An example part: BBa_K808000

This biobrick ("araC-Pbad") corresponds to an arabinose-inducible regulatory promoter/repressor unit. It contains the promoter as well as the coding sequence corresponding to the repressor AraC, which is transcribed in the opposite direction. Binding to L-arabinose causes AraC to change its conformation, relieving the repression and inducing transcription. Further details on the biology of this part, its characterisation and use by various teams are documented at http://parts.igem.org/Part:BBa_K808000.

12.2 CLASSIC CIRCUIT DESIGN EXPERIMENTS

The classic "circuit design" experiments played a key role in the early blooming of synthetic biology. They also set the stage for the key goal of synthetic biology—to design biological (gene regulatory) circuits that carry out a specific function, much like electronic circuits. Mathematical modelling also played a useful role here, enabling a model–driven circuit design approach. *E. coli* emerged as the front-runner for synthetic biology experiments given the ease of performing genetic manipulations and the wealth of knowledge accumulated on its biology over the years, including many gene regulatory systems—"parts" [6].

Below, we discuss two of the very first circuit design experiments, the repressilator and a toggle switch, both published in the same issue of *Nature*, on 20th January 2000. Interestingly, both circuits were built from similar parts, including an inducible promoter system, and a GFP-expression-based monitoring of circuit behaviour. Modelling was an integral component of both circuits too. Both studies used a workflow based on model-driven design, experimental measurements followed by hypothesis-driven debugging, quite closely mirroring the systems biology cycle (Figure 1.1).

12.2.1 Designing an oscillator: The repressilator

Elowitz and Liebler designed and constructed a novel synthetic network employing three transcriptional repressor systems—*lacI*, *λcI*, and *tetR*. In this circuit, the first repressor protein, LacI, inhibits the transcription of *tetR*. The protein product of *tetR*, in turn, inhibits the expression of *λcI*. The cycle is completed by the protein product of *λcI*, CI, which inhibits the expression of *lacI*.

These proteins were all derived from different organisms—LacI from *E. coli*, TetR from the tetracycline-resistance transposon Tn*10*, and the *cI* gene from λ phage. The readout for the oscillation was a green fluorescent protein (GFP). Figure 12.1a illustrates the architecture of this circuit. Most importantly, this network was the first example of "rational" (model-driven) network design. Let us take a look at the system of ODEs that were used to model this circuit:

$$\frac{dm_{lacI}}{dt} = -m_{lacI} + \frac{\alpha}{1 + [CI]^n} + \alpha_0 \tag{12.1a}$$

$$\frac{d[LacI]}{dt} = -\beta([LacI] - m_{lacI}) \tag{12.1b}$$

$$\frac{dm_{tetR}}{dt} = -m_{tetR} + \frac{\alpha}{1 + [LacI]^n} + \alpha_0 \tag{12.1c}$$

$$\frac{d[TetR]}{dt} = -\beta([TetR] - m_{tetR}) \tag{12.1d}$$

$$\frac{dm_{cI}}{dt} = -m_{cI} + \frac{\alpha}{1 + [TetR]^n} + \alpha_0 \tag{12.1e}$$

$$\frac{d[CI]}{dt} = -\beta([CI] - m_{cI}) \tag{12.1f}$$

The left-hand side of the equations denote the rate of change of the mRNA or protein concentrations, denoted by m_{xyz} or $[Xyz]$, respectively. α and $\alpha + \alpha_0$ are the number of protein copies per cell produced from a given promoter during growth in the presence ('promoter leakage') and absence of saturating amounts of the repressor. β denotes the ratio of the protein decay rate to the mRNA decay rate, n is a Hill coefficient. The protein concentrations are evidently written in terms of K_M, the concentration of repressors required to half-maximally repress a given promoter. Note that the model of the repressilator is available from the BioModels repository[2]. Depending on the parameters, this system can attain a stable steady-state or move towards sustained oscillations—this guided the design of appropriate promoters as well as the manipulation of protein half-lives. This

---

[2]https://www.ebi.ac.uk/biomodels/BIOMD0000000012

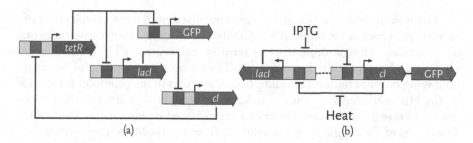

Figure 12.1 **Classic circuit designs.** (a) The repressilator, constructed from three repressor–promoter interactions, with TetR regulating a GFP 'reporter'. (b) A toggle switch, showing a pair of repressors wired to transcriptionally repress each other. Reprinted by permission from Springer Nature: D. E. Cameron, C. J. Bashor, and J. J. Collins, "A brief history of synthetic biology", *Nature Reviews Microbiology* 12:381–390 (2014). © (2014) Macmillan Publishers Limited.

study was the first step towards unravelling the design principles of biological circuits such as oscillators. The repressilator circuit exhibited periodic oscillations in GFP expression, persisting across a number of generations. The oscillations themselves were often noisy though, and also underwent dampening.

12.2.2   Toggle switch

In another classic study, Collins and co-workers [7] constructed a genetic toggle switch—a bi-stable gene regulatory network in *E. coli*. Much like the above study, they developed a mathematical model, and also theorised on the conditions necessary for bi-stability. They constructed the toggle switch from a pair of repressible promoters, arranged in a mutually inhibitory network. The circuit is flipped from one stable state to another by means of chemical or thermal induction. As a *part*, the toggle switch can be widely used, even as a 'cellular memory unit' [7]. Figure 12.1b illustrates the architecture of this circuit. CI inhibits transcription from *lacI* and is disengaged by heat. LacI inhibits transcription from *cI* and is disengaged by IPTG (isopropyl-β-D-thiogalactoside).

The behaviour of the toggle switch, including its bi-stable nature, are captured in the following simplified ODEs:

$$\frac{dm_{lacI}}{dt} = -m_{lacI} + \frac{\alpha_{lacI}}{1 + [CI]^{n_{cI}}} \tag{12.2}$$

$$\frac{dm_{cI}}{dt} = -m_{cI} + \frac{\alpha_{cI}}{1 + [LacI]^{n_{lacI}}} \tag{12.3}$$

In addition to the mRNA and protein symbols, which carry the same meanings as in Eq.12.1, we have α's and n's that capture the effective rate of synthesis of a repressor, and the co-operativity of the repression of a promoter, respectively.

The α parameters are 'lumped parameters' that describe the overall effect of a number of processes including RNA polymerase binding, open-complex formation, transcript elongation, transcript termination, repressor binding, ribosome binding, and polypeptide elongation [7]. The co-operative binding of repressors (following multimerisation) to multiple operator sites in the promoter is captured by the Hill coefficients n_{lacI} and n_{cI}. An analysis of the dynamics of these equations will reveal the relevant parameter regimes and conditions for bi-stability. The model of the toggle switch is available from the BioModels repository[3].

Several other circuits have since been designed and implemented [8–13], and have presented various interesting insights into the design principles of biological circuits, and are reviewed elsewhere [14–16].

12.3 DESIGNING MODULES

12.3.1 Exploring the design space

It is often helpful to view the design of a biological circuit in the context of the *design space*, comprising all possible circuits. In a classic study in 2005, Wagner enumerated 729 possible two-gene circuits and investigated their ability to produce circadian oscillations. This study underlined the importance of both topology and parameters in network function, and particularly studied the robustness of the circadian oscillations to perturbations in both topology and parameters. This study also had interesting implications for studying the evolvability of circadian oscillations, as we will see in §13.3.

Beyond oscillators and switches, another functionality that has attracted much attention is 'perfect adaptation'. In a seminal piece of work in 2009, Tang and co-workers [17] studied a very large number of possible alternate topologies of protein circuits comprising three enzymes. They considered circuits with three nodes (enzymes), A, B, and C, with the input stimulus at node A and the output at node C.

Combinatorially speaking, for a circuit containing three proteins, there are nine possible interactions (include self-interactions), each of which can be 'positive' (activating), 'negative' (repressing/inhibiting), or absent, leading to a total of $3^9 = 19,683$ possibilities. However, a number of these (3,645) are *disconnected*, *i.e.* lack a direct or indirect link from the input node to the output node. Removing these results in 16,038 connected topologies. The corresponding scenario for two-node networks is illustrated in Figure 12.2.

For each of the 16,038 topologies, Tang and co-workers explored 10^4 possible parameter sets, to investigate the ability of each of these enzyme networks to show perfect adaptation, as defined by conditions on the response (output) to a disturbance/stimulus (input). Interestingly, they found only two families of topologies that could show adaptation, viz. a negative feedback loop with a buffering node, and an incoherent feed-forward loop with a proportioner node. Most importantly, this study pointed towards important design principles of such

---

[3] https://www.ebi.ac.uk/biomodels/BIOMD0000000507

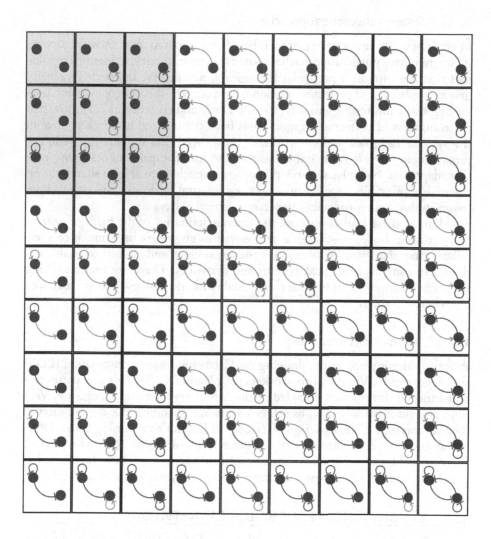

Figure 12.2 **Two-node topologies.** A total of 81 possible wirings exist, for two-node protein networks. The grey-shaded topologies are 'disconnected', lacking a link between the input and output nodes. The normal arrowheads indicate activation, and the bar-headed arrows indicate inhibition/repression.

networks, in terms of the minimally necessary network motifs, and also set the tone for other studies on adaptation [18].

12.3.2 Systems-theoretic approaches

In contrast to the above strategy that is based on enumeration of network topologies, others have pursued approaches rooted in systems theory, to identify topologies that can support a particular biological functionality. Interestingly, many groups have explored perfect adaptation from a control-theoretic viewpoint, presenting many interesting insights [19–22]. Doyle and co-workers [21] show that the robustness of perfect adaptation arises from the integral feedback control in the system. Tangirala and co-workers [22] showed that systems-theoretic approaches can provide comparable insights into motifs capable of achieving perfect adaptation. Notably, systems-theoretic approaches are able to shed light on the design principles of such networks, *e.g.* integral feedback, and other *circuit constructs* that can reproducibly generate a desired behaviour.

Many of these studies reiterate the interdisciplinarity of the fields of systems and synthetic biology—chemical and electrical engineers have much to contribute from their vast literature on controls, systems and circuits to enable the design of complex biological systems and networks. There are many exciting studies at the interface of systems theory and synthetic biology, and are reviewed elsewhere [23–26].

12.3.3 Automating circuit design

A very intriguing approach, drawing on Electronic Design Automation (EDA) was developed by Voigt and co-workers [27]. Named *Cello*, their tool parses circuit function information encoded in the hardware description language, Verilog, and automatically designs a DNA sequence encoding the desired circuit. The authors used Cello to design 60 circuits in *E. coli*. From Verilog codes, DNA sequences were built automatically, ultimately requiring 880,000 base pairs of DNA assembly. Overall, across circuits, 92% of 400+ output states functioned as predicted, with 45 circuits performing correctly in every output state.

12.4 DESIGN PRINCIPLES OF BIOLOGICAL NETWORKS

As we discussed very early on in §1.2, a lofty goal of modelling in systems biology is to be able to unravel the design principles of biological networks. The pursuit of synthetic biology, too, is marked by a desire to understand the underlying design principles of biological networks. What are these *design principles*? They could be thought of in the same way as motifs (see §3.4)—over-represented or recurring paradigms in biological networks, which are possibly associated with a specific biological functionality, *e.g.* oscillation or adaptation. In fact, Alves and co-workers define biological design principles as "*repeated qualitative and quantitative features of biological components and their interactions that are observed in molecular systems at high frequencies and improve the functional performance of a system that executes a specific process"* [28].

Many design principles pervade biological systems across different functionalities, *e.g.* redundancy, modularity, exaptation, and robustness. Note that these design principles are really not strategies that an organism employs to achieve a particular functionality (vis-à-vis a human who designs circuits or other things); rather, these are *post facto* observations of repeated paradigms that have emerged, possibly as a consequence of various evolutionary processes [29].

12.4.1 Redundancy

Redundancy, particularly functional redundancy, is a defining feature of biological networks, ranging from genetic networks to metabolic networks. A very useful approach to study redundancy in networks is the analysis of synthetic lethals (see §10.2). Synthetic lethals are essentially an extreme case of genetic interaction, where the combined effect of a pair of alterations (mutations, or deletions) differs drastically from the effect of individual alterations [30]. Functional redundancy is often achieved when genes or genomes duplicate. The duplicated set of genes results in redundancy; additionally, the duplicated genes may diversify (to varying extents) resulting in part conservation of function and part novel function.

Functionally redundant synthetic lethal reactions confer robustness against single mutations. Notably, genes involved in catabolic pathways, central metabolism, as well as those involved in reactions that carry a higher flux tend to have a higher rate of duplication, compared to genes that catalyse anabolic pathways [31, 32]. Sambamoorthy and Raman [33] performed a detailed analysis of synthetic lethals in metabolism, shedding light on the level of functional redundancy in metabolic networks of organisms such as *E. coli*.

12.4.2 Modularity

Another important design principle commonly found in metabolic networks is modularity. Some seminal articles by Wagner [34] and Hartwell [35] have highlighted the concept of modularity in biological networks. *In silico* simulations have also been employed to elucidate and illustrate the emergence of modularity in complex biological networks. In a classic study in 2005, Kashtan and Alon [36] showed that exposure to varying environmental pressures leads to the spontaneous evolution of modularity in network structure, and network motifs.

Modularity has also been studied in metabolic networks, by various groups [37–40]. Metabolic networks also are remarkably modular and evolve modularity in response to environmental pressures as well. Ruppin and co-workers have addressed how modularity changes as organisms evolve to incorporate newer reactions and specialise to different niches [39]. Despite the wealth of studies, the evolution of modularity in biological networks remains an exciting field with many interesting open questions. As we will see in §13.1.1.3, modularity is a key feature that also facilitates robustness.

12.4.3 Exaptation

Exaptations are pre-adaptations—these are traits initially naturally selected for a particular role, but later co-opted for an altogether different purpose [41]. While adaptation typically involves new features that arise as a consequence of natural selection, exaptations involve the co-option of an already selected feature. An interesting example of exaptation in metabolic networks was illustrated by Barve and Wagner [42]. Organisms may adapt to a given environment by acquiring new reactions—these very same reactions may become essential for survival in a different environment. Thus, this can be viewed as a pre-adaptation; the original purpose for which the reaction was naturally selected for is quite different from its utility for survival in a different environment. A number of complex innovations have also been shown to emerge from the addition of even single reactions to metabolic networks [43]. The bi-functionality of certain enzymes also serve as good examples of exaptation in metabolic networks [44]. Exaptation, particularly in metabolic networks, seems to provide novel capabilities to metabolic networks and paves the way for many evolutionary innovations.

12.4.4 Robustness

Robustness is another widely observed recurring design principle in biological networks [45]. Underlying robustness are also important circuit devices such as feedback loops or coherent feed-forward loops [46]. Various mechanisms facilitate robustness, such as modularity, decoupling, and redundancy, as we will discuss in §13.1.1. Robustness, and how it is intertwined with evolvability forms the central theme of the next chapter.

12.5 COMPUTING WITH CELLS

A very intriguing aspect of synthetic biology is the exploitation of biological macromolecules, or even cells, for computation. *Molecular computation* has enabled the solution of various computational problems, in many unconventional ways. One can argue that the first DNA computing experiments pre-date the emergence of synthetic biology as a discipline. Nevertheless, the last decade or so has seen synthetic biology and molecular computing grow together, contributing to one another's successes. Many exciting questions lie ahead of us, as we explore molecular computing paradigms. Notably, von Neumann's thought-provoking study as early as 1956 [47] on the *"synthesis of reliable organisms from unreliable components"* is all the more relevant, as scientists scamper to build algorithms and computers out of 'unreliable' processes such as DNA replication/hybridisation. The big question in front of us is whether these systems are useable and present important advantages despite their obvious shortcomings and disadvantages. In the following pages, we will briefly discuss some aspects of molecular computing, beginning with Adleman's classic experiment.

12.5.1 Adleman's classic experiment

In 1994, Leonard Adleman solved a Hamiltonian problem using DNA computing [48]. A Hamiltonian path[4] is a path that traverses all the nodes in a graph *exactly once* (not necessarily traversing all the edges). Adleman's algorithm, which is essentially brute force yet elegant at the same time, involves the following steps:

1. Generate a large number of random paths in the given (directed) network
2. Eliminate paths that do not start/end at the desired node
3. Eliminate paths that do not pass through exactly n nodes
4. Eliminate paths that pass through any node in the graph more than once
5. Any remaining paths are solutions to the Hamiltonian

While the above algorithm is computationally utterly infeasible on a silicon computer, it makes a lot of sense once you bring in the massive parallelism that is inherent in DNA. Of course, it is not obvious how to *implement* the above algorithm on DNA—which is where we defer to the genius of Adleman. Adleman encoded a seven-node directed graph onto DNA. Every node in the graph ($i = 0, 1, \ldots, 6$) was coded for by a random 20-mer DNA oligonucleotide sequence, denoted O_i. For every edge (i, j) in the graph, a new oligonucleotide $O_{i \to j}$ was synthesised as the 3' 10-mer of O_i followed by the 5' 10-mer of O_j. For $i = 0$ (starting vertex) and $j = 6$ (ending vertex), the entire 20-mer (O_1 or O_6) was used[5].

The novel encoding described above enabled the performance of the various steps of the algorithm using the *tools* of DNA computing—essentially Watson–Crick pairing, polymerases, nucleases, ligases, gel electrophoresis, and of course, the polymerase chain reaction (PCR). So, how do we implement the above algorithm's steps exactly? The answer is really simple—just throw all the 'DNA edges' into a test tube and let them react! The end result is a mixture of several trillion paths, including the solution to the Hamiltonian, if any. By using techniques such as gel electrophoresis and affinity purification, all the incorrect paths were weeded out to reveal the solution to the Hamiltonian—a 140 bp molecule.

Since Adleman's classic paper, many other hard computational problems have been solved using DNA computing, such as a 20-variable instance of the three-satisfiability[6] (3-SAT) problem [49]. Further, many intriguing molecular devices have been built, as we discuss in the following section.

12.5.2 Examples of circuits that can compute

A complex, distributed logic circuit using NOR gates has been implemented in *E. coli* [50]. The gates were wired together using quorum molecules produc-

---

[4]Recall the Eulerian path discussed in §2.1.1 and Figure 2.3, where every *edge* must be traversed instead.

[5]This encoding results in paths of six edges (that include either terminus) being encoded for by 140-bp oligonucleotides.

[6]Similar to the Hamiltonian path problem, this problem also falls under the class of "NP-complete" problems—essentially amongst the hardest computational problems, although any solution to such problems can be verified rapidly.

ing many different kinds of circuits. Shapiro and co-workers [51] built a programmable finite automaton comprising DNA, restriction nuclease, and ligase to solve computational problems autonomously. Qian and Winfree [52] have built a large-scale DNA circuit that is based on DNA strand displacement reactions. An excellent review that covers various aspects of molecular computation has been published by Benenson [53].

12.5.3 DNA data storage

An emerging area of interest is the storage of data in DNA. Even Adleman remarked about the information density of DNA in his classic paper mentioned above [48]. Despite the remarkable information density of DNA, it remains a challenge to effective encode, store, and accurately retrieve information from DNA. For a detailed review, which also interestingly discusses *in vivo* DNA data storage, see [54].

In a very interesting example, Church, Gao, and Kosuri encoded an HTML version of Church's very interesting book "Regenesis" [55] using DNA. The encoded data included 53,426 words, 11 JPG images, and one JavaScript program totalling 5.27 megabits. Hughes, Church, and co-workers have presented a very interesting commentary on the information retention, density, and energetics of DNA for a variety of applications.

12.6   CHALLENGES

As DNA synthesis technologies become increasingly cheaper and effective, and the reliance on traditional molecular cloning continues to diminish, synthetic biology will enter a *log phase* of even rapider growth. Nevertheless, the ability to predictively design biological circuits and systems will be limited by our ability to model circuit behaviour and iteratively improve the models based on experimental datasets. For greater success in synthetic biology approaches, it is important to be able to reliably compose parts into larger circuits—and handle the ensuing cross-talk with the existing complex cellular circuitry as well.

It will be very interesting to see what newer design principles can be uncovered by studying varied organisms and their circuitry. While many 'standard' circuits such as integral feedback loops and logic gate-based circuits have been designed, there remains much room to explore more novel circuit designs that do not borrow from our current knowledge of engineering design elements. The remarkable plasticity of biological systems and millions of years of evolution have resulted in many novel circuits that are capable of unique functionalities.

True modularity is a somewhat *ideal* concept—in every real system, there are always interactions between modules resulting in *retroactivity*—the phenomenon by which the behaviour of an upstream component is affected by the connection to one or more downstream components [56]. Retroactivity has major implications for the design and construction of biological circuits and has been the subject of many investigations [57–59]. Retroactivity plays a significant role in design failures; further, the *context* into which a circuit is integrated also has major implications for its function. Cardinale and Arkin present a very interesting discussion of sources of design failures in synthetic biological systems [60].

Noise is another major challenge in biological systems, both when we study them and when we attempt to engineer them [61]. Although noise has a generally more positive connotation in biological systems compared to engineering, where noise is almost always despised, it can prove to be challenging in the modelling and characterisation of circuit behaviours. Also, noise propagation can be another significant challenge to surmount during circuit design and implementation.

The field of molecular computation has witnessed several advances over the last two decades. Despite the inherent parallelism and energy efficiency, molecular computation presents many challenges, from the "unreliable" behaviour of biological macromolecules to the design of new algorithms to exploit the new computational paradigms that molecular computation brings forth. Several strides have been made in DNA data storage too, over the last few years, and it seems poised to go beyond an expensive lab experiment to an unusual yet effective form of data archival, sometime in the present decade.

12.7   SOFTWARE TOOLS

iBioSim [62] (https://async.ece.utah.edu/tools/ibiosim/) was primarily developed for the design, analysis, and modelling of genetic circuits. It can also analyse models representing metabolic networks, cell signalling pathways, and other biochemical systems.

Cello ⌊27⌋ (see §12.3.3) is available from https://github.com/CIDARLAB/cello.

EXERCISES

12.1 The iGEM website hosts extensive details on all the projects done till date. Identify any project of your interest from the iGEM (your country likely has one or more iGEM teams, wherever you are!) and study the various parts used by the team. For instance, the team from IIT Madras in 2019 used the following parts: BBa_K3163001, BBa_K3163002, BBa_K3163003, BBa_K3163004 (http://parts.igem.org/cgi/partsdb/pgroup.cgi?pgroup=iGEM2019&group=IIT-Madras).

12.2 Approximately how much data can you store in 1 ng of double-stranded DNA? Assume that the average weight of a DNA base-pair is 620 Da. State any assumptions you make.

12.3 Building on the above calculations, calculate the information density of DNA (in bits/mm^3) and compare it with that of a large capacity micro-SDXC card[7].

---

[7]See reference [63] for similar calculations for a variety of data storage devices.

[LAB EXERCISES]

12.4 Simulate the repressilator using the ODEs from Equation 12.1. You can obtain the model from the BioModels repository[8]. Can you tinker with the parameters α and α_0 and observe their effect on the oscillations.

12.5 Can you extend the above ODEs to 'larger ring' repressilators of $n = 4$ and $n = 5$ proteins, cyclically repressing one another's expression, in similar fashion? Do you still see oscillations? Are the parameter sets that display oscillation fewer in number or more in number?

12.6 Repeat the simulations of Ma *et al* [17] for a couple of example networks and measure the fraction of (randomly sampled) parameter sets for which each network exhibits perfect adaptation.

12.7 Can you design a set of (different) DNA sequences for solving Adleman's original Hamiltonian problem?

REFERENCES

[1] R. Kelwick, L. Bowater, K. H. Yeoman, and R. P. Bowater, "Promoting microbiology education through the iGEM synthetic biology competition", *FEMS Microbiology Letters* **362**:fnv129 (2015).

[2] R. P. Shetty, D. Endy, and T. F. Knight, "Engineering BioBrick vectors from BioBrick parts", *Journal of Biological Engineering* **2**:5 (2008).

[3] G. Røkke, E. Korvald, J. Pahr, O. Øys, and R. Lale, "BioBrick assembly standards and techniques and associated software tools", in *DNA Cloning and Assembly Methods*, Methods in Molecular Biology (Totowa, NJ, 2014), pp. 1–24.

[4] B. Canton, A. Labno, and D. Endy, "Refinement and standardization of synthetic biological parts and devices", *Nature Biotechnology* **26**:787–793 (2008).

[5] J. B. Lucks, L. Qi, W. R. Whitaker, and A. P. Arkin, "Toward scalable parts families for predictable design of biological circuits", *Current Opinion in Microbiology* **11**:567–573 (2008).

[6] D. E. Cameron, C. J. Bashor, and J. J. Collins, "A brief history of synthetic biology", *Nature Reviews Microbiology* **12**:381–390 (2014).

[7] T. S. Gardner, C. R. Cantor, and J. J. Collins, "Construction of a genetic toggle switch in *Escherichia coli*", *Nature* **403**:339–342 (2000).

[8] J. Stricker, S. Cookson, M. R. Bennett, W. H. Mather, L. S. Tsimring, and J. Hasty, "A fast, robust and tunable synthetic gene oscillator", *Nature* **456**:516–519 (2008).

---

[8] https://www.ebi.ac.uk/biomodels/BIOMD0000000012

[9] J. J. Tabor, H. M. Salis, Z. B. Simpson, et al., "A synthetic genetic edge detection program", *Cell* **137**:1272–1281 (2009).

[10] C. Liao, A. E. Blanchard, and T. Lu, "An integrative circuit-host modelling framework for predicting synthetic gene network behaviours", *Nature Microbiology* **2**:1658–1666 (2017).

[11] L. Potvin-Trottier, N. D. Lord, G. Vinnicombe, and J. Paulsson, "Synchronous long-term oscillations in a synthetic gene circuit", *Nature* **538**:514–517 (2016).

[12] J. Bonnet, P. Yin, M. E. Ortiz, P. Subsoontorn, and D. Endy, "Amplifying genetic logic gates", *Science* **340**:599–603 (2013).

[13] A. E. Friedland, T. K. Lu, X. Wang, D. Shi, G. Church, and J. J. Collins, "Synthetic gene networks that count", *Science* **324**:1199–1202 (2009).

[14] M. Xie and M. Fussenegger, "Designing cell function: assembly of synthetic gene circuits for cell biology applications", *Nature Reviews Molecular Cell Biology* **19**:507–525 (2018).

[15] P.-F. Xia, H. Ling, J. L. Foo, and M. W. Chang, "Synthetic genetic circuits for programmable biological functionalities", *Biotechnology Advances* **37**:107393 (2019).

[16] T. K. Lu, A. S. Khalil, and J. J. Collins, "Next-generation synthetic gene networks", *Nature Biotechnology* **27**:1139–1150 (2009).

[17] W. Ma, A. Trusina, H. El-Samad, W. A. Lim, and C. Tang, "Defining network topologies that can achieve biochemical adaptation", *Cell* **138**:760–773 (2009).

[18] R. P. Araujo and L. A. Liotta, "The topological requirements for robust perfect adaptation in networks of any size", *Nature Communications* **9**:1757 (2018).

[19] T. Drengstig, H. R. Ueda, and P. Ruoff, "Predicting perfect adaptation motifs in reaction kinetic networks", *The Journal of Physical Chemistry B* **112**:16752–16758 (2008).

[20] S. Waldherr, S. Streif, and F. Allgöwer, "Design of biomolecular network modifications to achieve adaptation", *IET Systems Biology* **6**:223–231 (2012).

[21] T. M. Yi, Y. Huang, M. I. Simon, and J. Doyle, "Robust perfect adaptation in bacterial chemotaxis through integral feedback control", *Proceedings of the National Academy of Sciences USA* **97**:4649–4653 (2000).

[22] P. Bhattacharya, K. Raman, and A. K. Tangirala, "A systems-theoretic approach towards designing biological networks for perfect adaptation", *IFAC-PapersOnLine*, 5th IFAC Conference on Advances in Control and Optimization of Dynamical Systems ACODS 2018 **51**:307–312 (2018).

[23] D. Del Vecchio, "A control theoretic framework for modular analysis and design of biomolecular networks", *Annual Reviews in Control* **37**:333–345 (2013).

[24] D. Del Vecchio, A. J. Dy, and Y. Qian, "Control theory meets synthetic biology", *Journal of the Royal Society Interface* **13**:20160380 (2016).

[25] V. Hsiao, A. Swaminathan, and R. M. Murray, "Control theory for synthetic biology: recent advances in system characterization, control design, and controller implementation for synthetic biology", *IEEE Control Systems Magazine* **38**:32–62 (2018).

[26] V. Kulkarni, G.-B. Stan, and K. Raman, eds., *A Systems Theoretic Approach to Systems and Synthetic Biology II: Analysis and Design of Cellular Systems* (Springer Netherlands, 2014).

[27] A. A. K. Nielsen, B. S. Der, J. Shin, et al., "Genetic circuit design automation", *Science* **352**:aac7341 (2016).

[28] B. Salvado, H. Karathia, A. U. Chimenos, et al., "Methods for and results from the study of design principles in molecular systems", *Mathematical Biosciences* **231**:3–18 (2011).

[29] G. Sambamoorthy, H. Sinha, and K. Raman, "Evolutionary design principles in metabolism", *Proceedings of the Royal Society B: Biological Sciences* **286**:20190098 (2019).

[30] R. Mani, R. P. St.Onge, J. L. Hartman, G. Giaever, and F. P. Roth, "Defining genetic interaction", *Proceedings of the National Academy of Sciences USA* **105**:3461–3466 (2008).

[31] E. Marland, A. Prachumwat, N. Maltsev, Z. Gu, and W.-H. Li, "Higher gene duplicabilities for metabolic proteins than for nonmetabolic proteins in yeast and *E. coli*", *Journal of Molecular Evolution* **59**:806–814 (2004).

[32] D. Vitkup, P. Kharchenko, and A. Wagner, "Influence of metabolic network structure and function on enzyme evolution", *Genome Biology* **7**:R39 (2006).

[33] G. Sambamoorthy and K. Raman, "Understanding the evolution of functional redundancy in metabolic networks", *Bioinformatics* **34**:i981–i987 (2018).

[34] G. P. Wagner and L. Altenberg, "Perspective: complex adaptations and the evolution of evolvability", *Evolution* **50**:967–976 (1996).

[35] L. H. Hartwell, J. J. Hopfield, S. Leibler, and A. W. Murray, "From molecular to modular cell biology", *Nature* **402**:C47–C52 (1999).

[36] N. Kashtan and U. Alon, "Spontaneous evolution of modularity and network motifs", *Proceedings of the National Academy of Sciences USA* **102**:13773–13778 (2005).

[37] R. Guimerà and L. A. Nunes Amaral, "Functional cartography of complex metabolic networks", *Nature* **433**:895–900 (2005).

[38] A. Hintze and C. Adami, "Evolution of complex modular biological networks", PLoS Computational Biology 4:e23 (2008).

[39] A. Kreimer, E. Borenstein, U. Gophna, and E. Ruppin, "The evolution of modularity in bacterial metabolic networks", Proceedings of the National Academy of Sciences USA 105:6976–6981 (2008).

[40] M. Parter, N. Kashtan, and U. Alon, "Environmental variability and modularity of bacterial metabolic networks", BMC Evolutionary Biology 7:169 (2007).

[41] S. J. Gould and E. S. Vrba, "Exaptation—a missing term in the science of form", Paleobiology 8:4–15 (1982).

[42] A. Barve and A. Wagner, "A latent capacity for evolutionary innovation through exaptation in metabolic systems", Nature 500:203–206 (2013).

[43] B. Szappanos, J. Fritzemeier, B. Csörgő, et al., "Adaptive evolution of complex innovations through stepwise metabolic niche expansion", Nature Communications 7:11607 (2016).

[44] M. G. Plach, B. Reisinger, R. Sterner, and R. Merkl, "Long-term persistence of bi-functionality contributes to the robustness of microbial life through exaptation", PLoS Genetics 12:e1005836 (2016).

[45] J. Stelling, U. Sauer, Z. Szallasi, F. J. Doyle, and J. Doyle, "Robustness of cellular functions", Cell 118:675–685 (2004).

[46] D.-H. Le and Y.-K. Kwon, "A coherent feedforward loop design principle to sustain robustness of biological networks", Bioinformatics 29:630–637 (2013).

[47] J. von Neumann, "Probabilistic logics and the synthesis of reliable organisms from unreliable components", in Automata studies. (AM-34) (Princeton, 1956), pp. 43–98.

[48] L. M. Adleman, "Molecular computation of solutions to combinatorial problems", Science 266:1021–1024 (1994).

[49] R. S. Braich, N. Chelyapov, C. Johnson, P. W. K. Rothemund, and L. Adleman, "Solution of a 20-variable 3-SAT problem on a DNA computer", Science 296:499–502 (2002).

[50] A. Tamsir, J. J. Tabor, and C. A. Voigt, "Robust multicellular computing using genetically encoded NOR gates and chemical 'wires'", Nature 469:212–215 (2011).

[51] Y. Benenson, T. Paz-Elizur, R. Adar, E. Keinan, Z. Livneh, and E. Shapiro, "Programmable and autonomous computing machine made of biomolecules", Nature 414:430–434 (2001).

[52] L. Qian and E. Winfree, "Scaling up digital circuit computation with DNA strand displacement cascades", Science 332:1196–1201 (2011).

[53] Y. Benenson, "Biomolecular computing systems: principles, progress and potential", Nature Reviews Genetics 13:455–468 (2012).

[54]   L. Ceze, J. Nivala, and K. Strauss, "Molecular digital data storage using DNA", *Nature Reviews Genetics* **20**:456–466 (2019).

[55]   G. M. Church and E. Regis, *Regenesis: how synthetic biology will reinvent nature and ourselves* (Basic Books, 2014).

[56]   D. Del Vecchio, A. J. Ninfa, and E. D. Sontag, "Modular cell biology: retroactivity and insulation", *Molecular Systems Biology* **4**:161 (2008).

[57]   A. Gyorgy and D. Del Vecchio, "Modular composition of gene transcription networks", *PLoS Computational Biology* **10**:e1003486 (2014).

[58]   L. Pantoja-Hernández, E. Álvarez-Buylla, C. F. Aguilar-Ibáñez, A. Garay-Arroyo, A. Soria-López, and J. C. Martínez-García, "Retroactivity effects dependency on the transcription factors binding mechanisms", *Journal of Theoretical Biology* **410**:77–106 (2016).

[59]   R. Shah and D. Del Vecchio, "Signaling architectures that transmit unidirectional information despite retroactivity", *Biophysical Journal* **113**:728–742 (2017).

[60]   S. Cardinale and A. P. Arkin, "Contextualizing context for synthetic biology – identifying causes of failure of synthetic biological systems", *Biotechnology Journal* **7**:856–866 (2012).

[61]   L. S. Tsimring, "Noise in biology", *Reports on Progress in Physics* **77**:026601 (2014).

[62]   L. Watanabe, T. Nguyen, M. Zhang, et al., "iBioSim 3: a tool for model-based genetic circuit design", *ACS Synthetic Biology* **8**:1560–1563 (2019).

[63]   G. M. Church, Y. Gao, and S. Kosuri, "Next-generation digital information storage in DNA", *Science* **337**:1628–1628 (2012).

FURTHER READING

P. E. M. Purnick and R. Weiss, "The second wave of synthetic biology: from modules to systems", *Nature Reviews Molecular Cell Biology* **10**:410–422 (2009).

O. Purcell, N. J. Savery, C. S. Grierson, and M. di Bernardo, "A comparative analysis of synthetic genetic oscillators", *Journal of the Royal Society Interface* **7**:1503–1524 (2010).

C. D. Smolke, "Building outside of the box: iGEM and the BioBricks Foundation", *Nature Biotechnology* **27**:1099–1102 (2009).

A. J. Ruben and L. F. Landweber, "The past, present and future of molecular computing", *Nature Reviews Molecular Cell Biology* **1**:69–72 (2000).

CHAPTER 13

Robustness and evolvability of biological systems

CONTENTS

W HAT is robustness? Broadly speaking, it is the ability of a system to continue to function in the face of perturbations [1]. However, one must take care to qualify this statement more carefully. It is not possible for any system to maintain *all* of its functions under the influence of *all* possible perturbations. Therefore, an operational definition of robustness [2] must specify (i) the specific characteristics/behaviours that remain unchanged and (ii) the disturbances/perturbations for which these hold. In the context of the model of a system, robustness can be quantified in terms of the parameter space in which a desired specific behaviour (*e.g.* oscillation, or adaptation) is observed. A more casual definition of robustness would be *the ability to remain unchanged in the face of perturbations*—this is again incorrect, because even a robust system maintains only *specific* functions, by altering various components or the underlying network to finally result in a system whose behaviour remains largely constant or undisturbed. No system can keep all of its components, networks and modes of operation unchanged in perturbation.

Another concept that is intimately connected with robustness is *evolvability*, again a distinguishing feature of biological systems. Evolvability is simply the ability of a biological system to produce heritable phenotypic variation [3]. Of course, these phenotypes are non–lethal and are produced via genetic variation, or mutations. More broadly, Andreas Wagner [3] defines evolvability as the *"ability to produce phenotypic diversity, novel solutions to the problems organisms face, and evolutionary innovations"*.

13.1 ROBUSTNESS IN BIOLOGICAL SYSTEMS

Robustness is a very defining characteristic of biological systems. Unlike most engineered systems, which are often designed to be robust, biological systems are typically *naturally* and inherently robust. Biological systems maintain many phenotypes in the face of various perturbations, both genetic and environmental. Notably, this robustness is exhibited at different levels—from that of simple cellular processes such as chemotaxis [4], to system-level homoeostasis.

In the context of the various systems we have seen thus far, metabolism is a classic example of a highly robust system. *E. coli* can accommodate several deletions to its genes, without majorly altering its growth rate[1]. Similarly, scale-free networks can tolerate the removal of multiple (non-hub) nodes, without a significant impact on network connectivity/communication. In the following section, we will look at some of the key mechanisms that facilitate robustness.

13.1.1 Key mechanisms

Biological systems exploit several key devices or mechanisms to achieve robustness; Kitano [5] presents an excellent account of these mechanisms, also juxtaposing them with those observed in engineered human-made systems.

13.1.1.1 Redundancy and diversity

A common and important device for robustness is redundancy, or the existence of multiple alternate components/fail-safe mechanisms to achieve a particular end. For instance, there could be a pair of isozymes in a cell, which can compensate for one another in the event of the loss of a single enzyme. In this case, the alternate components may be nearly identical. On the other hand, systems may also have diverse mechanisms, where components very different from one another can compensate for losses. Such "distributed robustness" [6] is a common occurrence in evolved biological systems, where there exist multiple heterogeneous components and modules with overlapping functions. This is also a consequence of events such as gene duplications in the evolutionary history of the organism.

Synthetic lethals observed in metabolic networks (§10.2) are nice examples of the remarkable redundancy in metabolism; for instance, ATP synthesis in the cell can be achieved by oxidative phosphorylation or substrate–level phosphorylation, and genes in these pathways will be synthetically lethal with one another. The ability to switch between these alternate pathways gives cells a greater ability to survive in a variety of conditions.

---

[1]Of course, *E. coli* achieves this by re-routing fluxes through alternate pathways. Thus, while 'growth' is robust to gene deletions, the pathway fluxes themselves cannot be!

13.1.1.2   Feedback control

Another classic device for robustness is the existence of control mechanisms, which typically rely on feedback. Negative feedback loops, commonly found in human-made control systems, enable the sensing of disturbances and adjusting the system to counteract. Positive feedback loops are also found in biological systems—these typically contribute to a switch-like behaviour, such as those observed in *Drosophila melanogaster* segment polarity network. Feedback loops also abound in metabolic networks, where the fluxes of metabolites are carefully controlled. System control through feedback is critical, even to leverage the compensatory functions of alternate components/mechanisms described above.

13.1.1.3   Modularity

Modularity is an effective mechanism, and a common engineering design strategy as well, to segregate functions and contain perturbations and damage locally, in complex systems. Modules are easier to define in the context of graphs (as in §3.3.1) but somewhat harder to define unambiguously in the context of complex cellular networks. Modularity can exist in different levels—spatial, temporal, and also functional. Compartmentalisation of cells in higher organisms is a classic example of modularity, particularly spatial isolation. Modularity has been studied in a variety of networks (see §12.4.2), and is a commonly observed design principle in metabolic networks [7].

13.1.1.4   Decoupling

Decoupling involves the isolation of low-level variation in a system from high-level functionalities. It is very common to find different levels of *buffering* in cells, where mutations or molecular-level perturbations and genetic variation [8] are decoupled from the phenotype. Decoupling and buffering ensure that slight changes in stimuli, noise, and small genetic perturbations do not produce significant impacts on cellular phenotypes.

13.1.2   Hierarchies and protocols

The mere existence of various mechanisms for facilitating robustness would be insufficient in the absence of organisational structures for embedding the different modules and efficient means for their reliable co-ordination. Hierarchies and protocols facilitate robustness by fulfilling this requirement.

Protocols encompass the set of rules that govern the exchange of information between different parts of a system. They facilitate co-ordination between diverse components. Much like Lego bricks that snap together (the patented 'snap connection' being the protocol), cells are connected by means of shared energy currencies and reduction equivalents, shared mechanisms of gene expression, phosphorylation, and so on [2, 9]. Notably, these systems are also organised hierarchically, to facilitate *layered* regulation.

13.1.3 Organising principles

Are there universal design principles underlying robust cellular systems? Do they come with inherent risks and limitations? Indeed, we have already studied such a design principle—*power laws*. Power-law systems that we studied in Chapter 3 exhibit robustness—to random failures. They have a peculiar organisation with a few hubs and many peripheral nodes. They also have the inherent risk of susceptibility to targeted attacks on the hub nodes, as we discussed in §3.5.2.

Kitano [5] argues that there are a few key architectural requirements for robustness: (i) the mechanisms that preserve the components and interactions in a network against perturbations (mutations) must be capable of generating genetic variation, (ii) modules must exist, which preserve their functions against mutations/perturbations, and (iii) building blocks of conserved core processes, which are modular and serve fundamental functions like metabolism, cell cycle or transcription; these modules can be wired together to create diverse phenotypes. One architectural paradigm that satisfies these requirements is a modularised nested *bow-tie*, or hour-glass structure (if you rotate it 90°), where input and output modules are connected to a conserved core.

13.1.3.1 Bow-tie structure

The complex network of biochemical conversions underlying metabolism can be viewed as a bunch of nested bow-ties, as illustrated in Figure 13.1. The bow-ties produce activated carriers and energy currencies, such as ATP, NAD, and NADP. The bow-tie for catabolism produces several key precursor metabolites, which form the starting points for the biosynthesis of carbohydrates, fatty acids, amino acids, nucleotides, and co-factor building blocks. The presence of a 'conserved core' at the 'knot' facilitates control, organisation, and management of the heterogeneities in metabolic networks, arising from enzyme specificity, action and regulation, as well as variations in the flux and concentration of various substrates.

Note that the input and output 'fans' have high variability and few constraints, and are general purpose and flexible. They mostly comprise linear pathways for catabolism or anabolism. These pathways are also dispensable and have many alternate routes, leading to robustness. Thus, complex macromolecules in the input are broken down into small building blocks (that are relatively fewer in number), which form the basis for the synthesis of various complex macromolecules, such as proteins, lipids, or carbohydrates in living cells. The core, on the other hand, is very specialised, rigid, efficient, and exhibits little variability. Notably, this hour-glass or bow-tie architecture is a distinctive feature of many modern technologies, from the power grid to the Internet [10]. Alon and co-workers have illustrated how bow-tie architectures can evolve in biological systems, by means of *in silico* simulations of systems evolving to fulfil a given input–output goal [11].

Bow-ties present many functional advantages. Imagine a *flat* architecture, where individual specialised pathways connect multiple substrates to multiple products. Such a system would be highly inefficient on the counts of high

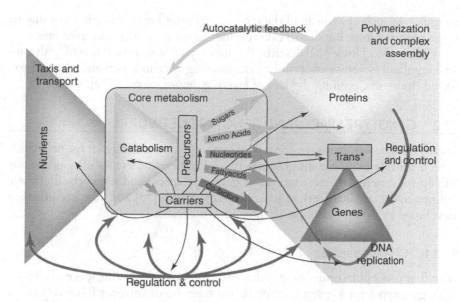

Figure 13.1 **The nested bow-tie architectures of metabolism.** The bow-ties take as input a wide range of nutrients and produce a large variety of products and complex macromolecules using a relatively few intermediate common currencies. The common currencies and their enzymes form the 'knot' of the bow-tie. The overall bow-tie can be decomposed into three principal subsidiary bow-ties, as discussed in reference [10]. Reprinted from M. Csete and J. Doyle, "Bow ties, metabolism and disease", *Trends in Biotechnology* **22**:446–450 (2004), © (2004), with permission from Elsevier.

enzyme complexity, rather difficult to regulate, and utterly unevolvable. On the other hand, bow-ties can accommodate divergent demands on metabolic systems, achieve high throughputs of metabolites with relatively few specialised enzymes and also facilitate regulation at various time-scales. Most importantly, the shared interfaces and protocols enable *plug-and-play*, where less central reactions and pathways (in the periphery) can be readily exchanged or added. The core, of course, is tightly regulated, highly conserved, and not very open to evolution.

Bow-ties also bring in their share of risks. Unlike simpler systems, complex systems are often associated with fragilities that result from specific faults. Much like scale-free networks that are vulnerable to targeted attacks on their hubs, bow-tie architectures are fragile at their core, and failures in the core will affect the entire system. The core is also not very variable; for instance, we have pathways such as the TCA cycle at the heart of catabolism, which is ubiquitous in all living cells. Thus, while such architectures contribute to remarkable robustness on certain axes, they do possess an Achilles' heel, namely specific vulnerabilities.

Fragility in such systems is all the more interesting because it arises not due to large perturbations, but due to catastrophic responses to small perturbations.

Carlson and Doyle [12] describe this *robust-yet-fragile* behaviours of both natural (evolved) systems and optimised engineering systems as consequences of profound trade-offs, positing the idea of *highly optimised tolerance* [12].

13.2 GENOTYPE SPACES AND GENOTYPE NETWORKS

In this section, we discuss the concept of genotype spaces and genotype networks, which provide excellent insights into the robustness and evolvability of various biological systems. These concepts of genotype spaces and networks find their origins in the seminal work of John Maynard-Smith [13] and have been further developed by various scientists in more recent decades [14, 15].

13.2.1 Genotype spaces

The first concept to understand is that of a genotype space—the space of all possible genotypes for a given system. To illustrate, let us consider RNA sequences of length 30. If we enumerate all possible 30-mer RNA molecules, there are obviously $4^{30} \approx 10^{18}$ of them, making up the *discrete* genotype space. This space can be thought to be connected, with edges connecting genotypes that vary in a single nucleotide, *i.e.* a point mutation. How many neighbours does every RNA sequence have? Every base in a sequence of length L can be replaced by the remaining three bases[2]; thus, an RNA sequence of length L will have 3L neighbours. Indeed, this network is a regular lattice, with each of the 4^L nodes connected to 3L neighbours. Once we have understood the genotype space, we now need to map phenotypes on to the genotypes. Figure 13.2a illustrates the genotype space for an arbitrary system.

13.2.2 Genotype–phenotype mapping

While the genotype denotes the genetic makeup, the phenotype refers to higher-level observable characteristics or traits of an organism. At the molecular level, features such as protein structures, RNA structures or folds can be regarded as molecular phenotypes. At a higher level of organisation, namely the network level, more complex phenotypes can be considered. Table 13.1 lists some examples of genotype–phenotype mapping from RNA to metabolic networks to artificial human-made systems such as digital circuits. Once we determine the mapping between genotype and phenotype—this could be via an experiment, or theoretically, via computation—we can unravel the various genotype networks that span the genotype space. Figure 13.2b shows phenotypes mapped on to genotypes using a variety of colours. Evidently, the genotype space is laced with numerous genotype networks, of different sizes and connectivities.

---

[2]A can be replaced by U, C, or G, for example.

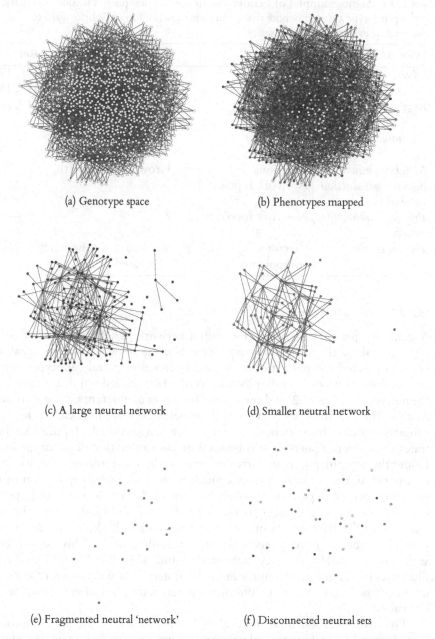

(a) Genotype space

(b) Phenotypes mapped

(c) A large neutral network

(d) Smaller neutral network

(e) Fragmented neutral 'network'

(f) Disconnected neutral sets

Figure 13.2 **A schematic representation of genotype space and neutral networks.** Every node in each of the networks represents a genotype, say RNA. Two nodes are connected by an edge if they differ by a single mutation. Nodes of the same colour, in panels b–f have the same phenotype, *e.g.* RNA structure. Note, however, that real genotype spaces harbour astronomical numbers of genotypes.

TABLE 13.1 **Some examples of genotype–phenotype mapping.** The question marks in the phenotype column denote that phenotypes are not too straightforward to characterise in some of these cases.

System	Genotype	Phenotype	Ref.
RNA	Sequence	Secondary structure	[16, 17]
Proteins	Sequence	Structure/fold	[15, 18]
Regulatory networks	Regulatory interactions	Gene expression patterns	[19, 20]
Circadian oscillators	Regulatory interactions	Oscillations	[21]
Metabolic networks	Reactions	Growth/No-growth	[22]
Signal transduction networks	Network topology	?	[23]
Protein interaction networks	Network topology	?	—
Digital circuits	Circuit wiring/ components	Boolean function computed	[24]

13.2.2.1 Neutral networks

A genotype network, also called a neutral network [13, 14], is a set of genotypes that share the same phenotype. Notably, the network (strictly speaking, every connected component of the network) contains several genotypes, which are reachable from one another by a series of mutations (edges) that do not leave the network. Figure 13.2c–f show neutral networks of different sizes and connectivities. The size and connectivity of these networks have implications for both robustness and evolvability of the system, as we will see shortly. Figure 13.3 illustrates the impact of phenotype robustness on the exploration of genotype space. Evidently, a population that starts evolving on the more robust network (bottom panels in the figure) achieves a much more widespread exploration of the genotype space, compared to those evolving on the less robust network (top panels). This also has an interesting consequence for evolvability; the population that has reached the far confines of the neutral network will likely have access to a greater number of novel phenotypes in its neighbourhood. This has also been well demonstrated by a variety of theoretical studies [17, 22]. Figure 13.4 further illustrates this concept, by illustrating neutral networks with different sizes and connectivities, and how they differ in their access to innovation in their neighbourhood.

Theoretically, robustness and evolvability need not co-exist. As it is apparent from Figure 13.4, robust neutral networks, depending on their structure, may or may not have access to novel phenotypes in the neighbourhood. A large connected neutral network is nevertheless robust—this is because a large number of genotypes have the same phenotype; further any genotype on the network can accommodate a series of mutations, without losing its phenotype (*i.e.* falling off

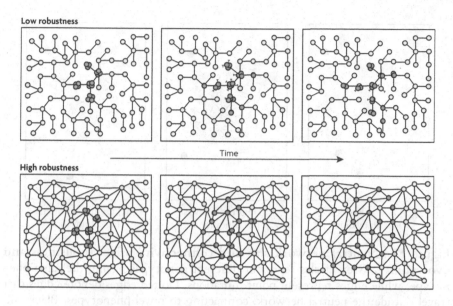

Figure 13.3 **Robust phenotypes facilitate widespread exploration of geno-type space.** Every node in the panels represents a genotype. Two genotypes are connected if they differ by a single mutation. The panels on the top correspond to a phenotype/neutral network with low robustness, while those on the bottom correspond to neutral networks of high robustness. The coloured circles represent individual members of a population evolving on these networks; the stubs correspond to hypothetical deleterious mutations, which cause the individual to fall off the neutral network (owing to a mutation to a genotype not on the present neutral network). From left to right, we see the population evolving over time. All panels contain the same number of genotypes, to facilitate comparison. Reprinted by permission from Springer Nature: A. Wagner, "Neutralism and selectionism: a network-based reconciliation", *Nature Reviews Genetics* 9:965–974 (2008), © Springer Macmillan Publishers Limited (2008).

the network). Thus, while robustness can be gauged from the size or density of the network, access to novel phenotypes is less obviously inferred—although, it is very likely that large neutral networks have greater access to novel phenotypes in their neighbourhood.

13.3   QUANTIFYING ROBUSTNESS AND EVOLVABILITY

The paradigms discussed above, namely that of genotype spaces and networks, and the expository toy examples in Figure 13.3 and Figure 13.4 provide interesting insights into the relationships between robustness and evolvability. However, for a better understanding of robustness and evolvability, as well as their inter-relationships, it is essential to *quantify* them. A quantitative understanding can also

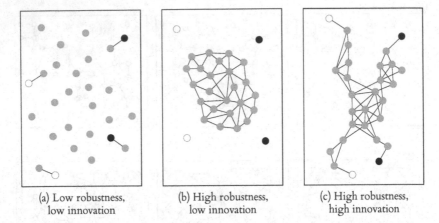

(a) Low robustness,　　　(b) High robustness,　　　(c) High robustness,
　low innovation　　　　　low innovation　　　　　high innovation

Figure 13.4 **Neutral network size and connectivity affect robustness and evolvability.** Grey circles represent different genotypes having the same phenotype, while edges represent point mutations. Dotted edges denote edges that travel outside the neutral network, connecting to novel phenotypes. Black and open circles represent genotypes with novel phenotypes. All panels contain the same number of nodes (24); the neutral networks in panels (b) and (c) also have the same number of edges, but vary markedly in their robustness and evolvability. Adapted with permission, from S. Ciliberti, O. C. Martin, and A. Wagner, "Innovation and robustness in complex regulatory gene networks", *Proceedings of the National Academy of Sciences USA* **104**:13591–13596 (2007). © (2007) National Academy of Sciences, U.S.A.

help resolve the apparent antagonism between robustness and evolvability. For, superficially, robustness refers to an ability to resist change, whereas evolvability involves the ability to produce phenotypic changes.

In particular, it is helpful to resolve the robustness and evolvability of the genotype and phenotype separately. Building on the concepts of genotype networks discussed above, Wagner defined robustness and evolvability for genotypes and phenotypes separately as follows [17]:

Genotype robustness The number of *neutral neighbours* of a genotype **G**; a neutral neighbour of a particular genotype has the same phenotype.

Phenotype robustness The number of neutral neighbours averaged over all genotypes, with a given phenotype **P** (average genotype degree in the neutral network)

Genotype evolvability The number of different phenotypes found in the *1-neighbourhood* of a genotype **G**. The 1-neighbourhood of a phenotype or, equivalently, the 1-neighbourhood of a neutral network is the set of

Quantifying model robustness

Often, the model of a system provides a window into system behaviour, which can then be examined for its robustness. Although there are experimental measures of robustness to perturbations for a few systems, a much larger number of theoretical studies have examined diverse systems such as regulatory, signalling, or metabolic networks for their robustness.

There are broadly two aspects to look at when studying the robustness of a model: (i) parametric robustness and (ii) topological robustness. In parametric robustness, we study the variation of model behaviour/output—this could potentially be measured in terms of a sum square error—with respect to changes in one or more parameters. This is similar to the sensitivity analyses described in §6.4.2. In the case of topological robustness, we study the variation of model behaviour to changes in model topology, *i.e.* the gain or loss of one or more interactions (see §6.4.2.4). One might also think of a topological change as an extreme variation of a parameter, all the way up to zero, but this is atypical; parameters are normally varied around a given point by a few per cent.

It is very interesting to study the region(s) in *parameter space*, or *topology space*, where a model shows similar behaviour. A model may exhibit one or more *viable* regions where it shows *acceptable* behaviour—typically, this would mean that the model fit, as estimated by an objective function that evaluates predictions against data, is at *acceptable* levels. Different models for a given system may vary in the size (volume) of the viable region, the number of such regions and how distributed they are in (the respective) space, as shown below. Again, this can be very useful for informing model selection.

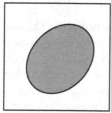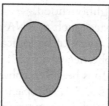

Consider each of the diagrams above to represent a region of (two-dimensional) parameter space, with the shaded regions representing viable regions of the space. While the viable volume, or area, in this case, is the same for all three models, the nature of their shapes, sizes, and connectivity vary. The model corresponding to the left-most panel is most robust, with a single region containing all viable parameter sets, while that corresponding to the right-most panel is least robust, with the viable parameter sets highly disconnected and fragmented in parameter space.

genotypes that differ from the genotypes that exhibit the same phenotype by a point mutation.

Phenotype evolvability The number of different phenotypes found in the 1-neighbourhood of a phenotype **P**, *i.e.* 1-neighbourhood of all genotypes exhibiting the phenotype **P**.

Wagner applied the above definitions to a system of RNA sequences (genotypes) and computationally predicted RNA secondary structures (phenotypes) and showed that genotype robustness and evolvability are negatively correlated. More importantly, he showed that phenotype robustness and evolvability are positively correlated. All of these definitions assume that all neighbours of a genotype are equally likely to be encountered as a result of a mutation. Partha and Raman [26] extended these definitions to scenarios where there are unequal probabilities of mutations to different genotypes, from a given genotype. Interestingly, they found that the correlations held well, despite the fact that assuming equal-likelihood mutations underestimated robustness and overestimated evolvability.

Why do robust phenotypes exhibit higher evolvability? Firstly, robust phenotypes are associated with vastly larger neutral networks, *i.e.* many more sequences in the genotype space fold into the structure corresponding to robust phenotypes. Secondly, the neighbourhood of the sequences on the neutral network contains very many different (novel) structures. This again reiterates the observations made from Figures 13.3 and 13.4c, where the synergism between robustness and evolvability could be observed in populations evolving on a neutral network. On large neutral networks, the populations can more often than not stay *on* the neutral network, suffering fewer losses through deleterious mutations. As a result, they accumulate much higher genotypic diversity, *i.e.* they are well-spread in the networks, which in turn allows them to access much greater phenotypic diversity (novel phenotypes) in their neighbourhood.

Beyond macromolecular spaces such as RNA and proteins [18], such studies have also been extended to a variety of biological networks, such as gene regulatory networks [20], metabolic networks [22], and signalling networks [24], demonstrating similar relationships between robustness and evolvability.

Further, as we discussed in §12.3.1, the space of possible network topologies can be studied to understand both the parametric robustness of a system, as well as its robustness in the face of topological perturbations. Wagner [21] has illustrated the evolution of robustness in two-gene circadian oscillators. By constructing a topology graph—essentially a neutral network—where each node represents a topology capable of showing circadian oscillation, and edges connect topologies that differ by a single regulatory interaction, Wagner showed that there are many topologies that can exhibit robust circadian oscillations. Furthermore, he showed that these topologies are highly connected; that is, it is possible to gradually 'evolve' to any circuit topology within this network[3] from any other topology within such networks.

---

[3]Strictly speaking, within each connected component of this network, the largest of which is most interesting.

Through this chapter, we have scratched the surface of the exciting field of *evolutionary systems biology*. This field is rapidly advancing through a combination of both *in silico* and *in vitro* experiments that are shedding fascinating insights on the organisation of various biological systems, relationships between their structure and function, and consequences for their robustness and evolvability.

13.4   SOFTWARE TOOLS

Genonets Server   [27], available at http://ieu-genonets.uzh.ch aids in performing a suite of analyses to characterise the genotype/phenotype space for different kinds of systems.

EXERCISES

13.1   Consider a 'genotype space' corresponding to peptides of length 50. Assume proteins to evolve by random changes of a single amino acid. What is the size of the genotype space? In the genotype space, what is the number of neighbours for any given genotype?

13.2   Consider the genotype space described in the previous question. Hypothetically, if arginine, alanine, glycine and lysine are present consecutively in positions 24, 25, 26, and 27, then the peptide exhibits a particular function, no matter what the rest of the amino acids are. However, if any of these four amino acids change, the function is lost completely. Based on this information, can you quantify the (genotype) robustness of the 50-mer peptide Gly_{23}-Arg-Ala-Gly-Lys_{24}?

[LAB EXERCISES]

13.3   Consider the Goodwin oscillator [28], also available from BioModels (https://www.ebi.ac.uk/biomodels/MODEL0911532520). This oscillator lies at the core of the study by Wagner [21]. There are $3^6 = 729$ different topologies considered by Wagner. Can you write a piece of code that generates a system of ODEs for any given topology. (Perhaps, an easy way to represent any topology is by a string of six ternary digits (*trits*), for each of the possible interactions). Once you have this piece of code, can you generate 20 topologies at random and quantify their ability to produce circadian oscillations?

13.4   Every topology in the above system has 12 neighbours (Why?). Can you compute the genotype robustness of some one topology? State any assumptions you have to make.

13.5   Write a piece of code that can generate 10^3 random RNA sequences of L = 10. Use the Genonets server to compute basic genotype robustness and evolvability.

REFERENCES

[1] A. Wagner, *Robustness and evolvability in living systems* (Princeton University Press, 2005).

[2] J. Stelling, U. Sauer, Z. Szallasi, F. J. Doyle, and J. Doyle, "Robustness of cellular functions", *Cell* **118**:675–685 (2004).

[3] A. Wagner, "Gene duplications, robustness and evolutionary innovations", *Bioessays* **30**:367–373 (2008).

[4] U. Alon, M. G. Surette, N. Barkai, and S. Leibler, "Robustness in bacterial chemotaxis", *Nature* **397**:168–171 (1999).

[5] H. Kitano, "Biological robustness", *Nature Reviews Genetics* **5**:826–837 (2004).

[6] A. Wagner, "Distributed robustness versus redundancy as causes of mutational robustness", *BioEssays: News and Reviews in Molecular, Cellular and Developmental Biology* **27**:176–188 (2005).

[7] G. Sambamoorthy, H. Sinha, and K. Raman, "Evolutionary design principles in metabolism", *Proceedings of the Royal Society B: Biological Sciences* **286**:20190098 (2019).

[8] J. L. Hartman, B. Garvik, and L. Hartwell, "Principles for the buffering of genetic variation", *Science* **291**:1001–1004 (2001).

[9] M. E. Csete and J. C. Doyle, "Reverse engineering of biological complexity", *Science* **295**:1664–1669 (2002).

[10] M. Csete and J. Doyle, "Bow ties, metabolism and disease", *Trends in Biotechnology* **22**:446–450 (2004).

[11] T. Friedlander, A. E. Mayo, T. Tlusty, and U. Alon, "Evolution of bow-tie architectures in biology", *PLoS Computational Biology* **11**:e1004055 (2015).

[12] J. M. Carlson and J. Doyle, "Highly optimized tolerance: a mechanism for power laws in designed systems", *Physical Review E* **60**:1412–1427 (1999).

[13] J. Maynard Smith, "Natural selection and the concept of a protein space", *Nature* **225**:563–564 (1970).

[14] P. Schuster, W. Fontana, P. F. Stadler, and I. L. Hofacker, "From sequences to shapes and back: a case study in RNA secondary structures", *Proceedings of the Royal Society B: Biological Sciences* **255**:279–284 (1994).

[15] A. Babajide, I. L. Hofacker, M. J. Sippl, and P. F. Stadler, "Neutral networks in protein space: a computational study based on knowledge-based potentials of mean force", *Folding and Design* **2**:261–269 (1997).

[16] L. W. Ancel and W. Fontana, "Plasticity, evolvability, and modularity in RNA", *Journal of Experimental Zoology* **288**:242–283 (2000).

[17] A. Wagner, "Robustness and evolvability: a paradox resolved", *Proceedings of the Royal Society B: Biological Sciences* **275**:91–100 (2008).

[18] E. Ferrada and A. Wagner, "Protein robustness promotes evolutionary innovations on large evolutionary time-scales", *Proceedings of the Royal Society B: Biological Sciences* **275**:1595–1602 (2008).

[19] S. Ciliberti, O. C. Martin, and A. Wagner, "Robustness can evolve gradually in complex regulatory gene networks with varying topology", *PLoS Computational Biology* **3**:e15 (2007).

[20] S. Ciliberti, O. C. Martin, and A. Wagner, "Innovation and robustness in complex regulatory gene networks", *Proceedings of the National Academy of Sciences USA* **104**:13591–13596 (2007).

[21] A. Wagner, "Circuit topology and the evolution of robustness in two–gene circadian oscillators", *Proceedings of the National Academy of Sciences USA* **102**:11775–11780 (2005).

[22] J. F. Matias Rodrigues and A. Wagner, "Evolutionary plasticity and innovations in complex metabolic reaction networks", *PLoS Computational Biology* **5**:e1000613+ (2009).

[23] K. Raman and A. Wagner, "The evolvability of programmable hardware", *Journal of the Royal Society Interface* **8**:269–281 (2010).

[24] K. Raman and A. Wagner, "Evolvability and robustness in a complex signalling circuit", *Molecular BioSystems* **7**:1081–1092 (2011).

[25] A. Wagner, "Neutralism and selectionism: a network-based reconciliation", *Nature Reviews Genetics* **9**:965–974 (2008).

[26] R. Partha and K. Raman, "Revisiting robustness and evolvability: evolution in weighted genotype spaces", *PLoS ONE* **9**:e112792+ (2014).

[27] F. Khalid, J. Aguilar-Rodríguez, A. Wagner, and J. L. Payne, "Genonets server—a web server for the construction, analysis and visualization of genotype networks", *Nucleic Acids Research* **44**:W70–W76 (2016).

[28] B. C. Goodwin, "Oscillatory behavior in enzymatic control processes", *Advances in Enzyme Regulation* **3**:425–438 (1965).

FURTHER READING

J. Masel and M. V. Trotter, "Robustness and evolvability", *Trends in Genetics* **26**:406–414 (2010).

M.-A. Félix and M. Barkoulas, "Pervasive robustness in biological systems", *Nature Reviews Genetics* **16**:483–496 (2015).

M. Khammash, "An engineering viewpoint on biological robustness", *BMC Biology* **14**:22 (2016).

CHAPTER 14

Epilogue: The road ahead

THIRTEEN CHAPTERS AGO, we set out to explore the exciting world of systems biology. Across the four parts of this book, we have forayed into many interesting topics, studied several new algorithms, mathematical tools, and techniques. I hope that, at the end of this exercise, your interest in computational systems biology has been sufficiently piqued, that you can now turn to the latest research papers and texts dedicated to the various modelling paradigms discussed in this book.

Beginning with a quick introduction to the philosophy of systems biology and mathematical modelling, we proceeded to study interaction-based models, *i.e.* static network models. We then studied mechanistic models of dynamical biological systems and constraint-based models of metabolic networks, finally reviewing some of the most cutting-edge advances in the field. Table 14.1 presents a quick overview of the various modelling paradigms discussed in this book, along with the model systems they are typically applicable to, how the methods are parametrised during reconstruction, the typical analyses one performs with these tools and the predictions that they typically turn up, as well as some key advantages and disadvantages of these methods.

Throughout this book, we have emphasised several challenges that a modeller faces in actual practice—these challenges also present many interesting research problems to solve, which can pave the way for an even better understanding of biological complexity. These and many other broader challenges need to be surmounted in the coming years, if we are to conquer biological complexity and predictively manipulate biological systems.

In every part of this book, we have reflected on some of the key applications of the methodologies that we have studied. Broadly, four classes of applications exist: (i) industrial applications, such as strain design for metabolic engineering, (ii) drug target identification—from infectious diseases to cancer, (iii) model-driven discovery of new biology, including prediction of enzyme function or new interactions, and filling knowledge gaps, and (iv) understanding cellular function and the organisational principles of the underlying networks. In all cases, the applications emerge from an analysis of various networks and their properties, typically

TABLE 14.1   An overview of various modelling paradigms discussed in this book.

Model Systems	Parametrisation	Typical Predictions	Advantages	Disadvantages
Static Network Models • Protein interaction networks • Gene networks • Large- and small-scale networks	• Edges representing relationships • Edge weights can quantify strengths/confidences	• Important nodes • Important edges, *i.e.* interactions • Important groups of nodes or sub-networks (clusters, modules, motifs)	• Nice synthesis of known information • Quick and easy	• No dynamics, typically
• Genome-scale metabolic networks	• Network topology	• Reachability	• Work with "draft" reconstructions	• No dynamics
Dynamic Kinetic Models • Small-scale biological processes • Signalling and regulatory networks	• Detailed kinetic parameters, *e.g.* v_{max}, K_M	• Concentrations of different species • Reaction rates • Sensitivity analyses	• Mechanistic insights • Capture dynamics—critical in biology	• Parameter estimation is challenging • Curse of dimensionality • Limited model size
Boolean Models • Small-scale biological processes • Signalling and regulatory networks	• Rules of interaction	• Regulatory states • Key interactions	• Mechanistic insights • Capture dynamics • Capture biological knowledge effectively	• Biological systems are seldom discrete
Constraint-based Models • Genome-scale metabolic networks	• Stoichiometry • Uptake rates, biomass composition	• Possible metabolic states • Key genes, proteins or reactions • Targets for manipulation	• No kinetics required • Enable true systems-level modelling • Mechanistic insights • Versatile	• Not easy to capture metabolite concentrations • More complex methods required to integrate/infer regulation/dynamics

followed by the prediction of cellular phenotypes and/or analyses of various perturbations to the network(s).

In particular, systems-level approaches are gaining importance in studying diseases. In the post-genomic era, it has become all the more apparent that diseases are not merely a result of something going wrong in a single gene, protein or pathway—while such an origin may exist for some diseases, most diseases *are system-level dysfunctions*. Perhaps no other application is as critical as the study of health and disease, using systems-level models.

SYSTEMS BIOLOGY IN HEALTH AND DISEASE

Systems biology of infectious disease

At the time of the writing of this book, the world continues to be ravaged by SARS-CoV-2, with over a million deaths and over 30 million infections. A number of studies have used SIR models (§1.5.2) and variants to model the spread of the disease [1]. Systems approaches have been applied to predict new targets based on constraint-based modelling [2], or possible drug re-purposings [3]. A very recent study in 2020 modelled the interplay between the human coronavirus–host interactome and drug targets in the human PPI network [3]. Three networks were built and integrated: (a) a network comprising 119 host proteins associated with four known human coronaviruses—these proteins were direct targets of the viral proteins or were involved in infection pathways, (b) a drug–target network comprising 2000+ drugs (both experimental and FDA-approved), and (c) the human interactome. They predicted both drugs and drug combinations that can be tested for efficacy: the combinations were captured by a *complementary exposure* pattern, where the targets of the drugs both hit the virus–host sub-network, but fall in separate neighbourhoods in the human interactome.

An interesting approach to study pathogenic organisms is the reconstruction of strain-specific models. Building on the same concepts of metabolic network reconstruction and analysis (Part III and Appendix B), Palsson and co-workers constructed genome-scale models for 64 different strains of *Staphylococcus aureus* [4] and 410 *Salmonella* strains spanning 64 serovars [5]. A comparative analysis of the models could point towards metabolic capabilities that were linked to pathogenic traits. Systems approaches have also been extensively used to study infectious diseases such as tuberculosis, as detailed in reference [6]. For an excellent review of systems biology and its application to infectious disease, see reference [7].

Cancer systems biology

Beyond infectious diseases, systems approaches are perhaps most valuable to dissect cancer—undeniably, cancer represents a very complex system-level dysfunction, with cross-talks between several signalling, metabolic, and regulatory pathways, interactions between various types of cells and the immune system, and so on. Indeed, cancer has been referred to as a "*systems biology disease*" [8, 9].

"The next generation" [10] of the seminal paper on the "hallmarks of cancer" [11] makes several more references to networks, including signalling and regulatory networks, and their cross-talk. Du and Elemento [12] too highlight the importance of a systems-level perspective to complement the reductionist approach, if we are to disentangle the complexity of cancer processes. Systems-level models will be indispensable to identify and validate targets, biomarkers, and other effective therapeutic strategies. Kitano [13] presented a systems perspective of cancer, as a *robust* system, emphasising the importance of modelling in understanding the underlying complex networks.

Metabolic modelling has also been widely used to study cancers, as reviewed in references [14, 15]. Typically, studies focus on capturing the metabolic state of cancer cells and comparing them with healthy cells—this enables the identification of key metabolic differences between healthy and cancerous cells [16, 17]. Models can also extrapolate the effect of drug inhibition of one or more proteins [16] and aid in predicting possible interventions. Algorithms such as mCADRE (see §9.4.3) enable us to leverage the large amounts of transcriptomic data that have been accumulated on cancer cells by mapping them onto tissue-specific metabolic networks, to ultimately unravel key metabolic signatures of cancer cells.

Dynamic models have provided insights into some of the key processes underlying cancer metastasis and progression [18, 19], viz. epithelial–mesenchymal transition (EMT) and the plasticity of these processes, which are central to cancer cell proliferation. The models provided key insights into the design principles of the underlying networks, especially the several feedback loops that promote plasticity. Another study [20] showed that cancer cells possess *hybrid* phenotypes where both oxidative and glycolytic metabolic states co-exist. This is unlike normal cells, which typically switch between the two states. The authors suggest that the hybrid state enhanced the metabolic plasticity of the cells, facilitating both tumorigenesis and metastasis. They also highlighted the role played by drugs in driving cancer cells into or away from the hybrid states, and the consequences for designing new therapies. Agent-based models have also been very useful to study cancer and other diseases at a systems-level, as reviewed in reference [21].

In another very interesting study, Kuenzi and Ideker [22] put together a "census" of pathway maps in cancer—comprising as many as 2,070 maps arising out of a variety of methods to map and model cancer pathways including: (a) genetic interaction networks, (b) protein–protein interaction networks, (c) inferred gene regulatory networks, (d) key sub-networks identified by mapping genomic alteration data onto a protein interaction network, (e) constraint-based models of metabolic networks integrated with gene expression, and (f) mechanistic ODE modelling of signalling networks. This paper presents a classic meta-analysis of the wealth of cancer systems biology literature.

Overall, the stage is set for large multi-scale personalised "whole-cell" models of cancer, which encompass various aspects, beginning with genomic alterations and their consequences, to the multitude of complex signalling networks involved, the metabolism of tumour cells, and their interactions with their microenvironment and also one another, as well as with the human immune system.

Systems biology in drug development: Quantitative Systems Pharmacology

Quantitative Systems Pharmacology (QSP) is the application of systems–level modelling approaches to extend classic pharmacokinetic (PK) and pharmacodynamic (PD) models. Pharmacokinetic models quantitatively describe *"what the body does to the drug"*, essentially the absorption, distribution, metabolism, and elimination (ADME) of a given drug. Pharmacodynamic models describe *"what the drug does to the body"* in terms of the receptors a drug binds to, as well as downstream effects. Overall, QSP brings a systems–level perspective to the classic PK/PD models, and enable the prediction of drug dosage regimens or drug–drug interactions. The PK/PD models are typically dynamic models (the central theme of Part II of this book), based on ODEs. QSP models have also been used by the USFDA (United States Food and Drug Administration) to recommend an alternate dosing regimen for a drug used to control low blood calcium in patients suffering from hypoparathyroidism [23].

QSP models have been developed for a wide variety of diseases, including cardiovascular diseases, infectious diseases, central nervous system diseases, and cancers [24]. The applications of QSP are ever–expanding and include the identification of novel targets for therapeutic interventions, identifying/evaluating potential biomarkers, evaluating combination strategies for new molecules, predicting the potential for human efficacy as well as safety of novel targets/compounds— through a mechanistic perspective, guiding the design of pre–clinical studies [25]. Most importantly, QSP models enable one to ask many 'what-if' questions, and ultimately cut down lengthy timelines for drug discovery and development. QSP models can also be integrated with other models, *e.g.* those of the gut microbiota, to address very interesting questions such as the effect of the gut microbiome on drug metabolism and efficacy [26].

KEY EMERGING TOPICS

Handling heterogeneity: From single systems to whole-cell models

A major challenge in biology, as we discussed in §1.3, is the remarkable heterogeneity of biological systems. Beyond this remarkable inherent heterogeneity of biological systems, we also need to reconcile the heterogeneity of models and methods, which span multiple (heterogeneous) time-scales, and perhaps even length-scales. Such a marriage of disparate models, which independently describe various aspects of a system can be a powerful device to understand biological complexity. Many approaches have been put forth in the past to integrate models of metabolic, regulatory, and signalling networks [27–29], but perhaps none as comprehensive and ambitious as the whole–cell model of *Mycoplasma genitalium*, developed by Covert and co-workers [30].

The whole-cell model of *M. genitalium* integrates as many as 28 sub-models, ranging from those of DNA replication and maintenance, to a genome-scale model of *M. genitalium* metabolism [31], as well as host interactions. Built on the basis of over 900 publications, the model contained over 1,900 experimentally observed parameters. For simulation, each process was independently modelled on a time-scale of 1s, followed by integrating the various sub-models at longer time-scales. The model contained 16 cell variables, which together represented the complete configuration of the modelled cell from metabolite, RNA, and protein copy numbers, to metabolic reaction fluxes, and even the host urogenital epithelium. The website, https://wholecell.org, provides many useful resources and perspectives on whole-cell modelling. We are likely to see more and more detailed models of complex cells, and perhaps even models that capture the interactions between whole-cell models in the coming years.

Another relevant challenge is the integration of heterogeneous multi-omic datasets—especially, transcriptomic, metabolomic, and proteomic (or even fluxomic) data to improve systems-level models, *e.g.* the genome-scale metabolic models (see also §9.4). An increasing amount of such data is becoming available owing to the proliferation of high-throughput technologies; it is easy to envisage that these datasets and their integration into large systems-level models will be a major aspect of systems biology in the coming decade.

Building large mechanistic models

In Part II, we studied both ODE-based dynamic modelling and Boolean network modelling. However, these models have been predominantly restricted to small networks, compared to the models we studied in Parts I and III. Large dynamic models can present insights not easily obtained from large static/steady-state models; still, there are serious challenges in the construction of large dynamic models. In particular, parameter estimation for such large systems becomes a near-impossible task. Whether such parameter sets exist in the first place is another very important question, and is driven by questions of system *identifiability* (see §6.1.2 and §6.5). Nevertheless, many valiant attempts have been made as yet, for organisms such as *E. coli* [32–34] and *M. tuberculosis* [35]. Of course, we must always remember that a dynamic model is not always necessary to draw valuable conclusions—every approximate[1] model helps us in ascertaining (part of) the truth about the real system!

Advances in computing and algorithms

The last decade has seen massive increases in processing power, as Moore's law has continued to hold its ground. Regular laptop computers are already capable of performing fairly heavy computational tasks, including a vast majority of the modelling exercises discussed in this book.

Unlike the field of molecular dynamics and molecular simulations, which continue to demand high computational resources, many seemingly difficult and computationally intensive tasks in systems-level modelling can be accomplished in reasonable time on regular desktop/laptop computers. Yet, there remain some

---

[1]Remember, "*all models are wrong*": see p. 11.

tasks that demand massive computational resources, especially (brute force) search and optimisation problems such as parameter estimation or combinatorial problems involved in strain optimisation and circuit design.

High-performance computing (HPC) has been leveraged widely in many interesting systems biology applications, such as the identification of synthetic lethal sets [36, 37], solving a large number of FBA problems [38], ODE integration for dynamic models [39], and parameter inference and model selection for dynamical models using Approximate Bayesian Computation [40]. Besozzi and co-workers present a nice review of a variety of HPC applications across bioinformatics, molecular dynamics/docking and systems biology [41].

Some software tools are also best used in a highly parallel environment, *e.g.* PyGMO (see §6.6). A large number of computationally intensive tasks can greatly benefit from the use of HPC, either CPU or GPGPU (general-purpose graphics processing units), including the reconstruction of metabolic networks (sequence comparisons), parameter estimation in large high-dimensional spaces, module identification in large biological networks, metagenomic data assembly (prior to reconstruction and microbiome modelling), the reconstruction of large number of tissue-specific metabolic networks, or the study of microbial interactions in large communities. Every deep learning or machine learning problem with large datasets will also benefit from HPC. Therefore, it is vital for students to acquire HPC skills as well, as we strive to build larger and more complex models spanning multiple scales. Wilson *et al* [42] provide a set of good scientific computing practices for systems biologists to adhere to (also see https://carpentries.org/).

Machine Learning in systems biology

In the last few years, machine learning (ML) and data science have begun to extend their influence ever so rapidly into most fields of science and engineering—and systems biology is no exception. Machine learning, both supervised and unsupervised, have many potent applications in systems biology, particularly to extract meaningful information out of multi-omic datasets coming from diverse experiments. For instance, Palsson and co-workers developed a robust model based on supervised ML, to accurately predict condition-specific gene expression for $> 1,110$ transcriptional units, and also identified ten regulatory modules [43]. Another interesting study [44] used unsupervised ML, including methods such as independent component analysis of 250+ RNAseq datasets, to shed further light on the architecture of the *E. coli* transcriptional regulatory network.

Collins and co-workers developed a white-box ML model to elucidate the metabolic aspects of antibiotic action [45]. By integrating biochemical screening, network modelling and ML, they showed how specific metabolic processes such as purine biosynthesis, are important in antibiotic lethality. In addition to large-scale omic datasets, network models, and metabolic model predictions can be integrated, for the ML model to learn and more reliably predict behaviours and unravel causal mechanisms. Reviews on the applications of ML in genome–scale modelling and metabolic engineering are available elsewhere [46, 47].

STANDARDS: ENSURING REPRODUCIBILITY OF BOTH DRY AND WET-LAB EXPERIMENTS

Community standards are a dire need to propel any field forward, more so for nascent and rapidly growing fields such as systems biology. With an explosion of data being generated from a variety of high-throughput experiments, it is important to have adequate annotation and metadata to scrupulously specify all the experimental details and conditions. This is important for both ensuring the veracity/fit of the data for a modelling purpose and the reproducibility of the experiment. The lack of reproducibility makes it very difficult to build new models on top of older ones, and is especially hard on young scientists and students making an entry into the field to build their skills by working with existing models.

Interestingly, the same challenges apply to *in silico* models as well. While one might naively expect *in silico* models to be far more reproducible than wet-lab experiments—at least the most complex issue of inherent biological variation is taken out of the equation—again, a lot of effort must be invested in exactly specifying the assumptions and exact conditions of model simulation.

In a rather revealing analysis, Malik-Sheriff and co-workers [48] attempted to reproduce as many as 455 kinetic models published in literature, and found that nearly half of the models (49%) could not be readily reproduced using just the information provided in the publications. Although part of these (12%) could be reproduced with empirical corrections or support from the original authors of the models, a staggering 37% remained non-reproducible as a result of missing values for parameters/initial conditions and/or inconsistencies in model structure. In over 40% of these non-reproducible models, the reasons for non-reproducibility of the results could not even be readily ascertained.

The above observations reinforce two very important issues: (i) the need for (and compliance to!) rigorous standards for encoding and reporting models and simulations, and (ii) the importance of independent curation/re-verification of published models in a resource such as BioModels [49]. Malik-Sheriff and co-workers have also proposed a reproducibility scorecard for models, which could be a useful check to ensure compliance.

While the above analyses specifically pertain to ODE-based models, similar issues exist with other kinds of models as well. We [50] and others [51] have previously bemoaned the issues with missing annotations and lack of standards for the exchange of genome-scale metabolic models. Many of these are being rapidly addressed by the community, through standards such as the SBML Level 3 Flux Balance Constraints (FBC) package [52]. A recent community effort has resulted in the development of a nice test suite named MEMOTE (for metabolic model tests) to assess the quality of genome-scale metabolic models [53]. Carey *et al* [54] have outlined an excellent set of best practices to create a 'gold standard' metabolic network reconstruction.

Another crucial aspect is the interoperability of models. It is easy to fall into the trap of hand-coding a model into a bunch of subroutines in your programming language of choice. Instead, it is important to ensure that the model is encoded in a form that is machine-readable, platform-independent, and inter-operable. There are hundreds of tools that can read and write models in formats such as SBML—it is always worth the effort to try and fit any new modelling paradigm or requirement to the existing standards, and use alternate approaches only under exceptional circumstances.

Despite the many issues discussed above, the community has been making rapid strides. It is heartening that a very large number of models are made available publicly in a variety of databases. Most publications these days include links to code repositories such as GitHub or general-purpose open-access repositories such as Zenodo[2]. Initiatives such as the "COmputational Modeling in BIology NEtwork" (COMBINE; http://co.mbine.org/) have been great enablers for the development of community standards and formats for a large variety of computational models. Several standards have already been defined, such as SBML [55] (see also Appendix D). On the wet-lab front too, journals such as *Cell Systems* have moved towards "STAR Methods" (STAR standing for Structured, Transparent, Accessible Reporting) from "Experimental Procedures" or "Materials and Methods" [56], with a view to improve the rigour and reproducibility of science.

A community of communities

We live in an era of (increasingly) community-driven open science. Data, publications, models, and just about everything else are more open than ever, bringing down both *physical* paywalls and *mental* silos that impeded the exchange and sharing of data. The present COVID-19 crisis has further expedited collaboration and data-sharing [57], as exemplified by the COVID-19 disease map [58], cataloguing the molecular processes of virus–host interactions. Indeed, such a rapid effort was possible only because of the groundwork laid by the community, through several such community-led initiatives such as Reactome ([59]; https://reactome.org/), WikiPathways ([60]; https://www.wikipathways.org/), IMEx Consortium (International Molecular Exchange Consortium [61]; http://www.imexconsortium.org/), Pathway Commons ([62]; https://www.pathwaycommons.org/), DisGeNET ([63]; https://www.disgenet.org/), ELIXIR ([64]; https://elixir-europe.org/), and the Disease Maps Project ([65]; https://disease-maps.org/)—to name a few.

Open data also has its challenges. The mushrooming of various general-purpose data repositories also has a flip side—the data may not be well-described or may be in arbitrary formats, which makes it harder for discovery and reuse by humans and machines alike. To mitigate this, a diverse group of stakeholders from across academia, industry, funding agencies, and scholarly publishers, have proposed Findable, Accessible, Interoperable, and Reusable *(FAIR) Data Principles* [66]. FAIR Data principles are a concise and measurable set of principles, complying with which can enable scientists (and others) to enhance the re-usability of their data. In particular, FAIR emphasises the need for enhancing the ability of machines to mine the data, over and above the re-use by

---

[2]See https://www.nature.com/sdata/policies/repositories for a list of data repositories recommended by the journal *Scientific Data*.

individuals. The FAIRDOMHub is a repository for publishing FAIR Data, Operating procedures, and Models (https://fairdomhub.org/) for the systems biology community [67].

All in all, one cannot exaggerate the importance of open data-sharing, interoperability, re-use, and reproducibility in all scientific research, and more so in systems biology, as we embark on integrating diverse models in our quest to answer incrementally harder questions on biological system structure and function.

LEARNING SYSTEMS BIOLOGY: BY DOING & COLLABORATING

Continuing from the discussion in the preface, this book on systems biology has emerged out of a course I have taught a variety of engineers at IIT Madras, with varied backgrounds and exposures to biology. Such interdisciplinary courses are tricky and difficult to *pitch* right, balancing the right amount of exposure to biological vs. mathematical and programming intricacies. This balance, which I am not sure I have yet achieved, improves as one iterates through the course; however, one thing is unequivocally clear—to get a solid understanding of computational systems biology, one must have hands-on lab sessions, to build models and simulate them.

Students, from across different years, and disciplines, have unanimously emphasised to me the value of lab sessions[3], and a seven-week long course project in honing their skills and improve their understanding of a variety of concepts covered in this course. This project, which I usually ask students to carry out in pairs, requires students to carry out a research project in the area of mathematical modelling of any biological system. I often try to pair up students in such a way that their biology and programming knowledge/skills are complementary, to the extent possible. This bringing together of 'unlike' minds fosters creativity, and the spirit of collaboration, which is indispensable for modern research, particularly in emergent interdisciplinary areas such as computational systems biology.

FINAL THOUGHTS

One of the loftiest goals of modelling is to understand the design principles of a system, even if it be a system shaped by millions of years of evolutionary forces. Unravelling this *operating system*, which translates the genotype of a system encoded in a few million (or billion) bases of DNA to a massively complex, robust, cellular machinery capable of exquisite functions, is perhaps one of the grandest challenges in modern biology. Ultimately, mirroring the behaviour of a prokaryotic cell, or full-blown virtual human beings—*silicon twins* of us carbon-based life-forms—remains one of the most exciting challenges ahead.

---

[3]This is another reason to emphasise "Lab exercises" at the end of every chapter.

REFERENCES

[1] W. C. Roda, M. B. Varughese, D. Han, and M. Y. Li, "Why is it difficult to accurately predict the COVID-19 epidemic?", *Infectious Disease Modelling* 5:271–281 (2020).

[2] A. Renz, L. Widerspick, and A. Dräger, *FBA reveals guanylate kinase as a potential target for antiviral therapies against SARS-CoV-2*, tech. rep. (2020).

[3] Y. Zhou, Y. Hou, J. Shen, Y. Huang, W. Martin, and F. Cheng, "Network-based drug repurposing for novel coronavirus 2019-nCoV /SARS-CoV-2", *Cell Discovery* 6:1–18 (2020).

[4] E. Bosi, J. M. Monk, R. K. Aziz, M. Fondi, V. Nizet, and B. Ø. Palsson, "Comparative genome-scale modelling of *Staphylococcus aureus* strains identifies strain-specific metabolic capabilities linked to pathogenicity", *Proceedings of the National Academy of Sciences USA* 113:E3801–E3809 (2016).

[5] Y. Seif, E. Kavvas, J.-C. Lachance, et al., "Genome-scale metabolic reconstructions of multiple *Salmonella* strains reveal serovar-specific metabolic traits", *Nature Communications* 9:3771 (2018).

[6] J. McFadden, D. J. V. Beste, and A. M. Kierzek, eds., *Systems biology of tuberculosis* (Springer-Verlag, New York, 2013).

[7] M. Eckhardt, J. F. Hultquist, R. M. Kaake, R. Hüttenhain, and N. J. Krogan, "A systems approach to infectious disease", *Nature Reviews Genetics* 21:339–354 (2020).

[8] J. J. Hornberg, F. J. Bruggeman, H. V. Westerhoff, and J. Lankelma, "Cancer: a systems biology disease", *BioSystems* 83:81–90 (2006).

[9] R. Laubenbacher, V. Hower, A. Jarrah, et al., "A systems biology view of cancer", *Biochimica et biophysica acta* 1796:129–139 (2009).

[10] D. Hanahan and R. A. Weinberg, "The hallmarks of cancer", *Cell* 100:57–70 (2000).

[11] D. Hanahan and R. A. Weinberg, "Hallmarks of cancer: the next generation", *Cell* 144:646–674 (2011).

[12] W. Du and O. Elemento, "Cancer systems biology: embracing complexity to develop better anticancer therapeutic strategies", *Oncogene* 34:3215–3225 (2015).

[13] H. Kitano, "Cancer as a robust system: implications for anticancer therapy", *Nature Reviews Cancer* 4:227–235 (2004).

[14] A. Nilsson and J. Nielsen, "Genome scale metabolic modeling of cancer", *Metabolic Engineering* 43:103–112 (2017).

[15] L. Jerby and E. Ruppin, "Predicting drug targets and biomarkers of cancer via genome-scale metabolic modeling", *Clinical Cancer Research* 18:5572–5584 (2012).

[16] O. Folger, L. Jerby, C. Frezza, E. Gottlieb, E. Ruppin, and T. Shlomi, "Predicting selective drug targets in cancer through metabolic networks", *Molecular Systems Biology* **7**:501 (2011).

[17] S. Sahoo, R. K. Ravi Kumar, B. Nicolay, et al., "Metabolite systems profiling identifies exploitable weaknesses in retinoblastoma", *FEBS Letters* **593**:23–41 (2019).

[18] K. Hari, B. Sabuwala, B. V. Subramani, et al., "Identifying inhibitors of epithelial–mesenchymal plasticity using a network topology-based approach", *npj Systems Biology and Applications* **6**:1–12 (2020).

[19] S. Sarkar, S. K. Sinha, H. Levine, M. K. Jolly, and P. S. Dutta, "Anticipating critical transitions in epithelial–hybrid–mesenchymal cell-fate determination", *Proceedings of the National Academy of Sciences USA* **116**:26343–26352 (2019).

[20] L. Yu, M. Lu, D. Jia, et al., "Modeling the Genetic Regulation of Cancer Metabolism: Interplay between Glycolysis and Oxidative Phosphorylation", *Cancer Research* **77**:1564–1574 (2017).

[21] G. An, Q. Mi, J. Dutta-Moscato, and Y. Vodovotz, "Agent-based models in translational systems biology", *WIREs Systems Biology and Medicine* **1**:159–171 (2009).

[22] B. M. Kuenzi and T. Ideker, "A census of pathway maps in cancer systems biology", *Nature Reviews Cancer* **20**:233–246 (2020).

[23] M. C. Peterson and M. M. Riggs, "FDA advisory meeting clinical pharmacology review utilizes a Quantitative Systems Pharmacology (QSP) model: a watershed moment?", *CPT: Pharmacometrics & Systems Pharmacology* **4**:189–192 (2015).

[24] S. G. Wicha and C. Kloft, "Quantitative systems pharmacology in model-informed drug development and therapeutic use", *Current Opinion in Systems Biology* **10**:19–25 (2018).

[25] K. Gadkar, D. Kirouac, N. Parrott, and S. Ramanujan, "Quantitative systems pharmacology: a promising approach for translational pharmacology", *Drug Discovery Today: Technologies* **21-22**:57–65 (2016).

[26] I. Thiele, C. M. Clancy, A. Heinken, and R. M. T. Fleming, "Quantitative systems pharmacology and the personalized drug–microbiota–diet axis", *Current Opinion in Systems Biology* **4**:43–52 (2017).

[27] J. M. Lee, E. P. Gianchandani, J. A. Eddy, and J. A. Papin, "Dynamic analysis of integrated signaling, metabolic, and regulatory networks", *PLoS Computational Biology* **4**:e1000086 (2008).

[28] T. Shlomi, Y. Eisenberg, R. Sharan, and E. Ruppin, "A genome-scale computational study of the interplay between transcriptional regulation and metabolism", *Molecular Systems Biology* **3**:101 (2007).

[29] M. W. Covert, N. Xiao, T. J. Chen, and J. R. Karr, "Integrating metabolic, transcriptional regulatory and signal transduction models in *Escherichia coli*", *Bioinformatics* **24**:2044–2050 (2008).

[30] J. R. Karr, J. C. Sanghvi, D. N. Macklin, et al., "A whole-cell computational model predicts phenotype from genotype", *Cell* **150**:389–401 (2012).

[31] P. F. Suthers, M. S. Dasika, V. S. S. Kumar, G. Denisov, J. I. Glass, and C. D. Maranas, "A genome-scale metabolic reconstruction of *Mycoplasma genitalium, i*PS189", *PLoS Computational Biology* **5**:e1000285+ (2009).

[32] A. Barve, A. Gupta, S. M. Solapure, et al., "A kinetic platform for *in silico* modeling of the metabolic dynamics in *Escherichia coli*", *Advances and Applications in Bioinformatics and Chemistry: AABC* **3**:97–110 (2010).

[33] K. Smallbone, E. Simeonidis, N. Swainston, and P. Mendes, "Towards a genome-scale kinetic model of cellular metabolism", *BMC Systems Biology* **4**:6 (2010).

[34] A. Khodayari and C. D. Maranas, "A genome-scale *Escherichia coli* kinetic metabolic model k-Ecoli457 satisfying flux data for multiple mutant strains", *Nature Communications* **7**:13806 (2016).

[35] D. A. Adiamah and J.-M. Schwartz, "Construction of a genome-scale kinetic model of *Mycobacterium tuberculosis* using generic rate equations", *Metabolites* **2**:382–397 (2012).

[36] C. S. Henry, F. Xia, and R. Stevens, "Application of high-performance computing to the reconstruction, analysis, and optimization of genome-scale metabolic models", *Journal of Physics: Conference Series* **180**:012025 (2009).

[37] A. Pratapa, S. Balachandran, and K. Raman, "Fast-SL: an efficient algorithm to identify synthetic lethal sets in metabolic networks", *Bioinformatics* **31**:3299–3305 (2015).

[38] L. Heirendt, I. Thiele, and R. M. Fleming, "DistributedFBA.jl: high-level, high-performance flux balance analysis in Julia", *Bioinformatics* (2017).

[39] L. A. Harris, M. S. Nobile, J. C. Pino, et al., "GPU-powered model analysis with PySB/cupSODA", *Bioinformatics* **33**:3492–3494 (2017).

[40] J. Liepe, C. Barnes, E. Cule, et al., "ABC-SysBio–Approximate Bayesian computation in Python with GPU support", *Bioinformatics* **26**:1797–1799 (2010).

[41] M. S. Nobile, P. Cazzaniga, A. Tangherloni, and D. Besozzi, "Graphics processing units in bioinformatics, computational biology and systems biology", *Briefings in Bioinformatics* **18**:870–885 (2017).

[42] G. Wilson, J. Bryan, K. Cranston, J. Kitzes, L. Nederbragt, and T. K. Teal, "Good enough practices in scientific computing", *PLoS Computational Biology* **13**:e1005510 (2017).

[43] X. Fang, A. Sastry, N. Mih, et al., "Global transcriptional regulatory network for *Escherichia coli* robustly connects gene expression to transcription factor activities", *Proceedings of the National Academy of Sciences USA* **114**:10286–10291 (2017).

[44] A. V. Sastry, Y. Gao, R. Szubin, et al., "The *Escherichia coli* transcriptome mostly consists of independently regulated modules", *Nature Communications* **10**:5536 (2019).

[45] J. H. Yang, S. N. Wright, M. Hamblin, et al., "A white-box machine learning approach for revealing antibiotic mechanisms of action", *Cell* **177**:1649–1661.e9 (2019).

[46] G. Zampieri, S. Vijayakumar, E. Yaneske, and C. Angione, "Machine and deep learning meet genome-scale metabolic modeling", *PLoS Computational Biology* **15**:e1007084 (2019).

[47] G. B. Kim, W. J. Kim, H. U. Kim, and S. Y. Lee, "Machine learning applications in systems metabolic engineering", *Current Opinion in Biotechnology* **64**:1–9 (2020).

[48] K. Tiwari, S. Kananathan, M. G. Roberts, et al., "Reproducibility in systems biology modelling", *Mol Syst Biol* **17**:e9982 (2021).

[49] R. S. Malik-Sheriff, M. Glont, T. V. N. Nguyen, et al., "BioModels—15 years of sharing computational models in life science", *Nucleic Acids Research* **48**:D407–D415 (2020).

[50] A. Ravikrishnan and K. Raman, "Critical assessment of genome-scale metabolic networks: the need for a unified standard", *Briefings in Bioinformatics* **16**:1057–1068 (2015).

[51] A. Ebrahim, E. Almaas, E. Bauer, et al., "Do genome-scale models need exact solvers or clearer standards?", *Molecular Systems Biology* **11**:831 (2015).

[52] B. G. Olivier and F. T. Bergmann, "SBML Level 3 Package: Flux Balance Constraints version 2", *Journal of Integrative Bioinformatics* **15**:20170082 (2018).

[53] C. Lieven, M. E. Beber, B. G. Olivier, et al., "MEMOTE for standardized genome-scale metabolic model testing", *Nature Biotechnology* **38**:272–276 (2020).

[54] M. A. Carey, A. Dräger, M. E. Beber, J. A. Papin, and J. T. Yurkovich, "Community standards to facilitate development and address challenges in metabolic modeling", *Molecular Systems Biology* **16**:e9235 (2020).

[55] S. M. Keating, D. Waltemath, M. König, et al., "SBML Level 3: an extensible format for the exchange and reuse of biological models", *Molecular Systems Biology* **16**:e9110 (2020).

[56] E. Marcus, "A STAR is born", *Cell* **166**:1059–1060 (2016).

[57] J. J. Lee and J. P. Haupt, "Scientific globalism during a global crisis: research collaboration and open access publications on COVID-19", *Higher Education* (2020).

[58] M. Ostaszewski, A. Mazein, M. E. Gillespie, et al., "COVID-19 Disease Map, building a computational repository of SARS-CoV-2 virus-host interaction mechanisms", *Scientific Data* **7**:136 (2020).

[59] B. Jassal, L. Matthews, G. Viteri, et al., "The reactome pathway knowledgebase", *Nucleic Acids Research* **48**:D498–D503 (2020).

[60] D. N. Slenter, M. Kutmon, K. Hanspers, et al., "WikiPathways: a multifaceted pathway database bridging metabolomics to other omics research", *Nucleic Acids Research* **46**:D661–D667 (2018).

[61] S. Orchard, S. Kerrien, S. Abbani, et al., "Protein interaction data curation: the International Molecular Exchange (IMEx) consortium", *Nature Methods* **9**:345–350 (2012).

[62] I. Rodchenkov, O. Babur, A. Luna, et al., "Pathway Commons 2019 update: integration, analysis and exploration of pathway data", *Nucleic Acids Research* **48**:D489–D497 (2020).

[63] J. Piñero, J. M. Ramírez-Anguita, J. Saüch-Pitarch, et al., "The DisGeNET knowledge platform for disease genomics: 2019 update", *Nucleic Acids Research* **48**:D845–D855 (2020).

[64] R. Drysdale, C. E. Cook, R. Petryszak, et al., "The ELIXIR Core Data Resources: fundamental infrastructure for the life sciences", *Bioinformatics* **36**:2636–2642 (2020).

[65] A. Mazein, M. Ostaszewski, I. Kuperstein, et al., "Systems medicine disease maps: community-driven comprehensive representation of disease mechanisms", *npj Systems Biology and Applications* **4**:1–10 (2018).

[66] M. D. Wilkinson, M. Dumontier, I. J. Aalbersberg, et al., "The FAIR guiding principles for scientific data management and stewardship", *Scientific Data* **3**:160018 (2016).

[67] K. Wolstencroft, O. Krebs, J. L. Snoep, et al., "FAIRDOMHub: a repository and collaboration environment for sharing systems biology research", *Nucleic Acids Research* **45**:D404–D407 (2017).

FURTHER READING

M. Cvijovic, J. Almquist, J. Hagmar, et al., "Bridging the gaps in systems biology", *Molecular Genetics and Genomics* **289**:727–734 (2014).

K. Schwenk, D. K. Padilla, G. S. Bakken, and R. J. Full, "Grand challenges in organismal biology", *Integrative and Comparative Biology* **49**:7–14 (2009).

I. Ajmera, M. Swat, C. Laibe, N. L. Novère, and V. Chelliah, "The impact of mathematical modeling on the understanding of diabetes and related complications", *CPT: Pharmacometrics & Systems Pharmacology* **2**:54 (2013).

B. Schoeberl, "Quantitative systems pharmacology models as a key to translational medicine", *Current Opinion in Systems Biology* **16**:25–31 (2019).

E. Barillot, L. Calzone, P. Hupe, J.-P. Vert, and A. Zinovyev, *Computational systems biology of cancer* (CRC Press, Boca Raton, FL, 2012).

J. R. Karr, K. Takahashi, and A. Funahashi, "The principles of whole-cell modeling", *Current Opinion in Microbiology* **27**:18–24 (2015).

B. Szigeti, Y. D. Roth, J. A. P. Sekar, A. P. Goldberg, S. C. Pochiraju, and J. R. Karr, "A blueprint for human whole-cell modeling", *Current Opinion in Systems Biology* **7**:8–15 (2018).

D. M. Camacho, K. M. Collins, R. K. Powers, J. C. Costello, and J. J. Collins, "Next-generation machine learning for biological networks", *Cell* **173**:1581–1592 (2018).

S. Srinivasan, W. R. Cluett, and R. Mahadevan, "Constructing kinetic models of metabolism at genome-scales: A review", *Biotechnology Journal* **10**:1345–1359 (2015).

P. Macklin, "Key challenges facing data-driven multicellular systems biology", *GigaScience* **8**:giz127 (2019).

P. Mendes, "Reproducible research using biomodels", *Bulletin of Mathematical Biology* **80**:3081–3087 (2018).

Online-only appendices

APPENDIX A: INTRODUCTION TO KEY BIOLOGICAL CONCEPTS

This appendix provides a quick introduction to biology, focussing on basic biological macromolecules such as nucleic acids, proteins, carbohydrates, lipids, and small molecules. It also provides a brief overview of information flow in biological systems, discussing aspects of DNA replication, mRNA transcription, protein translation, and post–translational modifications.

APPENDIX B: RECONSTRUCTION OF BIOLOGICAL NETWORKS

This appendix provides an overview of reconstruction of various types of biological networks, viz. gene regulatory networks, protein–protein interaction networks, signalling networks, and metabolic networks.

APPENDIX C: DATABASES FOR SYSTEMS BIOLOGY

This appendix presents a brief overview of databases of PPIs and functional associations, metabolic network databases, gene regulatory network databases, model databases, and certain miscellaneous databases.

APPENDIX D: SOFTWARE TOOLS COMPENDIUM

Beginning with a summary of standards and formats, this appendix details various software tools that are useful for all the modelling paradigms studied in the book, such as network biology, dynamic modelling and parameter estimation, Boolean networks, constraint-based modelling, and community modelling.

APPENDIX E: MATLAB FOR SYSTEMS BIOLOGY

This appendix provides a solid introduction to the basics of using MATLAB, followed by a discussion on how a variety of tools and toolboxes can be used for handling various modelling and simulation tasks in systems biology.

Appendices are located online–only at https://www.routledge.com/9781138597327.

Index

In addition to page numbers pointing to main text discussion, the suffixes f and T point to a relevant figure or table, respectively. Page numbers in **bold face** indicate equations or definitions, while those in *italics* indicate exercise questions.

Printed in the United States
by Baker & Taylor Publisher Services